306
Current Topics in Microbiology and Immunology

W. M. Shafer (Ed.)

Antimicrobial Peptides and Human Disease

With 12 Figures and 4 Tables

 Springer

William M. Shafer, Ph.D.
Department of Microbiology and Immunology
3001 Rollins Research Center
Emory University School of Medicine
Atlanta, GA 30322
USA

e-mail: wshafer@emory.edu

Cover Illustration by Dawn M.E. Bowdish, Donald J. Davidson and Robert E.W. Hancock
(Cover figure reproduced with kind permission of Leukemia Research) (this volume)

Library of Congress Catalog Number 72-152360

ISSN 0070-217X
ISBN-10 3-540-29915-7 Springer Berlin Heidelberg New York
ISBN-13 978-3-540-29915-8 Springer Berlin Heidelberg New York

Springer is a part of Springer Science+Business Media
springeronline.com
© Springer-Verlag Berlin Heidelberg 2006
Printed in Germany

Editor: Simon Rallison, Heidelberg
Desk editor: Anne Clauss, Heidelberg
Production editor: Nadja Kroke, Leipzig
Cover design: design & production GmbH, Heidelberg
Typesetting: LE-TEX Jelonek, Schmidt & Vöckler GbR, Leipzig
Printed on acid-free paper   SPIN 11332879   27/3150/YL – 5 4 3 2 1 0

Preface

Microbes are in our midst soon after birth. Thankfully, the number of harmless (and often beneficial) microbes far outnumber those that would do us harm. Our ability to ward off pathogens in our environment, including those that can colonize our exterior and/or interior surfaces, depends on the integrative action of the innate and adaptive immunity systems. This volume of CTMI, entitled *Antimicrobial Peptides and Human Disease*, is dedicated to the role of antimicrobial peptides (AMPs) in the innate host defense system of Homo sapiens.

The concept that oxygen-independent killing systems of phagocytic cells is in part attributable to the antibiotic-like action of AMPs (and antibacterial proteins) stored in cytoplasmic granules served as a stimulus for AMP research in humans. Unfortunately, this early work received little notice and was over-shadowed by investigations of the oxidative microbial killing components of phagocytes. Only a handful of laboratories were interested in these curious antimicrobial peptides and proteins. However, in 1980 Hans Boman's group [1] reported on the purification, characterization and antimicrobial action of an AMP from a non-human source – the moth *Hyalophora cecropia*, providing an important precedent and a catalyst for the field. Soon thereafter, other groups announced the purification of AMPs from a variety of sources including vertebrates, invertebrates and plants. It is now apparent that all living systems (including microbes) have the capacity to produce AMPs. It is also clear that AMPs evolved long before the development of adaptive immunity systems, and their induction following injury or infection is a highly conserved innate immune response to microbes. Indeed, they represent the earliest form of host defense. Thus, AMPs are important and understanding their contribution to host defense has promise for the advancement of medicine.

With respect to humans, the groundbreaking work of Lehrer's group in the early 1980s [2] that characterized the alpha-defensins from human polymorphonuclear leukocytes set the stage for the next 20 years of AMP research as it pertains to the human innate immune response. Two decades after the first reports of human neutrophil-derived defensins, we now have a better, but still incomplete, understanding of the AMP repertoire possessed by humans. Less clear, however, are their direct and/or indirect roles in host defense during infection. However, considerable progress has been made in this area and these advances are highlighted in several chapters of this book.

Using small intestinal Paneth cell alpha-defensins as model AMPs, A. Ouellette describes in detail the synthesis and function of AMPs. This system serves as a very nice model that has revealed fundamental information regarding the regulation of AMP gene expression, AMP activation and secretion to the extracellular fluid in response to microbes. While increasing evidence implicates the antibiotic-like action of AMPs as being fundamentally important in host defense against infection, the immunomodulatory activities of these peptides is being increasingly appreciated. Under certain circumstances, these AMP immunomodulatory activities may contribute more to overall host defense than their antimicrobial properties. It is becoming increasingly evident that through their immunomodulatory activities certain AMPs connect the innate and adaptive immune responses, providing what may be essential links for maintaining the overall fidelity of host defense. The wide range and importance of such activities displayed by members of the defensin and cathelicidin peptide families, the two main classes of human AMPs, are discussed in detail by D. Bowdish et al.

The contribution of AMPs to the ability of humans to defend themselves from infectious agents, especially on respiratory and alimentary mucosal surfaces, are placed in overall context by B. Agerberth and G. Gudmundsson. This chapter sets the stage for a series of chapters that review and highlight the contribution of AMPs to overall host defense at many different sites, beginning with the notion that AMPs are an essential component of the defensive barrier imposed by our largest organ (skin). As emphasized by M. Braff and R. Gallo, through their bactericidal action and immunomodulatory activities, AMPs have great promise as futuristic therapeutic agents. If this goal is realized, clinicians will have a new treatment option in combating common skin pathogens that are frequently resistant to multiple antibiotics. Invasive bloodstream infections are often caused by bacteria that resist multiple antibiotics. These infections are responsible for a significant number of deaths worldwide each year. M. Yeaman and A. Bayer describe how neutrophil- and platelet-derived AMPs combat invasive bloodstream infections, and they provide new insights as to the biologic importance of the immunomodulatory action of such peptides during bacteremic disease. Airborne transmission of pathogens is a major mechanism of spread of infectious diseases. As is highlighted by D. Laube et al., the presence and activity of AMPs are critical to the overall health of the respiratory tract. For instance, the presence of AMPs in airway surface fluid (ASF) and the role of these peptides in host defense has been a subject of great interest in that the ASF from patients with cystic fibrosis may be inhibitory for AMPs, which may explain these patients' frequent and often life-threatening infections. Strategies that optimize AMP activity or the application of AMPs to the respiratory surface epithelium have promise in human medicine for the treatment of lung infections that are refractory to other treatment options.

One recurring theme in many of the chapters in this volume is that AMPs have functions beyond antimicrobial action. In this respect, hepcidin, a remarkable AMP and a peptide hormone that is the homeostatic regulator of iron availability, is discussed by T. Ganz. During infection and inflammation, hepcidin synthesis is up-regulated and, because the availability of free iron is critical for efficient microbial growth, the presence and action of hepcidin can contribute to host defense. Sexually transmitted infections (STI), including transmission of the human immunodeficiency virus, continue to be a world-wide public health concern. The male and female genital tracts are sites for AMP production, due to the infiltration of phagocytic cells or the inducible synthesis of AMPs by epithelial cells. A. Cole describes their role in innate host defense against STIs with special emphasis on the contribution of genital tract AMPs to the overall host defensive strategy of protecting the vaginal mucosal surface from infection. Microbes do not stand idly by as they are assaulted by AMP. In fact, they have mechanisms to reduce or thwart the killing action of AMP. In this respect, the multiple mechanisms developed by Gram-negative and Gram-positive bacteria to escape or reduce the bactericidal action of AMP are reviewed by A. Peschel. In recent years, we have come to appreciate that many pathogens grow in communities and communicate via chemical signals as a population, and bacteria growing in one type of community (biofilms) differ significantly from those growing alone (planktonic growth). This and the recognition that microbial biofilms often are less susceptible to antimicrobials (including AMPs) have stimulated researchers to understand virulence mechanisms that operate during infection on host or implant surfaces. M. Otto reviews progress in this area of research and provides insights regarding new strategies to destroy biofilm formation during infection.

Those of us who study AMPs have benefited tremendously from the pioneering early workers who advanced fundamental knowledge and principles regarding critical topics of host defense, infectious diseases, cell biology, biochemistry and pathology. Those accomplishments often were made without the benefit of the instrumentation, techniques, and technologies in molecular biology, biochemistry and cell biology that enabled the field to progress to its current level. We owe them much for their achievements, insights and prescience. This book is dedicated to those insightful scientists and their accomplishments.

Atlanta, 2005   *William M. Shafer*

References

1. Hultmark D, Steiner H, Rasmuson T, Boman HG (1980) Insect immunity. Purification and properties of three inducible bactericidal proteins from hemolymph of immunized pupae of *Hyalophora cecropia*. Eur J Biochem 106:7–16
2. Selsted ME, Harwig SS, Ganz T, Schilling JW, Lehrer RI (1985) Primary structure of three human neutrophil defensins. J Clin Invest 76:1436–1439

List of Contents

List of Contributors

(Addresses stated at the beginning of respective chapters)

Agerberth, B.   67

Bayer, A. S.   111
Bowdish, D. M. E.   27
Braff, M. H.   91

Cole, A. M.   199

Davidson, D. J.   27
Diamond, G.   153

Gallo, R. L.   91
Ganz, T.   183
Guðmundsson, G. H.   67

Hancock, R. E. W.   27

Kisich, K. O.   153
Kraus, D.   235

Laube, D. M.   153

Otto, M.   251
Ouellette, A. J.   1

Peschel, A.   235

Ryan, L. K.   153

Yeaman, M. R.   111
Yim, S.   153

CTMI (2006) 306:1–25

Paneth Cell α-Defensin Synthesis and Function

A. J. Ouellette (✉)

Department of Pathology & Laboratory Medicine, Med Sci D440, School of Medicine, College of Health Sciences, University of California, Irvine, CA 92697-4800, USA
aouellet@uci.edu

Abstract Endogenous antimicrobial peptides (AMPs) mediate innate immunity in every species in which they have been investigated. Cathelicidins and defensins are the two major AMP families in mammals, and they are abundant components of phagocytic leukocytes and are released by epithelial cells at mucosal surfaces. In the small intestine, Paneth cells at the base of the crypts of Lieberkühn secrete α-defensins and additional AMPs at high levels in response to cholinergic stimulation and when exposed to bacterial antigens. Paneth cell α-defensins evolved to function in the extracellular environment with broad-spectrum antimicrobial activities, and they constitute the majority of bactericidal peptide activity secreted by Paneth cells. The release of Paneth cell products into the crypt lumen is inferred to protect mitotically active crypt cells from colonization by potential pathogens and confers protection from enteric infection, as is evident from the immunity of mice expressing a human Paneth cell α-defensin transgene to oral infection by *Salmonella enterica* serovar Typhimurium. α-Defensins in Paneth cell secretions also may interact with bacteria in the intestinal lumen above the crypt–villus boundary and influence the composition of the enteric microbial flora. Mutations that cause defects in the activation, secretion, dissolution, and bactericidal effects of Paneth cell AMPs may alter crypt innate immunity and contribute to immunopathology.

1
Antimicrobial Peptides and Innate Immunity

Originally characterized in plants and insects, gene-encoded antimicrobial peptides (AMPs) are now recognized as a widely deployed mechanism of biochemical defense against potential pathogens [12, 152, 153]. Although exceptions exist, AMPs characteristically are 100 amino acids or less in chain length, cationic at neutral pH, and have broad-spectrum antibiotic activity against a wide range of microbes and certain viruses at low micromolar concentrations [11]. AMPs have been isolated from all phyla examined, and their expression and release is observed in diverse host defense settings [43].

In mammals, AMPs are produced by phagocytes and by diverse mucosal epithelia [62]. In cells of myeloid origin, AMPs are stored in granules and contribute to nonoxidative bacterial cell killing after phagocytosis and phagolysosomal fusion [27, 60, 71, 74, 94, 109, 114]. In the airway, skin, oropharynx, gingival crevice, ocular tissues, urogenital epithelium, and gastrointestinal tract, constituent epithelial cells release AMPs, and convincing evidence implicates the peptides as mediators of innate immunity [53, 58, 62, 63, 75, 105]. AMPs also accumulate in secretory granules for later release as components of regulated exocytotic pathways, as Paneth cells in the small intestine exemplify [9, 85].

AMPs are highly varied in structure. They range from linear, random coil molecules that adopt α-helical structures in hydrophobic environments to highly constrained, β-sheet-containing peptides that contain multiple disulfide bonds and have no α-helical component. Despite their diverse primary, secondary, and tertiary structures, most AMPs are amphipathic [50, 72, 143, 150], and it is amphipathicity and selective affinity for anionic phospholipid bilayers that enables peptide-mediated binding and disruption of microbial cell membranes [48, 49, 116, 140]. Most AMPs kill their microbial targets by peptide-mediated membrane disruption, creating defects that dissipate cellular electrochemical gradients, leading to microbial cell death [12, 152]. Two large AMP families are found in mammals, the cathelicidins and defensins.

2
Cathelicidins

The cathelicidins are AMPs that exhibit a phylogenetically conserved precursor structure from which varied AMPs are derived [133]. Cathelicidins occur widely, being found in primates, ungulates, and rodents, but also in non-mammals including chicken, trout, and the Atlantic hagfish, a primitive vertebrate [133]. Cathelicidin precursors contain an N-terminal "Cathelin" domain, named as such for its resemblance to an apparent cysteine protease inhibitor, and a C-terminal AMP domain. The relatedness of varied cathelicidins often is evident only from the Cathelin prodomains of the precursors [151]. The genomes of certain species such as cattle and swine contain numerous and diverse cathelicidin genes, but in mice, rats, rhesus macaques, and humans, only a single gene is expressed from this diverse family [133]. C-terminally derived cathelicidin peptides are highly variable in primary, secondary, and tertiary structures, ranging from α-helical peptides and peptides with very high proline content to approximately 2-kDa peptides that contain two disulfide bonds. In cells of myeloid lineage, procathelicidins are activated by serine proteinases, including elastase or proteinase 3 [86, 107, 117, 121, 122].

3
Defensins: A Cysteine-Rich AMP Family

Defensins were among the first AMPs to be described [59, 108], consisting of three subfamilies of cationic, Cys-rich AMPs, the α-, β-, and θ-defensins (Fig. 1). Defensins have broad-spectrum antimicrobial activities [115]. The α-defensins are major granule constituents of mammalian phagocytic leukocytes and of small intestinal Paneth cells [32]. α-Defensins have been estimated to reach concentrations of approximately 10 mg/ml in neutrophil phagolysosomes, accounting for 5%–18% of total cellular protein [27], and Paneth cells secrete α-defensins at local concentrations of 25–100 mg/ml in mouse small intestinal crypts [4, 34]. The β-defensins, discovered in cattle as AMPs of airway and lingual epithelial cells and in bovine neutrophil granules [22, 103, 112], exist in varied species and are expressed by a great variety of epithelial cell types and at many more sites than the α-defensins [105]. For example, β-defensin peptides or transcripts have been detected in human kidney, skin, pancreas, gingiva, tongue, esophagus, salivary gland, cornea, and airway epithelium and in epithelial cells of several species [105]. The θ-defensins are approximately 2-kDa macrocyclic peptides from rhesus macaque neutrophils and monocytes that, like all defensins,

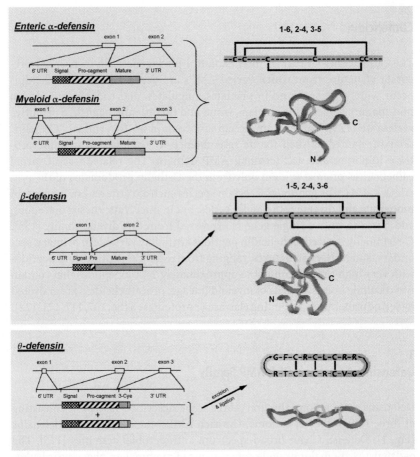

Fig. 1 α-Defensin gene and precursor structure. Schematics of the myeloid and Paneth cell α-defensin genes and precursors are aligned at *left*. The signal peptide and prore-gions are *crosshatched* differentially, and the *gray* C-terminal region of the precursors denote the residues that constitute the mature α-defensin peptide. The 3D structure shown is that of rabbit α-defensin RK-1. (Reprinted from [115], Fig. 1)

are stabilized by three disulfide bonds (Fig. 1) [65, 132, 135]. The ligation mechanisms that circularize the closed θ-defensin polypeptide chain remain unknown. The biology of defensins as chemotactic stimuli in the innate im-mune response and their influences on adaptive immune function have been reviewed [146] and will not be considered here.

4
Defensin Structures

The three-dimensional structures of several α- and β-defensins have been determined by both NMR and X-ray crystallography techniques [115, 140], and the 3D structures contain a canonical triple-stranded antiparallel β-sheet motif. The 3- to 4-kDa α- and β-defensins both contain six cysteine residues, which form invariant disulfide bond pairings that distinguish the individual defensin subfamilies [110, 130]. Although the spacing of cysteine residues and Cys-Cys pairings of disulfide bonds in α- and β-defensins differ (Fig. 1), the peptides have remarkably similar folded conformations [87, 120, 158]. The crystal structure of the human neutrophil α-defensin HNP-3 is a non-covalent, amphipathic dimer in which arginine side chains lie equatorially above a hydrophobic surface consisting of apolar monomer side chains [39]. On the other hand, α-defensins from rabbit neutrophils, mouse Paneth cells, and rabbit kidney are monomeric in solution. Solution and crystallographic analyses of β-defensins have shown that the α- and β-defensin folds are similar as is the amphipathicity produced by the distribution of polar and hydrophobic side chains on the peptide surfaces. The folds of hBD-1 and -2 are similar to each other and to bovine β-defensin-12 [45, 46, 101, 158], all three of which are predominated by a triple-stranded β-sheet. However, both β-defensins possess short α-helical segments that α-defensins lack. The crystal structure of hBD-2 was found to be dimeric, but the interfacial contacts of the hBD-2 and HNP-3 dimers differ [39]. Certain hBD-2 crystal forms have been reported to be octameric, organized through interactions of monomer N-termini [45]. Solution structural studies of hBD-2, though, showed it to be a monomer at acidic and near-neutral pH [101]. There is no evident relationship between β-sheet content and antimicrobial activity, and knowledge of the general structural factors that modulate the antimicrobial spectrum and activity of defensins is for the most part lacking.

The solution structures of closed circular rhesus θ-defensin-1 (RTD-1) and its open chain analog oRTD-1 have been determined by two-dimensional NMR [134]. RTD-1 and oRTD-1 adopt very similar structures in water, containing an extended β-hairpin, structure with turns at one end in oRTD-1 or in both ends in circular RTD-1. The double stranded β-sheet region of the two molecules are flexible, and, because the structures and flexibilities of RTD-1 and oRTD-1 are similar, the reduced antimicrobial activity of oRTD-1 relative to circular RTD-1 are attributable to the charged N- and C-termini of the oRTD-1 molecule [134]. In contrast to many AMPs, RTD-1 has no amphiphilic character, even though surface models of RTD-1 exhibit a certain clustering of positive charges.

5
Mechanisms of Defensin Bactericidal Action

The α-defensins are microbicidal against Gram-positive and Gram-negative bacteria, certain fungi, spirochetes, protozoa, and enveloped viruses [62], and they kill bacteria by distinctive membrane disruptive mechanisms [140]. Human neutrophil α-defensins, in concert with diverse effectors stored in azurophilic granules mediate nonoxidative microbial cell killing by sequential permeabilization of the outer and inner membranes following phagolysosomal fusion [61]. The peptides induce the formation of ion channels in lipid bilayers [61], and peptide-elicited effects are influenced by cell and membrane energetics [27]. As noted previously, in contrast to human HNP-3, a noncovalent dimer, rabbit neutrophil α-defensins (NPs), and α-defensins from rabbit kidney and mouse Paneth cells are monomeric in solution [39, 54, 87, 128], and these distinct molecular forms differ with regard to the mechanisms by which they lyse model membranes in vitro. Although dimeric HNP-2 forms stable, 20-Å multimeric pores after insertion in model phospholipid bilayers [143], rabbit NP-1 monomers create short-lived defects [48]. The mouse Paneth cell α-defensin cryptdin-4 (Crp4) exhibits strong interfacial binding to model phospholipid membranes [100], and it induces graded fluorophore leakage from model membranes in a manner that is dependent on peptide concentration and vesicular phospholipid composition [99, 100]. Thus, even though the tertiary structure and topology of α-defensins are conserved, microbial cell killing by α-defensins of differing primary structure may vary markedly in their membrane disruptive mechanisms. Functional analysis of Crp4 disulfide mutants has shown that the canonical α-defensin disulfide arrangement provides critical peptide protection during precursor activation, but it is not a determinant of α-defensin bactericidal activity [69]. Disulfide bond disruption does not induce loss of function: the in vitro bactericidal activities of certain α-defensin disulfide variants with disordered structures equal or exceed those of the native Crp4 peptide. However, although native Crp4 is completely resistant to proteolysis by matrix metalloproteinase-7, the mouse pro-α-defensin convertase, all disulfide variants of Crp4 were cleaved extensively, eliminating peptide bactericidal activity [69]. Studies of model membrane interactions with RTD-1 and its open chain analog, oRTD-1, have shown that RTD-1 displays increased selectivity for anionic phospholipids rather than zwitterionic lipids. RTD-1 and oRTD-1 induced similar effects on vesicles, although cyclic RTD-1 induced stabilized lipid-peptide domains more efficiently than oRTD-1 [1].

6
Sites of Defensin Expression

6.1
β-Defensins

β-Defensins are widely expressed by varied epithelia, in bovine leukocytes, and certain β-defensin genes are expressed constitutively, while others may be inducible depending on the site of expression. For example, β-defensins of cattle tracheal and lingual epithelia [22] are constitutively expressed but also inducible by LPS or by wounding [23, 103]. Approximately 30 β-defensin genes have been identified in the human genome, and approximately 45 have been identified in the mouse [102, 106]. Many human β-defensin genes are adjacent to the α-defensin locus on 8p23, but additional novel β-defensin genes are located on chromosomes 6 and 20, and many β-defensin ORFs on chromosome 20 contain long C-terminal extensions [106]. In situ hybridization studies reveal that several of these genes are expressed at distinct sites along the epididymis [95], and evidence suggests a role for these β-defensins in sperm maturation [147, 149, 178]. hBD-2 and hBD-3 mRNAs are differentially expressed in numerous tissues, and the peptide levels are markedly increased in bronchoalveolar inflammation [40–42, 96] and in certain skin diseases such as psoriasis [37]. In contrast to α- and θ-defensins, β-defensin expression is inducible at the transcriptional level in many tissues, often involving a response to TLR-mediated production of proinflammatory cytokines.

6.2
α-Defensins

α-Defensins were identified initially as antimicrobial activities in cytoplasmic granules of neutrophils [154–156], and subsequent biochemical analyses showed that the molecules were cationic peptides with a tridisulfide array [27]. α-Defensins have been isolated from primate leukocytes and neutrophils of several rodents including rats, rabbits, guinea pigs, and hamsters [115]. Myeloid α-defensin mRNAs are expressed almost exclusively in the bone marrow, where they occur at highest levels in promyelocytes and at lower levels in myeloblasts and myelocytes [148]. Although neutrophils contain high levels of α-defensin peptides, defensin mRNAs are degraded during neutrophil differentiation. In contrast, circulating monocytes contain both α-defensin mRNAs and peptides [131, 132, 148].

Species may differ regarding the leukocyte lineages that express α-defensins, as exemplified by the appreciable levels of α-defensins in rabbit alveolar macrophages [28, 59, 108] and the curious lack of α-defensins in mouse neutrophils [24]. Human monocytes [68] and NK cells [14] express myeloid α-defensins HNP 1–3 constitutively, and microarray experiments also have provided evidence of HNP expression in NK cells [77]. Human CD56+ lymphocytes accumulate HNPs and the human cathelicidin LL-37 [2], and CD56+ NK cell exposure to *Klebsiella* species' outer membrane protein A and to *Escherichia coli* flagellin induces IFN-γ production and NK cell release of α-defensins [14]. θ-Defensin expression in rhesus monkeys has only been reported in neutrophils and monocytes [132].

Enteric α-defensins are components of Paneth cell secretory granules [20, 83, 84, 91, 111], and expression of other AMP classes, β-defensins, or cathelicidins, for example, has not been detected in the small bowel. In human, rat, mouse, and monkey small intestine, α-defensin transcripts have been localized exclusively to Paneth cells and the peptides to Paneth cell secretory granules [19, 55, 84, 91, 111]. Paneth cells from mice and rhesus macaques express numerous α-defensin peptide isoforms, yet only two Paneth cell α-defensins, HD5 and HD6, exist in humans [55, 56, 80, 85, 129]. Paneth cell α-defensins have been recovered from luminal rinses of human ileum [34] and washings of mouse jejunum and ileum [84, 111].

In humans and mice, Paneth cell α-defensin genes may be expressed in reproductive epithelium, as shown by the presence of HD5 in female genital tract epithelia [93, 127] and HD5 mRNA in normal vagina, ectocervix, endo-cervix, endometrium, and fallopian tube specimens. The highest endometrial HD5 levels occurred during the early secretory phase of the menstrual cycle, localized to the upper half of the stratified squamous epithelium of the vagina and ectocervix [93]. In endocervix, endometrium, and fallopian tube, HD5 was found in apically oriented granules and on the apical surface of some columnar epithelial cells. In addition, Paneth cell α-defensins have been de-tected in human oropharyngeal mucosa [26], and mouse Crps have been immunolocalized to Sertoli cells and Leydig cells of the testis [18, 36]. Rabbit kidney appears to express a subset of α-defensins distinct from the rabbit myeloid peptides [6, 144]. α-Defensins occur only in Paneth cells in the small intestine, despite their expression at these other sites [5, 83, 91, 111].

α-Defensin mRNAs accumulate in small bowel during intestinal develop-ment, coinciding with differentiation of the Paneth cell lineage during intesti-nal crypt ontogeny [13, 21, 70]. Four primary epithelial cell lineages, includ-ing absorptive enterocytes, goblet cells, enteroendocrine cells, and Paneth cells, derive from mitotically active progenitors that occupy the crypts of Lieberkühn [35, 119]. Paneth cells differentiate as they descend to the base of the crypt from the proliferative compartment [10, 15], and Paneth cells populate the base of small intestinal crypts in most mammals [92]. The HD5 and HD6 genes are expressed in the developing fetus as early as 13.5 weeks of gestation, as detected by RT-PCR, and their appearance coincides with Paneth cell ontogeny during gestation [70, 97]. The expression of HD5 and HD6 genes in utero shows that human Paneth cell α-defensin gene activation is independent of infectious stimuli, as was found in mice [5, 13]. In mice, certain Crp genes are expressed at a low level in the maturing epithelial cell monolayer of fetal and newborn small intestine in goblet-like cells that are scattered throughout the epithelial sheet [21]. Normally, the distribution of Paneth cells is restricted to the small intestine, but they may appear ectopically during episodes of inflammation and in Barrett's esophagus, Crohn's disease, gastritis, and ulcerative colitis [85]. The large, eosinophilic secretory granules of Paneth cells are rich in lysozyme [89], secretory phospholipase A_2 (sPLA$_2$), angiogenin-4, and α-defensins [92], all potent microbicides. Additional Paneth cell gene products have been identified using microarray approaches [124–126]. Because Paneth cells secrete many peptides and proteins with antimicrobial or immunoregulatory roles, hypotheses that Paneth cells are effectors of innate immunity have been subjected to analysis [92].

In mice, two α-defensin-related gene families that code for unusual, non-defensin, cysteine-rich sequence peptides, CRS1C and CRS4C, are also expressed specifically by Paneth cells [47, 51, 66, 79, 81]. Overall, levels of CRS and Crp mRNAs in Paneth cells are approximately the same, and the concentrations of both mRNA subfamilies increase coordinately during postnatal Paneth cell differentiation [79]. Like the α-defensins [52], CRS peptides are Paneth cell-specific gene products and coded by linked genes with conserved two-exon structures. The first exons of CRS genes and Crp genes are highly conserved in that their nucleotide sequences and deduced prepro regions are almost identical; however, the peptide-coding CRS gene second exons diverge markedly from the α-defensin-coding Crp genes [51, 66, 79]. Crp, CRS1C, and CRS4C precursors have nearly identical signal peptides and proregions, but the deduced CRS1C and CRS4C peptides contain 11 or 9 Cys residues, respectively, in a series of nine [C]-[X]-[Y] (CRS1C) or seven [C]-[P]-[Y] (CRS4C) triplet repeats (Fig. 2) [51, 79]. Like the α-defensins, CRS1C and CRS4C are cationic and cysteine-rich, and several isoforms of each occur in mouse Paneth cells [51]. The Cys distribution in CRS peptides differs markedly from the canonical α-defensin Cys spacing and tridisulfide array. To date, seven CRS1C isoform mRNAs (not shown), and eight CRS4C mRNA variants have been identified in C3H/HeJ, 129/J, and outbred Swiss mice [47, 51]. CRS1C, but not CRS4C, mRNA has been detected in mouse testis [18]. Several CRS4C isoforms were identified genetically and at the peptide level, and CRS4C monomers and dimers have been reported in extracts of mouse small intestine [47]. Homodimers and heterodimeric forms of CRS4C isoforms were detected in protein extracts, and both CRS4C forms behaved similarly in in vitro bactericidal peptide assays [47]. The biochemistry of CRS peptides and their role in the small bowel are largely unknown, but their genetic relation to α-defensins, their high level of Paneth cell expression, and their in vitro bactericidal activities implicate them in host defense.

7
Regulation of α-Defensin Biosynthesis

7.1
Transcriptional Regulation

α-Defensin genes map to 8p21–8pter through 8p23 in human are syntenic in mice [78, 88, 123] and are expressed predominantly in myeloid cells or in Paneth cells [113]. Myeloid α-defensin mRNAs are expressed almost exclusively in the bone marrow, where they are found at highest levels in promyelocytes and at lower levels in myeloblasts and myelocytes [148]. Enteric α-defensins occur exclusively in Paneth cells in normal small bowel [20, 83, 84, 91, 111]. Myeloid and Paneth cell α-defensin genes differ in that genes expressed in cells of myeloid origin consist of three exons, but those expressed in Paneth cells have two [7, 30, 52, 55, 56, 67]. In Paneth cell α-defensin genes, the 5′-untranslated region and the preprosegment are coded by exon 1 (Fig. 1, top), but an additional intron interrupts the 5′-untranslated region of myeloid α-defensin gene transcripts [82]. Regardless of the site of expression, the 5′- distal exon codes for the α-defensin peptide moiety [7, 31, 52, 55, 62, 67]. Studies of cis-acting DNA elements involved in myeloid α-defensin gene regulation identified a short region containing a CAAT box and a putative polyoma enhancer-binding protein 2/core-binding factor (PEBP2/CBF) site approximately 90 bp upstream of the transcription start site as important for defensin promoter activity [145]. In electrophoretic mobility shift assays, PEBP2/CBF or a PEBP2/CBF-related protein formed specific protein–DNA interactions, identifying trans- and cis- regulatory elements of rat α-defensin gene expression in high α-defensin-expressing promyelocytic cells [145]. Despite these advances, the mechanisms that specify myeloid α-defensin gene expression still remain obscure.

Paneth cell differentiation is determined by continuous Wnt signaling via the frizzled-5 receptor, and transcription of the Paneth cell α-defensin genes is mediated by β-catenin/TCF-4 recognition sites in the 5′- upstream regions of the gene transcription start sites as well as upstream of the gene coding for matrix metalloproteinase-7 (MMP-7), the mouse α-defensin convertase [3, 38, 90, 138]. Monocytes and NK cells also contain α-defensin

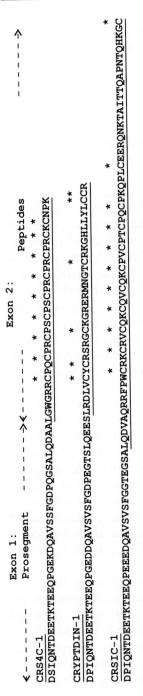

```
                Exon 1:                              Exon 2:
                Prosegment      - - - - - ->< - - - - - -       Peptides
<- - - - - - -                                     *   *   *   *   *   *
CRS4C-1                                       *   *   *   *   *   *   *   *
DSIQNTDEETKTEEQPGEKDQAVSSFGDPQGSALQDAALGWRRCPQCPRCPSCPSCPRCPRCPRCKCNPK

CRYPTDIN-1                    *       *       *                         **
DPIQNTDEETKTEEQPGEDDQAVSVSFGDPEGTSLQEESLRDLVCYCRSRGCKGRERMNGTCRKGHLLYLCCR

                                                                        *
CRS1C-1                       *       *   *   *   *   *   *   *   *   *
DPIQNTDEETKTEEQPEEEDQAVSVSFGGTEGSALQDVAQRRFPWCRKCRVCQKCRVCQKCPVCPTCPQCPKQPLCEERQNKTAITTQAPNTQHKGC
```

Fig. 2 Alignment of CRS1C-1, CRS4C-1, and Crp1 precursors. The deduced proforms of the precursors are aligned, beginning at residue position 20, following removal of the signal sequence. Natural proCrp1 has been isolated and the N-terminus confirmed [118]. *Asterisks above lines of sequence denote Cys residue positions, and the determined or deduced products of these precursors are underlined*

```
Pro-Crp1:
                                           43 44             53 54    58 59
20
DPIQNTDEETKTEEQPGEDDQAVS↓VSFGDPEGTS↓LQEES↓LRDLVCYCRSRGCKGRERMNGTCRKGHLLYLCCR

Pro-HD5:
                                      55 56            62 63
20
ESLQERADEATTQKQSGEGNQDLAISFAGNGLSALR↓TSGSQAR↓ATCYYRTGRCATRESLSGVCEISGRLYRLCCR
```

Fig. 3 Distinct mechanisms of pro-α-defensin activation in mouse and human Paneth cells. Human pro-HD5 is activated by anionic and meso-trypsins, which cleave the precursor at the sites depicted by *arrows* in pro-HD5. In contrast, MMP-7 cleaves all known mouse Paneth cell pro-α-defensins intracellularly at the sites depicted by *arrows* in pro-Crp1. In both sequences, proregions are underlined and the mature α-defensin peptides are italicized. NB: Mouse proCrps are resistant to in vitro proteolysis by trypsins (not shown)

mRNAs and peptides [14, 131, 132, 148], but regulatory elements equivalent to β-catenin/TCF-4 sites in Paneth cell α-defensin genes remain to be found in myeloid α-defensin gene promoters.

7.2
Posttranslational Activation

Rhesus θ-defensin peptides assemble from two distinct precursor molecules with each hemi-precursor contributing a nine-amino acid moiety to the final RTD-1 peptide [132], although the molecular mechanisms that catalyze or facilitate θ-defensin assembly in primates are not understood [132]. Rhesus pro-RTDs are products of different genes that resemble the three-exon myeloid α-defensin genes, except that they are truncated by stop codons in exon 3. In addition to heterodimeric RTD-1, homodimeric θ-defensins RTD-2 and 3 also have been isolated from monkey neutrophils [65, 135]. α-Defensin gene mutations that gave rise to the θ-defensin genes (*DEFT*) apparently arose in Old World monkeys, because rhesus *DEFT* homologs have not been found in prosimians or in New World monkeys [64]. Humans, chimpanzees, and gorillas lack θ-defensins, because the *DEFT* genes of those species harbor mutation(s) that create premature stop codons in the prepro regions of the precursors. However, at least one mutant human *DEFT* gene still is actively transcribed, and its nonfunctional mRNA accumulates to high abundance at several sites of expression [17].

The biosynthesis of Paneth cell pro-α-defensins involves posttranslational proteolytic activation. Although the enzymes that mediate pro-α-defensin processing in myeloid and epithelial cells are likely to differ, the overall processing schemes are similar in that all are processed from proforms by specific proteolytic cleavage steps. Both myeloid and Paneth cell α-defensins derive from approximately 10-kDa prepropeptides that contain canonical signal sequences, usually acidic proregions, and a approximately 3.5-kDa mature α-defensin peptide in the C-terminal portion of the precursor. In human and rabbit neutrophils, α-defensins are almost fully processed by primary cleavage steps that leave major intermediates of 75 and 56 amino acids and the mature HNP-1 peptide [29, 73, 136, 137].

The details of human and mouse Paneth cell α-defensin processing provide interesting contrasts and comparisons (Fig. 3). Human Paneth cells store unprocessed α-defensin precursors, e.g., proHD5$_{(20-94)}$ [34], which are cleaved after secretion by anionic and meso isoforms of trypsin at R62↓A63 to produce the predominant form of HD5 (Fig. 3) [34]. Additional, perhaps alternative, pro-HD5 processing sites include an HD5$_{(37-94)}$ variant that was isolated from secretions of human small intestinal crypts stimulated with carbamyl choline [20]. Trypsin may also process monkey Paneth cell pro-α-defensins after secretion [129], because all rhesus macaque Paneth cell α-defensin precursors deduced from cDNA sequences contain canonical trypsin cleavage sites at residue position 62, the junction of the proregion and the α-defensin N-terminus [129]. The rhesus Paneth cell α-defensin RED-4 has alternative N-termini that apparently result from activating cleavage steps at Arg62↓Arg63 and at Arg63↓Thr64.

In contrast to trypsin-mediated α-defensin activation of human Paneth cell α-defensins, mouse pro-α-defensin processing is mediated by MMP-7 [142]. In mice, MMP-7 is the sole pro-α-defensin activating convertase in Paneth cells [5, 142], and Paneth cells are the only mouse small intestinal epithelial cell lineage that expresses MMP-7, where it co-localizes with α-defensins and pro-α-defensins in dense secretory granules [141]. MMP-7 is essential for posttranslational processing of mouse pro-α-defensins [142], producing active 3.5-kDa α-defensins by cleaving precursors in vitro at conserved sites in the proregion and at the junction of the propeptide and the α-defensin [118] (Fig. 3). Enteric innate immunity to oral infection is defective in MMP-7-null mice, which are impaired in clearing of orally-administered E. coli from small intestine, and they are approximately ten times more susceptible than wild-type mice to systemic infection by S. enterica serovar Typhimurium via the oral route [142]. In vivo, mouse pro-α-defensin activation takes place intracellularly and prior to secretion [5], and MMP-7 gene disruption ablates proCrp processing and impairs innate immunity to oral bacterial infection [142]. MMP-7 proteolysis relieves the inhibition of these activities imposed by the covalently linked proregion.

8
α-Defensins Confer Innate Immunity

α-Defensin secretion by Paneth cells constitutes a key source of AMP activity in the crypt lumen [4, 34]. Normally, Paneth cells occur only in small bowel, occupying the base of the crypts of Lieberkühn in most mammals, although they may appear in esophagus, stomach, pancreas, and colon during episodes of inflammation [85]. In most mammals, Paneth cells populate every small intestinal crypt, ranging from 2–3 cells per crypt in proximal small intestine to approximately 10 cells per crypt distally [8]. The Paneth cell lineage differentiation program does not require contact with bacteria or bacterial antigens [5, 25], although some Paneth cell gene products, e.g., angiogenin-4, may be upregulated when germ-free mice are colonized [44].

The contribution of Paneth cell α-defensins to enteric mucosal immunity is evident from the phenotype of mice transgenic for the human Paneth cell α-defensin, HD5 (tg-HD5) [98]. These tg-HD5 mice express a human mini-gene, consisting of 3 kb of genomic DNA encompassing the two exons of the *DEFA5* HD5 gene, specifically in Paneth cells and at levels comparable to endogenous cryptdin expression. Given that Paneth cell α-defensin gene transcription is mediated by β-catenin/TCF-4 consensus sequences in the regions 5'- upstream of the MMP-7 and Paneth cell α-defensin gene transcription start sites [3, 38, 90, 138], these findings provide evidence that the fidelity of HD5 transgene expression in tg-HD5 mice is directed by *cis*-acting TCF-4 sites conserved in the human Paneth cell α-defensin genes. When challenged with *S. enterica* serovar Typhimurium (*S. typhimurium*) by oral inoculation, the tg-HD5 mice were immune to infection; inocula that were lethal to the control mice caused little perceptible disease in tg-HD5 mice. The numbers of *S. typhimurium* recovered from the intestinal lumen of tg-HD5 mice were much lower than those recovered from the wild type mice, suggesting that transgenic expression of HD5 led to increased killing of the bacteria. Because unpublished comparisons of HD5 and mouse Crps have shown that their bactericidal activities are similar in vitro against *S. typhimurium* (H. Tanabe, C.L. Bevins, A.J. Ouellette, unpublished data), immunity to *Salmonella* may result from interactions more complex than the introduction of a superior bactericidal peptide to the milieu of the mouse intestinal lumen. Perhaps the presence of HD5 in mouse Paneth cell secretions contributes to changes in the small intestinal microflora, creating a less hospitable environment for *S. typhimurium*.

9
Consequences of Paneth Cell Defects on Host Defense

In mice, crypt intermediate or granulo-mucous cells accumulate Paneth cell gene products in electron dense granules under conditions of disrupted crypt cell biology. For example, in mice expressing attenuated diphtheria toxin A fragment or SV40 large T antigen transgenes under control of the Crp2 gene promoter, a transient Paneth cell deficiency occurs in crypts of 3- to 4-week-old transgenic mice [33]. Intermediate cells and granule-containing goblet cells increase in number and accumulate electron-dense secretory granules that contain both Crps and $sPLA_2$ during the Paneth-cell-deficient period [33]. Normally, Paneth cells are found only in the small intestine, but in Barrett's esophagus, Crohn's disease, gastritis, and ulcerative colitis the cells appear ectopically along with lineage-specific markers [85]. Collectively, these disruptions of crypt cell biology apparently modify Wnt-regulated epithelial lineage differentiation programs, resulting in increased production of Paneth cell antimicrobial peptides for secretion.

Defects in Paneth cell physiology and α-defensin biology may predispose individuals to infectious challenges and perhaps to inflammatory bowel diseases. For example and as noted previously, MMP-7 null mice are deficient in functional α-defensins due to defective pro-Crp processing, and their host defense to oral enteric infection is compromised in vivo [142]. Cystic fibrosis mice, null for the cystic fibrosis transmembrane conductance regulator, accumulate undissolved Paneth cell secretory granules in mucus-occluded crypts, resulting in decreased resistance to bacterial colonization of the small intestine, bacterial overgrowth of small bowel and mucus colonization, and an overwhelming predominance of Enterobacteriaceae in the microflora [16, 76]. In human small intestinal epithelium, NOD2, an intracellular peptidoglycan pattern recognition receptor and a susceptibility gene for Crohn's ileitis [104], is expressed in the Paneth cell [57]. Although proinflammatory cytokine levels are unaffected by NOD2 status, ileal HD-5 and HD-6 mRNA levels decline in Crohn's ileum, especially in patients with NOD2 mutations [139]. Possibly, NOD2 mutations alter human Paneth cell biology and the composition of their secretions, leading to diminished innate mucosal immunity. Thus, defects in activation, secretion, dissolution, and function of Paneth cell α-defensins all may affect crypt innate immunity adversely. Innate mucosal immunity may become deficient due to altered α-defensin levels or defects in Paneth cell or crypt cell biology that impair delivery of functional α-defensins and additional AMPs to the microenvironment of the crypt lumen.

Acknowledgements Supported by NIH Grant DK044632, the Human Frontiers Science Program, and the United States-Israel Binational Science Foundation.

References

1. Abuja PM, Zenz A, Trabi M, Craik DJ, Lohner K (2004) The cyclic antimicrobial peptide RTD-1 induces stabilized lipid-peptide domains more efficiently than its open-chain analogue. FEBS Lett 566:301–306
2. Agerberth B, Charo J, Werr J, Olsson B, Idali F, Lindbom L, Kiessling R, Jornvall H, Wigzell H, Gudmundsson GH (2000) The human antimicrobial and chemotactic peptides LL-37 and alpha-defensins are expressed by specific lymphocyte and monocyte populations. Blood 96:3086–3093
3. Andreu P, Colnot S, Godard C, Gad S, Chafey P, Niwa-Kawakita M, Laurent-Puig P, Kahn A, Robine S, Perret C, Romagnolo B (2005) Crypt-restricted proliferation and commitment to the Paneth cell lineage following Apc loss in the mouse intestine. Development 132:1443–1451

4. Ayabe T, Satchell DP, Wilson CL, Parks WC, Selsted ME, Ouellette AJ (2000) Secretion of microbicidal alpha-defensins by intestinal Paneth cells in response to bacteria. Nat Immunol 1:113–118

5. Ayabe T, Satchell DP, Pesendorfer P, Tanabe H, Wilson CL, Hagen SJ, Ouellette AJ (2002) Activation of Paneth cell alpha-defensins in mouse small intestine. J Biol Chem 277:5219–5228

6. Bateman A, MacLeod RJ, Lembessis P, Hu J, Esch F, Solomon S (1996) The isolation and characterization of a novel corticostatin/defensin-like peptide from the kidney. J Biol Chem 271:10654–10659

7. Bevins CL, Jones DE, Dutra A, Schaffzin J, Muenke M (1996) Human enteric defensin genes: chromosomal map position and a model for possible evolutionary relationships. Genomics 31:95–106

8. Bevins CL, Martin-Porter E, Ganz T (1999) Defensins and innate host defence of the gastrointestinal tract. Gut 45:911–915

9. Bevins CL (2004) The Paneth cell and the innate immune response. Curr Opin Gastroenterol 20:572–580

10. Bjerknes M, Cheng H (1981) The stem-cell zone of the small intestinal epithelium. I. Evidence from Paneth cells in the adult mouse. Am J Anat 160:51–63

11. Boman HG (1995) Peptide antibiotics and their role in innate immunity. Annu Rev Immunol 13:61–92

12. Brogden KA (2005) Antimicrobial peptides: pore formers or metabolic inhibitors in bacteria? Nat Rev Microbiol 3:238–250

13. Bry L, Falk P, Huttner K, Ouellette A, Midtvedt T, Gordon JI (1994) Paneth cell differentiation in the developing intestine of normal and transgenic mice. Proc Natl Acad Sci U S A 91:10335–10339

14. Chalifour A, Jeannin P, Gauchat JF, Blaecke A, Malissard M, N'Guyen T, Thieblemont N, Delneste Y (2004) Direct bacterial protein PAMPs recognition by human NK cells involves TLRs and triggers alpha-defensin production. Blood 104:1778–1783

15. Cheng H (1974) Origin, differentiation and renewal of the four main epithelial cell types in the mouse small intestine. IV. Paneth cells. Am J Anat 141:521–535

16. Clarke LL, Gawenis LR, Bradford EM, Judd LM, Boyle KT, Simpson JE, Shull GE, Tanabe H, Ouellette AJ, Franklin CL, Walker NM (2004) Abnormal Paneth cell granule dissolution and compromised resistance to bacterial colonization in the intestine of CF mice. Am J Physiol Gastrointest Liver Physiol 286:G1050–G1058

17. Cole AM, Hong T, Boo LM, Nguyen T, Zhao C, Bristol G, Zack JA, Waring AJ, Yang OO, Lehrer RI (2002) Retrocyclin: a primate peptide that protects cells from infection by T- and M-tropic strains of HIV-1. Proc Natl Acad Sci U S A 99:1813–1818

18. Com E, Bourgeon F, Evrard B, Ganz T, Colleu D, Jegou B, Pineau C (2003) Expression of antimicrobial defensins in the male reproductive tract of rats, mice, and humans. Biol Reprod 68:95–104

19. Condon MR, Viera A, D'Alessio M, Diamond G (1999) Induction of a rat enteric defensin gene by hemorrhagic shock. Infect Immun 67:4787–4793

20. Cunliffe RN, Rose FR, Keyte J, Abberley L, Chan WC, Mahida YR (2001) Human defensin 5 is stored in precursor form in normal Paneth cells and is expressed by some villous epithelial cells and by metaplastic Paneth cells in the colon in inflammatory bowel disease. Gut 48:176–185

21. Darmoul D, Brown D, Selsted ME, Ouellette AJ (1997) Cryptdin gene expression in developing mouse small intestine. Am J Physiol 272:G197–G206

22. Diamond G, Zasloff M, Eck H, Brasseur M, Maloy WL, Bevins CL (1991) Tracheal antimicrobial peptide, a cysteine-rich peptide from mammalian tracheal mucosa: peptide isolation and cloning of a cDNA. Proc Natl Acad Sci U S A 88:3952–3956

23. Diamond G, Bevins CL (1994) Endotoxin upregulates expression of an antimicrobial peptide gene in mammalian airway epithelial cells. Chest 105:51S–52S

24. Eisenhauer PB, Lehrer RI (1992) Mouse neutrophils lack defensins. Infect Immun 60:3446–3447

25. Falk PG, Hooper LV, Midtvedt T, Gordon JI (1998) Creating and maintaining the gastrointestinal ecosystem: what we know and need to know from gnotobiology. Microbiol Mol Biol Rev 62:1157–1170

26. Frye M, Bargon J, Dauletbaev N, Weber A, Wagner TO, Gropp R (2000) Expression of human alpha-defensin 5 (HD5) mRNA in nasal and bronchial epithelial cells. J Clin Pathol 53:770–773

27. Ganz T, Selsted ME, Szklarek D, Harwig SS, Daher K, Bainton DF, Lehrer RI (1985) Defensins. Natural peptide antibiotics of human neutrophils. J Clin Invest 76:1427–1435

28. Ganz T, Sherman MP, Selsted ME, Lehrer RI (1985) Newborn rabbit alveolar macrophages are deficient in two microbicidal cationic peptides, MCP-1 and MCP-2. Am Rev Respir Dis 132:901–904

29. Ganz T, Liu L, Valore EV, Oren A (1993) Posttranslational processing and targeting of transgenic human defensin in murine granulocyte, macrophage, fibroblast, and pituitary adenoma cell lines. Blood 82:641–650

30. Ganz T (1994) Biosynthesis of defensins and other antimicrobial peptides. Ciba Found Symp 186:62–71; discussion 71–76

31. Ganz T (1999) Defensins and host defense. Science 286:420–421

32. Ganz T (2003) Defensins: antimicrobial peptides of innate immunity. Nat Rev Immunol 3:710–720

33. Garabedian EM, Roberts LJ, McNevin MS, Gordon JI (1997) Examining the role of Paneth cells in the small intestine by lineage ablation in transgenic mice. J Biol Chem 272:23729–23740

34. Ghosh D, Porter E, Shen B, Lee SK, Wilk D, Drazba J, Yadav SP, Crabb JW, Ganz T, Bevins CL (2002) Paneth cell trypsin is the processing enzyme for human defensin-5. Nat Immunol 3:583–590

35. Gordon JI, Hermiston ML (1994) Differentiation and self-renewal in the mouse gastrointestinal epithelium. Curr Opin Cell Biol 6:795–803

36. Grandjean V, Vincent S, Martin L, Rassoulzadegan M, Cuzin F (1997) Antimicrobial protection of the mouse testis: synthesis of defensins of the cryptdin family. Biol Reprod 57:1115–1122

37. Harder J, Schroder JM (2005) Psoriatic scales: a promising source for the isolation of human skin-derived antimicrobial proteins. J Leukoc Biol 77:476–486

38. He XC, Zhang J, Tong WG, Tawfik O, Ross J, Scoville DH, Tian Q, Zeng X, He X, Wiedemann LM, Mishina Y, Li L (2004) BMP signaling inhibits intestinal stem cell self-renewal through suppression of Wnt-beta-catenin signaling. Nat Genet 36:1117–1121

39. Hill CP, Yee J, Selsted ME, Eisenberg D (1991) Crystal structure of defensin HNP-3, an amphiphilic dimer: mechanisms of membrane permeabilization. Science 251:1481–1485

40. Hiratsuka T, Nakazato M, Date Y, Ashitani J, Minematsu T, Chino N, Matsukura S (1998) Identification of human beta-defensin-2 in respiratory tract and plasma and its increase in bacterial pneumonia. Biochem Biophys Res Commun 249:943–947

41. Hiratsuka T, Nakazato M, Ihi T, Minematsu T, Chino N, Nakanishi T, Shimizu A, Kangawa K, Matsukura S (2000) Structural analysis of human beta-defensin-1 and its significance in urinary tract infection. Nephron 85:34–40

42. Hiratsuka T, Mukae H, Iiboshi H, Ashitani J, Nabeshima K, Minematsu T, Chino N, Ihi T, Kohno S, Nakazato M (2003) Increased concentrations of human beta-defensins in plasma and bronchoalveolar lavage fluid of patients with diffuse panbronchiolitis. Thorax 58:425–430

43. Hoffmann JA (2004) Primitive immune systems. Immunol Rev 198:5–9

44. Hooper LV, Stappenbeck TS, Hong CV, Gordon JI (2003) Angiogenins: a new class of microbicidal proteins involved in innate immunity. Nat Immunol 4:269–273

45. Hoover DM, Rajashankar KR, Blumenthal R, Puri A, Oppenheim JJ, Chertov O, Lubkowski J (2000) The structure of human beta-defensin-2 shows evidence of higher order oligomerization. J Biol Chem 275:32911–32918

46. Hoover DM, Chertov O, Lubkowski J (2001) The structure of human beta-defensin-1. New insights into structural properties of beta-defensins. J Biol Chem 276:39021–29026

47. Hornef MW, Putsep K, Karlsson J, Refai E, Andersson M (2004) Increased diversity of intestinal antimicrobial peptides by covalent dimer formation. Nat Immunol 5:836–843

48. Hristova K, Selsted ME, White SH (1996) Interactions of monomeric rabbit neutrophil defensins with bilayers: comparison with dimeric human defensin HNP-2. Biochemistry 35:11888–11894

49. Hristova K, Selsted ME, White SH (1997) Critical role of lipid composition in membrane permeabilization by rabbit neutrophil defensins. J Biol Chem 272:24224–24233

50. Huang HW (1999) Peptide-lipid interactions and mechanisms of antimicrobial peptides. Novartis Found Symp 225:188–200; discussion 200–206

51. Huttner KM, Ouellette AJ (1994) A family of defensin-like genes codes for diverse cysteine-rich peptides in mouse Paneth cells. Genomics 24:99–109

52. Huttner KM, Selsted ME, Ouellette AJ (1994) Structure and diversity of the murine cryptdin gene family. Genomics 19:448–453

53. Iimura M, Gallo RL, Hase K, Miyamoto Y, Eckmann L, Kagnoff MF (2005) Cathelicidin mediates innate intestinal defense against colonization with epithelial adherent bacterial pathogens. J Immunol 174:4901–4907

54. Jing W, Hunter HN, Tanabe H, Ouellette AJ, Vogel HJ (2004) Solution structure of cryptdin-4, a mouse Paneth cell alpha-defensin. Biochemistry 43:15759–15766

55. Jones DE, Bevins CL (1992) Paneth cells of the human small intestine express an antimicrobial peptide gene. J Biol Chem 267:23216–23225

56. Jones DE, Bevins CL (1993) Defensin-6 mRNA in human Paneth cells: implications for antimicrobial peptides in host defense of the human bowel. FEBS Lett 315:187–192

57. Lala S, Ogura Y, Osborne C, Hor SY, Bromfield A, Davies S, Ogunbiyi O, Nunez G, Keshav S (2003) Crohn's disease and the NOD2 gene: a role for Paneth cells. Gastroenterology 125:47–57

58. Lee PH, Ohtake T, Zaiou M, Murakami M, Rudisill JA, Lin KH, Gallo RL (2005) Expression of an additional cathelicidin antimicrobial peptide protects against bacterial skin infection. Proc Natl Acad Sci U S A 102:3750–3755

59. Lehrer RI, Selsted ME, Szklarek D, Fleischmann J (1983) Antibacterial activity of microbicidal cationic proteins 1 and 2, natural peptide antibiotics of rabbit lung macrophages. Infect Immun 42:10–14

60. Lehrer RI, Ganz T, Selsted ME (1988) Oxygen-independent bactericidal systems. Mechanisms and disorders. Hematol Oncol Clin North Am 2:159–169

61. Lehrer RI, Barton A, Daher KA, Harwig SS, Ganz T, Selsted ME (1989) Interaction of human defensins with *Escherichia coli*. Mechanism of bactericidal activity. J Clin Invest 84:553–561

62. Lehrer RI, Ganz T (2002) Defensins of vertebrate animals. Curr Opin Immunol 14:96–102

63. Lehrer RI, Ganz T (2002) Cathelicidins: a family of endogenous antimicrobial peptides. Curr Opin Hematol 9:18–22

64. Lehrer RI (2004) Primate defensins. Nat Rev Microbiol 2:727–738

65. Leonova L, Kokryakov VN, Aleshina G, Hong T, Nguyen T, Zhao C, Waring AJ, Lehrer RI (2001) Circular minidefensins and posttranslational generation of molecular diversity. J Leukoc Biol 70:461–464

66. Lin MY, Munshi IA, Ouellette AJ (1992) The defensin-related murine CRS1C gene: expression in Paneth cells and linkage to Defcr, the cryptdin locus. Genomics 14:363–368

67. Linzmeier R, Michaelson D, Liu L, Ganz T (1993) The structure of neutrophil defensin genes. FEBS Lett 321:267–273

68. Mackewicz CE, Yuan J, Tran P, Diaz L, Mack E, Selsted ME, Levy JA (2003) Alpha-defensins can have anti-HIV activity but are not CD8 cell anti-HIV factors. Aids 17:F23–F32

69. Maemoto A, Qu X, Rosengren KJ, Tanabe H, Henschen-Edman A, Craik DJ, Ouellette AJ (2004) Functional analysis of the alpha-defensin disulfide array in mouse cryptdin-4. J Biol Chem 279:44188–44196

70. Mallow EB, Harris A, Salzman N, Russell JP, DeBerardinis RJ, Ruchelli E, Bevins CL (1996) Human enteric defensins. Gene structure and developmental expression. J Biol Chem 271:4038–4045

71. Mambula SS, Simons ER, Hastey R, Selsted ME, Levitz SM (2000) Human neutrophil-mediated nonoxidative antifungal activity against Cryptococcus neoformans. Infect Immun 68:6257–6264

72. Matsuzaki K, Mitani Y, Akada KY, Murase O, Yoneyama S, Zasloff M, Miyajima K (1998) Mechanism of synergism between antimicrobial peptides magainin 2 and PGLa. Biochemistry 37:15144–15153

73. Michaelson D, Rayner J, Couto M, Ganz T (1992) Cationic defensins arise from charge-neutralized propeptides: a mechanism for avoiding leukocyte autocytotoxicity? J Leukoc Biol 51:634–639

74. Newman SL, Gootee L, Gabay JE, Selsted ME (2000) Identification of constituents of human neutrophil azurophil granules that mediate fungistasis against Histoplasma capsulatum. Infect Immun 68:5668–5672

75. Nizet V, Ohtake T, Lauth X, Trowbridge J, Rudisill J, Dorschner RA, Pestonjamasp V, Piraino J, Huttner K, Gallo RL (2001) Innate antimicrobial peptide protects the skin from invasive bacterial infection. Nature 414:454–457

76. Norkina O, Burnett TG, De Lisle RC (2004) Bacterial overgrowth in the cystic fibrosis transmembrane conductance regulator null mouse small intestine. Infect Immun 72:6040–6049

77. Obata-Onai A, Hashimoto S, Onai N, Kurachi M, Nagai S, Shizuno K, Nagahata T, Mathushima K (2002) Comprehensive gene expression analysis of human NK cells and CD8(+) T lymphocytes. Int Immunol 14:1085–1098

78. Ouellette AJ, Pravtcheva D, Ruddle FH, James M (1989) Localization of the cryptdin locus on mouse chromosome 8. Genomics 5:233–239

79. Ouellette AJ, Lualdi JC (1990) A novel mouse gene family coding for cationic, cysteine-rich peptides. Regulation in small intestine and cells of myeloid origin. J Biol Chem 265:9831–9837

80. Ouellette AJ, Hsieh MM, Nosek MT, Cano-Gauci DF, Huttner KM, Buick RN, Selsted ME (1994) Mouse Paneth cell defensins: primary structures and antibacterial activities of numerous cryptdin isoforms. Infect Immun 62:5040–5047

81. Ouellette AJ, Lauldi JC (1994) A novel gene family coding for cationic, cysteine-rich peptides. Regulation in mouse small intestine and cells of myeloid origin. J Biol Chem 269:18702

82. Ouellette AJ, Selsted ME (1996) Paneth cell defensins: endogenous peptide components of intestinal host defense. FASEB J 10:1280–1289

83. Ouellette AJ, Darmoul D, Tran D, Huttner KM, Yuan J, Selsted ME (1999) Peptide localization and gene structure of cryptdin 4, a differentially expressed mouse Paneth cell alpha-defensin. Infect Immun 67:6643–6651

84. Ouellette AJ, Satchell DP, Hsieh MM, Hagen SJ, Selsted ME (2000) Characterization of luminal Paneth cell alpha-defensins in mouse small intestine. Attenuated antimicrobial activities of peptides with truncated amino termini. J Biol Chem 275:33969–33973

85. Ouellette AJ, Bevins CL (2001) Paneth cell defensins and innate immunity of the small bowel. Inflamm Bowel Dis 7:43–50

86. Panyutich A, Shi J, Boutz PL, Zhao C, Ganz T (1997) Porcine polymorphonuclear leukocytes generate extracellular microbicidal activity by elastase-mediated activation of secreted proprotegrins. Infect Immun 65:978–985

87. Pardi A, Zhang XL, Selsted ME, Skalicky JJ, Yip PF (1992) NMR studies of defensin antimicrobial peptides. 2. Three-dimensional structures of rabbit NP-2 and human HNP-1. Biochemistry 31:11357–11364

88. Patil A, Hughes AL, Zhang G (2004) Rapid evolution and diversification of mammalian alpha-defensins as revealed by comparative analysis of rodent and primate genes. Physiol Genomics 20:1–11

89. Peeters T, Vantrappen G (1975) The Paneth cell: a source of intestinal lysozyme. Gut 16:553–558

90. Pinto D, Clevers H (2005) Wnt control of stem cells and differentiation in the intestinal epithelium. Exp Cell Res 306:357–363

91. Porter EM, Liu L, Oren A, Anton PA, Ganz T (1997) Localization of human intestinal defensin 5 in Paneth cell granules. Infect Immun 65:2389–2395

92. Porter EM, Bevins CL, Ghosh D, Ganz T (2002) The multifaceted Paneth cell. Cell Mol Life Sci 59:156–170

93. Quayle AJ, Porter EM, Nussbaum AA, Wang YM, Brabec C, Yip KP, Mok SC (1998) Gene expression, immunolocalization, and secretion of human defensin-5 in human female reproductive tract. Am J Pathol 152:1247–1258

94. Rice WG, Ganz T, Kinkade JM Jr, Selsted ME, Lehrer RI, Parmley RT (1987) Defensin-rich dense granules of human neutrophils. Blood 70:757–765

95. Rodriguez-Jimenez FJ, Krause A, Schulz S, Forssmann WG, Conejo-Garcia JR, Schreeb R, Motzkus D (2003) Distribution of new human beta-defensin genes clustered on chromosome 20 in functionally different segments of epididymis. Genomics 81:175–183

96. Ross DJ, Cole AM, Yoshioka D, Park AK, Belperio JA, Laks H, Strieter RM, Lynch JP 3rd, Kubak B, Ardehali A, Ganz T (2004) Increased bronchoalveolar lavage human beta-defensin type 2 in bronchiolitis obliterans syndrome after lung transplantation. Transplantation 78:1222–1224

97. Salzman NH, Polin RA, Harris MC, Ruchelli E, Hebra A, Zirin-Butler S, Jawad A, Martin Porter E, Bevins CL (1998) Enteric defensin expression in necrotizing enterocolitis. Pediatr Res 44:20–26

98. Salzman NH, Ghosh D, Huttner KM, Paterson Y, Bevins CL (2003) Protection against enteric salmonellosis in transgenic mice expressing a human intestinal defensin. Nature 422:522–526

99. Satchell DP, Sheynis T, Kolusheva S, Cummings JE, Vanderlick TK, Jelinek R, Selsted ME, Ouellette AJ (2003) Quantitative interactions between cryptdin-4 amino terminal variants and membranes. Peptides 24:1793–1803

100. Satchell DP, Sheynis T, Shirafuji Y, Kolusheva S, Ouellette AJ, Jelinek R (2003) Interactions of mouse Paneth cell alpha-defensins and alpha-defensin precursors with membranes: prosegment inhibition of peptide association with biomimetic membranes. J Biol Chem 278:13838–13846

101. Sawai MV, Jia HP, Liu L, Aseyev V, Wiencek JM, McCray PB Jr, Ganz T, Kearney WR, Tack BF (2001) The NMR structure of human beta-defensin-2 reveals a novel alpha-helical segment. Biochemistry 40:3810–3816

102. Scheetz T, Bartlett JA, Walters JD, Schutte BC, Casavant TL, McCray PB Jr (2002) Genomics-based approaches to gene discovery in innate immunity. Immunol Rev 190:137–145

103. Schonwetter BS, Stolzenberg ED, Zasloff MA (1995) Epithelial antibiotics induced at sites of inflammation. Science 267:1645–1648

104. Schreiber S, Rosenstiel P, Albrecht M, Hampe J, Krawczak M (2005) Genetics of Crohn disease, an archetypal inflammatory barrier disease. Nat Rev Genet 6:376–388

105. Schutte BC, McCray PB Jr (2002) Beta-defensins in lung host defense. Annu Rev Physiol 64:709–748

106. Schutte BC, Mitros JP, Bartlett JA, Walters JD, Jia HP, Welsh MJ, Casavant TL, McCray PB Jr (2002) Discovery of five conserved beta -defensin gene clusters using a computational search strategy. Proc Natl Acad Sci U S A 99:2129–2133
107. Scocchi M, Skerlavaj B, Romeo D, Gennaro R (1992) Proteolytic cleavage by neutrophil elastase converts inactive storage proforms to antibacterial bactenecins. Eur J Biochem 209:589–595
108. Selsted ME, Brown DM, DeLange RJ, Lehrer RI (1983) Primary structures of MCP-1 and MCP-2, natural peptide antibiotics of rabbit lung macrophages. J Biol Chem 258:14485–14489
109. Selsted ME, Harwig SS, Ganz T, Schilling JW, Lehrer RI (1985) Primary structures of three human neutrophil defensins. J Clin Invest 76:1436–1439
110. Selsted ME, Harwig SS (1989) Determination of the disulfide array in the human defensin HNP-2. A covalently cyclized peptide. J Biol Chem 264:4003–4007
111. Selsted ME, Miller SI, Henschen AH, Ouellette AJ (1992) Enteric defensins: antibiotic peptide components of intestinal host defense. J Cell Biol 118:929–936
112. Selsted ME, Tang YQ, Morris WL, McGuire PA, Novotny MJ, Smith W, Henschen AH, Cullor JS (1993) Purification, primary structures, and antibacterial activities of beta-defensins, a new family of antimicrobial peptides from bovine neutrophils. J Biol Chem 268:6641–6648
113. Selsted ME, Ouellette AJ (1995) Defensins in granules of phagocytic and nonphagocytic cells. Trends Cell Biol 5:114–119
114. Selsted ME, Tang YQ, Morris WL, McGuire PA, Novotny MJ, Smith W, Henschen AH, Cullor JS (1996) Purification, primary structures, and antibacterial activities of beta-defensins, a new family of antimicrobial peptides from bovine neutrophils. J Biol Chem 271:16430
115. Selsted ME, Ouellette AJ (2005) Mammalian defensins in the antimicrobial immune response. Nat Immunol 6:551–557
116. Shai Y (1999) Mechanism of the binding, insertion and destabilization of phospholipid bilayer membranes by alpha-helical antimicrobial and cell non-selective membrane-lytic peptides. Biochim Biophys Acta 1462:55–70
117. Shinnar AE, Butler KL, Park HJ (2003) Cathelicidin family of antimicrobial peptides: proteolytic processing and protease resistance. Bioorg Chem 31:425–436
118. Shirafuji Y, Tanabe H, Satchell DP, Henschen-Edman A, Wilson CL, Ouellette AJ (2003) Structural determinants of procryptdin recognition and cleavage by matrix metalloproteinase-7. J Biol Chem 278:7910–7919
119. Simon TC, Gordon JI (1995) Intestinal epithelial cell differentiation: new insights from mice, flies and nematodes. Curr Opin Genet Devel 5:577–586
120. Skalicky JJ, Selsted ME, Pardi A (1994) Structure and dynamics of the neutrophil defensins NP-2, NP-5, and HNP-1: NMR studies of amide hydrogen exchange kinetics. Proteins 20:52–67
121. Sorensen OE, Follin P, Johnsen AH, Calafat J, Tjabringa GS, Hiemstra PS, Borregaard N (2001) Human cathelicidin, hCAP-18, is processed to the antimicrobial peptide LL-37 by extracellular cleavage with proteinase 3. Blood 97:3951–3959
122. Sorensen OE, Gram L, Johnsen AH, Andersson E, Bangsboll S, Tjabringa GS, Hiemstra PS, Malm J, Egesten A, Borregaard N (2003) Processing of seminal plasma hCAP-18 to ALL-38 by gastricsin: a novel mechanism of generating antimicrobial peptides in vagina. J Biol Chem 278:28540–28546

123. Sparkes RS, Kronenberg M, Heinzmann C, Daher KA, Klisak I, Ganz T, Mohandas T (1989) Assignment of defensin gene(s) to human chromosome 8p23. Genomics 5:240–244

124. Stappenbeck TS, Hooper LV, Gordon JI (2002) Developmental regulation of intestinal angiogenesis by indigenous microbes via Paneth cells. Proc Natl Acad Sci U S A 99:15451–15455

125. Stappenbeck TS, Hooper LV, Manchester JK, Wong MH, Gordon JI (2002) Laser capture microdissection of mouse intestine: characterizing mRNA and protein expression, and profiling intermediary metabolism in specified cell populations. Methods Enzymol 356:167–196

126. Stappenbeck TS, Mills JC, Gordon JI (2003) Molecular features of adult mouse small intestinal epithelial progenitors. Proc Natl Acad Sci U S A 100:1004–1009

127. Svinarich DM, Wolf NA, Gomez R, Gonik B, Romero R (1997) Detection of human defensin 5 in reproductive tissues. Am J Obstet Gynecol 176:470–475

128. Tanabe H, Ouellette AJ, Cocco MJ, Robinson WE Jr (2004) Differential effects on human immunodeficiency virus type 1 replication by alpha-defensins with comparable bactericidal activities. J Virol 78:11622–11631

129. Tanabe H, Yuan J, Zaragoza MM, Dandekar S, Henschen-Edman A, Selsted ME, Ouellette AJ (2004) Paneth cell alpha-defensins from rhesus macaque small intestine. Infect Immun 72:1470–1478

130. Tang YQ, Selsted ME (1993) Characterization of the disulfide motif in BNBD-12, an antimicrobial beta-defensin peptide from bovine neutrophils. J Biol Chem 268:6649–6653

131. Tang YQ, Yuan J, Miller CJ, Selsted ME (1999) Isolation, characterization, cDNA cloning, and antimicrobial properties of two distinct subfamilies of alpha-defensins from rhesus macaque leukocytes. Infect Immun 67:6139–6144

132. Tang YQ, Yuan J, Osapay G, Osapay K, Tran D, Miller CJ, Ouellette AJ, Selsted ME (1999) A cyclic antimicrobial peptide produced in primate leukocytes by the ligation of two truncated alpha-defensins. Science 286:498–502

133. Tomasinsig L, Zanetti M (2005) The cathelicidins—structure, function and evolution. Curr Protein Pept Sci 6:23–34

134. Trabi M, Schirra HJ, Craik DJ (2001) Three-dimensional structure of RTD-1, a cyclic antimicrobial defensin from Rhesus macaque leukocytes. Biochemistry 40:4211–4221

135. Tran D, Tran PA, Tang YQ, Yuan J, Cole T, Selsted ME (2002) Homodimeric theta-defensins from rhesus macaque leukocytes: isolation, synthesis, antimicrobial activities, and bacterial binding properties of the cyclic peptides. J Biol Chem 277:3079–3084

136. Valore EV, Ganz T (1992) Posttranslational processing of defensins in immature human myeloid cells. Blood 79:1538–1544

137. Valore EV, Martin E, Harwig SS, Ganz T (1996) Intramolecular inhibition of human defensin HNP-1 by its propiece. J Clin Invest 97:1624–1629

138. Van Es JH, Jay P, Gregorieff A, van Gijn ME, Jonkheer S, Hatzis P, Thiele A, van den Born M, Begthel H, Brabletz T, Taketo MM, Clevers H (2005) Wnt signalling induces maturation of Paneth cells in intestinal crypts. Nat Cell Biol 7:381–386

139. Wehkamp J, Harder J, Weichenthal M, Schwab M, Schaffeler E, Schlee M, Herrlinger KR, Stallmach A, Noack F, Fritz P, Schroder JM, Bevins CL, Fellermann K, Stange EF (2004) NOD2 (CARD15) mutations in Crohn's disease are associated with diminished mucosal alpha-defensin expression. Gut 53:1658–1664

140. White SH, Wimley WC, Selsted ME (1995) Structure, function, and membrane integration of defensins. Curr Opin Struct Biol 5:521–527

141. Wilson CL, Heppner KJ, Rudolph LA, Matrisian LM (1995) The metalloproteinase matrilysin is preferentially expressed by epithelial cells in a tissue-restricted pattern in the mouse. Mol Biol Cell 6:851–869

142. Wilson CL, Ouellette AJ, Satchell DP, Ayabe T, Lopez-Boado YS, Stratman JL, Hultgren SJ, Matrisian LM, Parks WC (1999) Regulation of intestinal alpha-defensin activation by the metalloproteinase matrilysin in innate host defense. Science 286:113–117

143. Wimley WC, Selsted ME, White SH (1994) Interactions between human defensins and lipid bilayers: evidence for formation of multimeric pores. Protein Sci 3:1362–1373

144. Wu ER, Daniel R, Bateman A (1998) RK-2: a novel rabbit kidney defensin and its implications for renal host defense. Peptides 19:793–799

145. Yamamoto CM, Banaiee N, Yount NY, Patel B, Selsted ME (2004) Alpha-defensin expression during myelopoiesis: identification of cis and trans elements that regulate expression of NP-3 in rat promyelocytes. J Leukoc Biol 75:332–341

146. Yang D, Biragyn A, Hoover DM, Lubkowski J, Oppenheim JJ (2004) Multiple roles of antimicrobial defensins, cathelicidins, and eosinophil-derived neurotoxin in host defense. Annu Rev Immunol 22:181–215

147. Yenugu S, Hamil KG, Radhakrishnan Y, French FS, Hall SH (2004) The androgen-regulated epididymal sperm-binding protein, human beta-defensin 118 (DEFB118) (formerly ESC42), is an antimicrobial beta-defensin. Endocrinology 145:3165–3173

148. Yount NY, Wang MS, Yuan J, Banaiee N, Ouellette AJ, Selsted ME (1995) Rat neutrophil defensins. Precursor structures and expression during neutrophilic myelopoiesis. J Immunol 155:4476–4484

149. Yudin AI, Tollner TL, Li MW, Treece CA, Overstreet JW, Cherr GN (2003) ESP13.2, a member of the beta defensin family, is a macaque sperm surface coating protein involved in the capacitation process. Biol Reprod 69:1118–1128

150. Zanetti M, Gennaro R, Romeo D (1997) The cathelicidin family of antimicrobial peptide precursors: a component of the oxygen-independent defense mechanisms of neutrophils. Ann N Y Acad Sci 832:147–162

151. Zanetti M, Gennaro R, Scocchi M, Skerlavaj B (2000) Structure and biology of cathelicidins. Adv Exp Med Biol 479:203–218

152. Zasloff M (2002) Antimicrobial peptides of multicellular organisms. Nature 415:389–395

153. Zasloff M (2002) Antimicrobial peptides in health and disease. N Engl J Med 347:1199–1200

154. Zeya HI, Spitznagel JK (1963) Antibacterial and enzymic basic proteins from leukocyte lysosomes: separation and identification. Science 142:1085–1087

155. Zeya HI, Spitznagel JK (1966) Cationic proteins of polymorphonuclear leukocyte lysosomes. I. Resolution of antibacterial and enzymatic activities. J Bacteriol 91:750–754

156. Zeya HI, Spitznagel JK (1966) Cationic proteins of polymorphonuclear leukocyte lysosomes. II. Composition, properties, and mechanism of antibacterial action. J Bacteriol 91:755–762

157. Zhou CX, Zhang YL, Xiao L, Zheng M, Leung KM, Chan MY, Lo PS, Tsang LL, Wong HY, Ho LS, Chung YW, Chan HC (2004) An epididymis-specific beta-defensin is important for the initiation of sperm maturation. Nat Cell Biol 6:458–464

158. Zimmermann GR, Legault P, Selsted ME, Pardi A (1995) Solution structure of bovine neutrophil beta-defensin-12: the peptide fold of the beta-defensins is identical to that of the classical defensins. Biochemistry 34:13663–13671

CTMI (2006) 306:27–66

Immunomodulatory Properties
of Defensins and Cathelicidins

D. M. E. Bowdish · D. J. Davidson · R. E. W. Hancock (✉)

Centre for Microbial Diseases and Immunity Research,
University of British Columbia, 232 Lower Mall Research Station,
Vancouver BC, V6T 1Z4, Canada
bob@cmdr.ubc.ca

Abstract Host defence peptides are a conserved component of the innate immune response in all complex life forms. In humans, the major classes of host defence peptides include the α- and β-defensins and the cathelicidin, hCAP-18/LL-37. These peptides are expressed in the granules of neutrophils and by a wide variety of tissue types. They have many roles in the immune response including both indirect and direct antimicrobial activity, the ability to act as chemokines as well as induce chemokine production leading to recruitment of leukocytes to the site of infection, the promotion of wound healing and an ability to modulate adaptive immunity. It appears that many of these properties are mediated though direct interaction of peptides with the cells of the innate immune response including monocytes, dendritic cells, T cells and epithelial cells. The importance of these peptides in immune responses has been demonstrated since animals defective in the expression of certain host defence peptides show greater susceptibility to bacterial infections. In the very few instances in which human patients have been demonstrated to have defective host defence peptide expression, these individuals suffer from frequent infections. Although studies of the immunomodulatory properties of these peptides are in their infancy, there is a growing body of evidence suggesting that the immunomodulatory properties of these small, naturally occurring molecules might be harnessed for development as novel therapeutic agents.

Abbreviations

PBMC	Peripheral blood mononuclear cells
HNP	Human neutrophil peptide
HBD	Human beta defensin
TLR	Toll-like receptor
LPS	Lipopolysaccharide
LTA	Lipoteichoic acid
TNF-α	Tumour necrosis factor alpha
MHC1	Major histocompatibility complex class 1
BAL	Bronchoalveolar lavage

1
Overview

1.1
Host Defence Peptides in Humans

Host defence peptides are small (generally less than 50 amino acids), positively charged peptides that are an evolutionarily conserved component of the innate immune response. Originally characterised as natural antimicrobial agents, it is becoming increasingly apparent that these peptides have a wide range of immunomodulatory properties that are either complementary to, or independent of, antimicrobial activity. Interest in the immunomodulatory functions of these peptides is increasing, and indeed many peptides and proteins with similar characteristics to host defence peptides have been found to have either antimicrobial or immunomodulatory properties in addition to their primary functions.

In humans the best-characterised host defence peptides are the defensins and the sole cathelicidin, hCAP-18/LL-37. The amino acid sequences of these peptides are summarised in Table 1. In general, the defensins are between 29 and 30 amino acids long (approximately 3.5 kDa) and contain three conserved

Table 1 Sequences of major human host defence peptides

		Sequence
α-Defensins	HNP1	ACYCRIPACIAGERRYGTCIYQGRLWAFCC
	HNP2	CYCRIPACIAGERRYGTCIYQGRLWAFCC
	HNP3	DCYCRIPACIAGERRYGTCIYQGRLWAFCC
	HNP4	VCSCRLVFCRRTELRVGNCLIGGVSFTYCCTRV
	HD5	ATCYCRHGRCATRESLSGVCEISGRLYRLCCR
	HD6	AFTCHCRRSCYSTEYSYGTCTVMGINHRFCCL
β-Defensin	HBD1	DHYNCVSSGGQCLYSACPIFTKIQGTCYRGKACCK
	HBD2	TCLKSGAICHPVFCPRRYKQIGTCGLPGTKCCKKP
	HBD3	GIINTLQKYYCRVRGGRCAVLSCLPKEEQIGKCSTRGRKCCRRKK
	HBD4	MQRLVLLLAVSLLLYQDLPVRSEFELDRICGYGTARCRKKCRSQE YRIGRCPNTYACCLRKWDESLLNRTKP
Cathelicidin	hCAP-18/ LL-37	LLGDFFRKSKEKIGKEFKRIVQRIKDFLRNLVPRTES

disulphide bridges (White et al. 1995). The genes for all human defensins are clustered on chromosome 8 (Sparkes et al. 1989; Maxwell et al. 2003). They are further subdivided to include the α- and β-defensins, a distinction based on the organisation of the three characteristic cystine disulphide bonds.

The canonical sequence of the α-defensins is $x_{1-2}CxCRx_{2-3}Cx_3Ex_3GxCx_3$ Gx_5CCx_{1-4}, where x represents any amino acid residue. These peptides are rich in cysteine, arginine, and aromatic residues (Selsted et al. 1985). The cysteines are linked 1–6, 2–4, and 3–5. Initially these peptides were isolated from the neutrophils and are thus called human neutrophil peptides (HNP)-1 to -3 (Ganz et al. 1985). The HNPs are expressed at the transcriptional level in the bone marrow, spleen and thymus, where they co-localise with peripheral blood leukocytes (Zhao et al. 1996). The three HNPs are highly homologous, differing by only one amino acid at the NH_2 terminus. Because of the high sequence similarity and difficulties in purifying the individual peptides as well as the high degree in functional similarity, the HNP1–3 are often studied as a group, although certain studies have demonstrated differences in their antimicrobial (Lehrer et al. 1988) and immunomodulatory activities (Chertov et al. 1996). HNP4 was identified as a HNP due to its structural homology to the HNP1–3 (Wilde et al. 1989). This gene differs from the other genes of this family by an extra 83-base pair segment that is apparently the result of a recent duplication within the coding region (Palfree et al. 1993). As with the other HNPs, HNP4 is found in the neutrophils, but is also called corticostatin because it exhibits corticostatic activity and inhibits corticotrophin-stimulated corticosterone production (Singh et al. 1988). Despite its variation from the conserved sequences of HNP1–3, it appears to have much more potent antimicrobial activity (Wilde et al. 1989).

Two other α-defensins, HD5 and HD6, are found solely in the intestinal tract. HD5 and HD6 were found to be expressed at the transcriptional level solely in the small intestine and in situ hybridisation demonstrated that this expression occurs in the Paneth cells (Jones and Bevins 1992, 1993). Southern blot analysis using a nucleotide probe for the conserved signal sequence of the defensins indicated that a number of genes with high homology to HNPs exist within the human genome.

The β-defensins are expressed in a variety of tissue types, including epithelial cells from the trachea and lung, in the salivary and mammary glands, in a variety of organs such as in the plasma and skin (Bensch et al. 1995; Zhao et al. 1996; Harder et al. 1997; reviewed in Lehrer and Ganz 2002). The expression of certain β-defensins is inducible upon stimulation with bacterial components or pro-inflammatory cytokines and thus these peptides are presumed to be an important component of host defence to infection or inflammation. The canonical sequence for the beta defensins is $x_{2-10}Cx_{5-6}(G/A)_{x}CX_{3-4}Cx_{9-13}C_{x4-7}CCx_{n}$. The best characterised members of the β-defensin family are HBD1–3; however, the antimicrobial properties of HBD4 have been recently published (Garcia et al. 2001) and over 20 potential β-defensin homologues have been identified in the human genome based on sequence similarity to HBD1–4 (Schutte et al. 2002).

The cathelicidins are an evolutionarily conserved family of host defence peptides which are found in cows, sheep, guinea pigs, rabbits, mice, and primates (reviewed in Zanetti 2004) and are characterised by having an evolutionarily conserved N-terminal domain called the cathelin domain. In addition, these peptides have a signal sequence, which is believed to target their delivery to the secondary granules of neutrophils. The C terminal domain, which is released by cleavage of proteases, has both antimicrobial and immunomodulatory properties. Despite the conserved nature of the cathelin domain, its function remains unclear, although it has been proposed to block the antimicrobial activity of the cleaved product, presumably as a mechanism which allows storage of the peptide in its inactive form (Zaiou et al. 2003), and there is some evidence that it has anti-protease activity (Zaiou et al. 2003). The sole human cathelicidin LL-37/hCAP-18 is found at high concentrations in its unprocessed form in the granules of neutrophils and is processed upon degranulation and release (Sorensen et al. 2001). Consequently, it is found at sites of neutrophil degranulation. It is also produced by epithelial cells and is found in a number of tissues and bodily fluids, including gastric juices, saliva, semen, sweat, plasma, airway surface liquid and breast milk (Bals et al. 1998c; Murakami et al. 2002b; Hase et al. 2003; Murakami et al. 2005). Generally epithelial cells have been shown to produce the hCAP-18 form. Although the hCAP-18 has been shown to be cleaved by the neutrophil protease, protease 3, when released from neutrophils, it is not entirely clear how or when hCAP-18 is cleaved when it is produced by epithelial cells. There are a variety of processed forms of hCAP-18 that result from as-yet uncharacterised cleavage processes. For example, a 6-kDa form is found in gastric juice (Hase et al. 2003), while numerous cleavage products are found in the sweat (Murakami et al. 2002b). As well, hCAP-18 from semen is cleaved to a 38-amino acid antimicrobial peptide ALL-38 in the vagina, thus potentially providing some antimicrobial protection after sexual intercourse (Sorensen et al. 2003). There appears to be some overlapping, complementary and possibly even enhanced antimicrobial activity of these isoforms (Murakami et al. 2004); however, to date there is no information about their immunomodulatory properties.

1.2
Induction of Host Defence Peptides

Due to the high homology of HNP1–3, they are often classed as a group. The HNPs are found exclusively in leukocytes and at their highest concentrations in neutrophils where they are localised to azurophilic granules. The recently described HNP4 is also found in azurophilic granules but at much lower concentrations (Wilde et al. 1989). To date there has been no indication that their expression levels can vary substantially. Neutrophils stimulated with IL-8, FMLP or phorbol 12-myristate 13-acetate causes degranulation and release of HNP1–3 (Chertov et al. 1996), thus it is believed that relatively high concentrations of these peptides are present at sites of infection or inflammation. Interestingly, in the course of infection HNP levels increase systemically. Although the mechanism is not entirely clear, the plasma levels of defensins increase significantly in patients experiencing septicaemia or meningitis (from approximately 40 ng/ml to as much as 170,000 ng/ml). The concentration of neutrophil defensins generally correlates with that of IL-8, a potent chemoattractant for neutrophils (Ashitani et al. 2002), as well as the presence of other neutrophil components such as elastase (Zhang et al. 2002). Presumably this occurs due to neutrophil degranulation in response to bacterial or pro-inflammatory stimuli as concentrations decrease after antibiotic treatment (Panyutich et al. 1993).

HD5 and HD6 are expressed throughout the gastrointestinal tract, with the highest levels of expression occurring in the jejunum and ileum (Dhaliwal et al. 2003). Specifically, immunohistochemical studies have shown that

HD5 is expressed by Paneth cells, some villous epithelial cells and in the terminal ileal mucosa (Cunliffe et al. 2001). Although there is great variability between individuals, levels of these defensins are increased in certain disease states such as acute coeliac sprue (Frye et al. 2000b) and are decreased in others such as HIV-related cryptosporidiosis (Kelly et al. 2004). In patients with Crohn's disease or ulcerative colitis, but not in healthy individuals, HD5 immunoreactive cells are present in the crypt region of a large proportion of colonic samples, indicating that expression in these disease states might be dysregulated (Cunliffe et al. 2001). Limited studies have demonstrated that HD5 and HD6 may also be present at low levels in airway epithelial cells (Frye et al. 2000a) or the female reproductive tract (Quayle et al. 1998).

The patterns of expression of the β-defensins are markedly different. In general HBD1 is not up-regulated during the course of infection or inflammation or by stimulation with pro-inflammatory cytokines or bacterial components and in many cases the presence of HBD1 is detectable at the transcriptional level but is undetectable at the protein level (O'Neil et al. 1999; Frye et al. 2000b). However its constitutive presence in airway surface liquid, in intestinal and colon cell lines and in other tissues and bodily fluids implies that it may be involved in maintenance and homeostasis of these areas (Salzman et al. 2003). The expression levels of HBD1 appear to be quite low, ranging from the pg/ml to ng/ml levels in most bodily fluids.

HBD2 is an inducible host defence peptide whose expression is altered under both infectious and inflammatory conditions. It has been found to be up-regulated by pro-inflammatory stimuli in oral epithelial cells and keratinocytes (Krisanaprakornkit et al. 2000), in intestinal and colonic epithelial cell lines (O'Neil et al. 1999; Ogushi et al. 2001; Vora et al. 2004) and in various lung epithelial cell lines (Singh et al. 1998; Becker et al. 2000). HBD2 is believed to be an important component of host defences, since HBD2 expression is depressed in patients with atopic dermatitis who often present with cases of acute and chronic colonisation by *Staphylococcus aureus* (Ong et al. 2002) but is increased in psoriatic skin, a disease in which patients are fairly resistant to bacterial infection. It has been demonstrated to be up-regulated by both commensal and pathogenic bacteria in the oral mucosa and keratinocytes, although by different mechanisms (Krisanaprakornkit et al. 2002; Chung and Dale 2004). HBD2 is inducible during the course of inflammation and infection in the gastrointestinal system, as observed at both the mRNA and protein levels. HBD2 expression in intestinal and colonic epithelial cell lines is increased upon stimulation with IL-1α, flagellin or bacteria, in an NF-κB-dependent manner (O'Neil et al. 1999; Ogushi et al. 2001), and by either lipopolysaccharide (LPS) or lipoteichoic acid (LTA) in a TLR4 and TLR2-dependent manner (Vora et al. 2004). Interestingly, other inflammatory mediators such as TNF-α and LPS do not induce HBD2 up-regulation (O'Neil et al. 1999). This may reflect a predominantly intracellular expression pattern of TLR4 in these cells, which has been suggested to be an evolutionary adaptation to the high bacterial load in the intestine (Naik et al. 2001; Hornef et al. 2002). Increases in HBD2 expression have been detected in inflamed intestinal and colonic tissues by RT-PCR and immunohistochemistry, in Crohn's disease and ulcerative colitis (Fahlgren et al. 2003), and in the stomach of patients suffering from *Helicobacter pylori*-induced gastritis (Wehkamp et al. 2003). In the lung, HBD2 has been found to be down-regulated in a variety of infectious and inflammatory diseases including cystic fibrosis and infections (Chen et al. 2004). In lung transplant patients, it has been found to be present at greater than ten times the concentration in patients suffering from bronchiolitis obliterans syndrome, a consequence of rejection of the transplant, compared to transplant patients without signs of rejection (Ross et al. 2004).

Less is known about the newly characterised HBD3. HBD3 expression and activity is not well characterised; however, it has been demonstrated to be inducibly expressed at the transcriptional level in bronchial epithelial cell lines stimulated with TNF-α, bacteria, or live rhinovirus (Harder et al. 2001; Duits et al. 2003). In addition, this peptide can be induced in amnion cells in response to bacterial components and is found at high concentrations in human fetal membranes, by immunohistology, indicating that it may be involved in maintaining the sterility of the intra-amniotic environment (Buhimschi et al. 2004).

The pro-cathelicidin hCAP18 is found at high concentrations in the granules of neutrophils and is produced by epithelial cells. The CAMP-gene promoter that directs the expression of this peptide contains many transcription factor binding sites (Larrick et al. 1996), including a vitamin D response element (Wang et al. 2004). Consistent with this, hCAP18, and/or its processed product LL-37, has been shown to be up-regulated in sinus epithelial cells, at the transcriptional level, by the bacterial products LPS and LTA (Nell et al. 2004). It is also increased in bronchial airway cells by IL-α (Erdag and Morgan 2002). In other cell types, pro-inflammatory cytokines do not increase the expression of this peptide, implying that other signalling pathways may be involved (Hase et al. 2003). Its expression is increased in gastric epithelial cells upon stimulation with a wild-type strain of *H. pylori*, but not a type IV secretion mutant (Hase et al. 2003), and is found at increased levels in other forms of bacterial infection. It is not entirely clear under what circumstances neutrophils release LL-37.

1.3
Antimicrobial Properties In Vitro

The antimicrobial activity of all host defence peptides is highest in media of low ionic strength, and the activity of most peptides is sensitive to the presence of physiological concentrations of ions such as Na^+, Mg^{2+} and Ca^{2+}. The antimicrobial properties of the β-defensins, for example demonstrate profound salt sensitivity, and in some cases their antimicrobial activity is completely lost at concentrations of 100 mM NaCl (Bals et al. 1998a, 1998b; Garcia et al. 2001). For example, HBD1 is antimicrobial towards Gram-negative bacteria at concentrations of 1–10 µg/ml (Singh et al. 1998), but its antimicrobial activity is almost completely abrogated by the presence of 100 mM sodium ions. Although HBD2 is slightly less sensitive to the presence of sodium ions, its antimicrobial activity is reduced from 0.001–0.1 µg/ml in conditions of low ionic strength to 0.5 µg/ml or more in the presence of 100 mM sodium (Singh et al. 1998). Of the human β-defensins, HBD3 is the most potent antimicrobial. This peptide is more basic, has a broader spectrum and stronger bactericidal activity against Gram-positive and Gram-negative bacteria, as well as yeast, and is salt-insensitive at concentrations less than 200 mM Na^+ ions (Harder et al. 2001).

Most biological fluids, including sputum (Halmerbauer et al. 2000), airway surface liquid (Baconnais et al. 1999) and serum/plasma (Hoshino et al. 2003) contain Mg^{2+} and Ca^{2+} at free concentrations between 1 and 2 mM, and the presence of these ions is generally more detrimental to antimicrobial activity than Na^+ alone. The α-defensins are susceptible to concentrations of Ca^{2+} and Mg^{2+} as low as 0.5 mM (Lehrer et al. 1988). In the presence of 100 mM Na^+ ions, the antimicrobial activity of LL-37 is decreased two- to eightfold (Turner et al. 1998), and in the presence of standard tissue culture media, which contains 150 mM NaCl and 1–2 mM of Mg^{2+} and Ca^{2+}, LL-37 has no killing activity against *S. aureus* or *Salmonella typhimurium* even at concentrations as high as 100 µg/ml (Bowdish et al. 2004).

In some cases, for example in the granules of neutrophils, the concentrations of host defence peptides are estimated to be as great as 10 mg/ml, and there is no doubt that upon ingestion of bacteria these concentrations are sufficient to cause direct antimicrobial activity, despite the presence of divalent cations or other inhibitory substances. However, it is questionable whether these concentrations are reached at mucosal surfaces. For example, in patients suffering from inflammatory lung disease or infection, the concentration of HNPs in the bronchoalveolar lavage (BAL) has been estimated to be 0.7–1.2 and 10 µg/ml in two different studies (Cole et al. 2001; Spencer et al. 2003). This contrasts with antimicrobial activity in low salt medium in vitro at greater than 10 µg/ml for *S. aureus* and *Escherichia coli* (Nagaoka et al. 2000), and reduction of the infectivity of adenoviruses in an airway epithelial model at concentrations of between 8–50 µg/ml (Spencer et al. 2003).

Interestingly, the antimicrobial activity of the host defence peptides may also be inhibited by components of serum. For example, HNP1 has also been demonstrated to possess antiviral activity towards enveloped viruses, but this activity is abrogated by the presence of serum or albumin (Daher et al. 1986). It is believed but has not been conclusively shown that the high concentrations of α-defensins in neutrophils would overcome any localised serum effects (Daher et al. 1986). Moreover, the antibacterial activity of LL-37 has been demonstrated to be abrogated by the presence of apolipoprotein (Wang et al. 1998).

In contrast to the antimicrobial activity of these peptides, their immunomodulatory properties are generally studied in the presence of standard tissue culture media, which contains physiologically relevant concentrations of ions and serum proteins. Under these conditions, host defence peptides have been demonstrated to induce chemokine production, reduce pro-inflammatory cytokine production, alter transcription, induce proliferation and angiogenesis, induce chemotaxis and alter dendritic cell differentiation. Thus the immunomodulatory properties of host defence peptides are unaffected by physiological ion concentrations and it seems possible that they may be the predominant function of these peptides in vivo. This perspective is quite controversial, and will remain so since it is extremely difficult to discriminate between direct and indirect (i.e. through stimulation of innate immunity) mechanisms of killing. One often used argument for the likely antimicrobial function of these peptides is their substantial variation over evolution (e.g. the mouse CRAMP and human LL-37 peptides share only 67% homology), which could have arisen from the evolutionary pressure of dealing with different pathogens. However, not all antimicrobial proteins are as divergent, and we note that a number of proteins involved in immunity and reproduction show similar "rapid evolution" to the antimicrobial peptides (Emes et al. 2003).

1.4
Evidence for Antimicrobial and Immunomodulatory Properties In Vivo

There has been some debate about whether host defence peptides might be directly antimicrobial in vivo. Certainly, neutrophil granules and the crypts of the lumen contain sufficiently high concentrations of peptides to ensure substantial antimicrobial activity; however, it is less clear that antimicrobial activity occurs at the lower concentrations in such sites as mucosal surfaces, and it is worth noting that such sites are often heavily colonised by a rich and diverse collection of commensal bacteria. The evidence for antimicrobial activity in certain body sites is in our opinion inconclusive. On the one hand, certain bodily fluids such as sinus fluid (Cole et al. 1999) and gastric fluids (Hase et al. 2003) can directly kill certain micro-organisms, and this antimicrobial activity is ablated or reduced by removal of proteins or immunodepletion with a peptide-specific antibody. However, in certain animal models in which peptides and bacteria are instilled simultaneously, bacterial counts are often not significantly different from mice treated with bacteria alone, despite improved outcome or reduced pro-inflammatory responses (Sawa et al. 1998; Giacometti et al. 2004). The difficulties in assessing the role of host defence peptides in vivo are profound, as it is almost impossible to account for synergistic interactions between peptides and other factors, to assess the actual concentrations at the sites of infection and to discriminate the direct antimicrobial activity of peptides from other less direct effects such as enhancement of inflammatory mechanisms (chemotaxis and recruitment of effector cells, enhancement of nonopsonic phagocytosis, etc.). Nonetheless, creative experiments and animal models have begun to elucidate the roles of these peptides in vivo.

In transgenic mouse model studies in which the expression of certain host defence peptides is ablated, these mice are somewhat more susceptible to infection and carry increased bacterial loads when challenged (Wilson et al. 1999; Nizet et al. 2001). Although this was interpreted as being due to direct antimicrobial activity, other components of host defences must be considered. For example, in a mouse model of peritoneal *Klebsiella pneumoniae* infection, small doses of HNP1 (4 ng–4 µg) caused an increase in leukocyte accumulation. In this model, it was the leukocyte accumulation which was linked to HNP1 induced antimicrobial activity, as the reduction in bacterial counts was significantly diminished in leukocytopenic mice. Similar results were observed in *S. aureus* thigh infections (Welling et al. 1998).

Gain of function studies have found that introducing or increasing the expression of a host defence peptide can reduce bacterial loads in certain animal models of infection. For example, adenovirus transfer of LL-37/hCAP-18 into the lungs of mice that were subsequently challenged with *Pseudomonas aeruginosa* led to a reduction in both the bacterial load and in production of the pro-inflammatory cytokine, TNF-α (Bals et al. 1999), and intriguingly, similar gene therapy decreased susceptibility to sepsis induced by LPS in the complete absence of bacterial infection. In other models, the simultaneous instillation into the mouse lung of *P. aeruginosa* and either of HBD2 or a LL-37 derivative led to reduced lung damage and pro-inflammatory cytokine production, but did not affect bacterial counts (Sawa et al. 1998).

There are very few human diseases that are characterised by defects in host defence peptide production, perhaps emphasizing their importance. However, the neutrophils of individuals with specific granule deficiency, a disease characterised by frequent and severe infections, have a reduction in the size of the peroxidase positive, defensins-containing granules (Parmley et al. 1989) and are deficient in defensins (Ganz et al. 1988). However, it is difficult to assess the extent to which these infections result from the lack of defensins, as these patients are also deficient in other neutrophil components. It is believed that the constitutive production and deposition of neutrophils is of crucial importance to maintaining the immunological balance of the mouth. Patients who suffer from morbus Kostman, a severe congenital neutropenia, and are treated with G-CSF to restore neutrophil level, do not express LL-37 in these cells. One of the manifestations of this disease is frequent and severe infections and periodontal disease (Putsep et al. 2002). It has been proposed that the absence of LL-37 may give a selective advantage to bacteria that at low levels are commensals but at higher levels are responsible for periodontal disease. It is unclear, however, whether LL-37 is directly microbicidal towards common pathogens of the mouth or marshals other defences. Although a number of oral bacteria are susceptible to LL-37 (<10 µg/ml) at 10 mM NaCl in vitro, far fewer bacteria are susceptible in physiologically more relevant isotonic environments (Tanaka et al. 2000). Although LL-37 has been detected in saliva, the actual concentration was not determined (Murakami et al. 2002a).

Other indirect evidence for the in vivo antimicrobial activity of host defence peptides is that a decreased level of expression often correlates with frequency or severity of disease. For example, HBD2 and LL-37 expression is depressed in patients with atopic dermatitis who often present with cases of acute or chronic colonisation by *S. aureus* (Ong et al. 2002). In contrast to atopic dermatitis, HBD2 expression is increased in psoriatic skin, a disease in which patients are fairly resistant to bacterial infections (Harder et al. 2001; Nomura et al. 2003).

Whether host defence peptides are directly or indirectly antimicrobial, it is apparent that it is of advantage for bacterial pathogens to subvert their expression or activity. For example, *Streptococcus pyogenes* binds to α_2-microglobin and secretes a small proteinase which inhibits LL-37 from interacting with the bacteria and thus prevents LL-37 mediated killing (Nyberg et al. 2004). LL-37 expression has been shown to be decreased in *Shigella* infection, consistent with a proposed mechanism of evasion by this bacterium (Islam et al. 2001). However, it is not clear whether this is a direct down-regulation of expression, or a consequence of denuding the epithelium, with reduced expression in the replacement cells.

2
Host Defence Peptides in the Innate Immune Response

2.1
Role of Host Defence Peptides in Wound Healing

2.1.1
Re-epithelisation and Proliferation

Early experiments with host defence peptides demonstrated that many of these peptides have mitogenic effects on a variety of cells and cell lines. Since modest to high concentrations of host defence peptides are found at sites of infection and inflammation, it has been hypothesised that this pro-liferative effect might be involved in wound healing and re-epithelisation. Consistent with this hypothesis, both the human and mouse cathelicidins are up-regulated at sites of incision or wounding, even if the wound is sterile. The appearance of cathelicidins in the skin has been ascribed to both synthesis within epidermal keratinocytes, and deposition from granulocytes that migrate to the site of injury (Dorschner et al. 2001). Upon incision, hCAP-18 (the precursor to LL-37) has been shown to be up-regulated in the epidermis bordering the wound. This increase in expression at both the RNA and protein levels was clearly evident at the migrating front of the wound during re-epithelialisation. Levels of hCAP-18 decreased following wound closure and eventually returned to baseline levels when the wound was intact and re-epithelisation was complete. hCAP-18 was found to be an active component in the process of re-epithelisation since antibodies specific for the peptide decreased the rate of re-epithelisation in a concentration-dependent manner (Heilborn et al. 2003). Consistent with this observation, low levels of LL-37 (as low as 50 ng/ml) have been demonstrated to increase proliferation in an endothelial cell line (Koczulla et al. 2003). The importance of this peptide in re-epithelisation has been further inferred from its presence in wounds which are healing normally, but its absence in chronic ulcers (Heilborn et al. 2003).

HNPs are potent mitogens for epithelial cells, squamous cell carcinoma cell lines and fibroblasts in vitro at low concentrations (Murphy et al. 1993; M. Nishimura et al. 2004). Interestingly, in one of the earliest studies of these effects it was demonstrated that the HNPs acted synergistically with insulin to induce proliferation (Murphy et al. 1993). In general it has been hypothesised that the mitogenic properties of the neutrophil defensins on non-myeloid cells is an important component of the healing process. However, certain tumours and tumour cell lines have been demonstrated to inappropriately express neutrophil defensins, and in such cases it is believed that this expression might lead to inappropriate proliferation. For example, in a renal carcinoma cell line, the α-defensins HNP1–3 are expressed at both the transcriptional and protein levels. At moderate levels (i.e. ≤12 μg/ml), the defensins had mitogenic activity on a subset of these cell lines. By influencing tumour cell proliferation, α-defensins could potentially modulate tumour progression of renal carcinoma cells (Muller et al. 2002).

Moderate concentrations (e.g. ≤10 μg/ml) of neutrophil defensins (HNP1–3) induce proliferation of a lung epithelial cell line in vitro (Aarbiou et al. 2002). Consistent with these observations, a combination of HNP1–3 caused a dose- and time-dependent increase in cell migration and wound closure of an airway epithelial cell line, possibly due to an ability to induce the expression of genes involved in proliferation (Aarbiou et al. 2004). The mitogenic activity of the HNPS and cathelicidins does not appear to be shared by the β-defensins. Although HBD2 has been demonstrated to be up-regulated in chronic ulcers (Butmarc et al. 2004), it has not been demonstrated to be involved in re-epithelisation. In addition, the β-defensins investigated did not increase the proliferation of epithelial cells, squamous cell carcinoma cell lines or fibroblasts (M. Nishimura et al. 2004).

2.1.2
Angiogenesis and Vasculogenesis

An interesting phenomenon which has been observed to occur in response to two cathelicidins, human LL-37 and its mouse homologue CRAMP, is the induction of angiogenesis, which is the process of blood vessel formation and/or growth. The formation of new blood vessels results in restoration of tissues increasing the oxygen supply and the provision of blood substances and cells to these tissues. As such it is a requirement for tissue repair and wound healing as well as for the marshalling of innate immunity. Thus this function is consistent with a role for host defence peptides in the maintenance and repair of tissues. In a chorioallantoic membrane assay, 5 μg of LL-37 induced an increase in blood vessel growth, while in a rabbit hind limb model of angiogenesis, collateral vessel growth and blood flow were increased (Koczulla et al. 2003). Interestingly however, despite the known chemotactic properties of this peptide, no inflammatory infiltrate was detected. The angiogenic properties of LL-37 appear to stem from its direct interaction with endothelial cells rather than induction of growth factors. These data are consistent with the observation that CRAMP knockout mice have reduced vascular structures at the wound edge at the site of injury (Koczulla et al. 2003).

A mouse β-defensin, DefBD29, has been shown to be involved in vasculogenesis, which is the differentiation of endothelial cells from progenitor cells during blood vessel development, leading to the de novo formation of blood vessels and tubes. Tumours expressing DefBD29 recruit dendritic cell (DC) precursors via CCR6 and result in enhanced vascularisation and growth in the presence of the cytokine Vegf-A (Conejo-Garcia et al. 2004). Interestingly, these DCs differentiate to express both DC and endothelial cell markers in response to Vegf, indicating that these cells undergo endothelial cell-like specialisation after or during migration to newly formed vessels. This implies that host defence peptides may play important roles in vascular development.

2.2
Role of Host Defence Peptides in Chemokine Production and Chemotaxis

It has been observed that there are similarities between chemokines and host defence peptides. Indeed, many chemokines have modest antimicrobial activity (Hieshima et al. 2003; Yang et al. 2003), while a derivative of the highly active antimicrobial peptide, horseshoe crab polyphemusin is a potent antagonist of CXCR4 (Tamamura et al. 1998). Indeed it has been proposed that certain host defence peptides have evolved from duplication of chemokine genes, although this connection is controversial (Durr and Peschel 2002; Yang et al. 2002); consistent with this, certain peptides have chemotactic activity. Interestingly, unlike the chemokines characterised to date, many host defence peptides appear to have chemotactic activity over a wide range of species, and generally speaking these activities are often observed at concentrations 100-fold or more higher than observed with the classical chemokines.

HNP1 and -2 have been demonstrated to induce chemotaxis of T cells in vitro at concentrations of between 0.1–100 ng/ml, with maximal activity occurring at less than 10 ng/ml (Chertov et al. 1996). HNP1 is a more potent chemoattractant of monocytes that HNP2, with optimal activity at concentrations of 10^{-8}–10^{-9} M, while HNP3 failed to induce significant chemotaxis (Territo et al. 1989). Conversely, these peptides were not chemotactic for neutrophils (Territo et al. 1989), and indeed a subsequent study demonstrated that HNP1 actually suppressed polymorphonuclear (PMN) migration to formyl-methionyl-leucyl-phenylalanine but not to interleukin 8 (Grutkoski et al. 2003). In BALB/c mice, 4 h after subcutaneous injection, a mixture of HNP1–3 was demonstrated to induce infiltration of PMNs and mononuclear cells, while in huPBL-SCID mice the defensins-induced infiltrate consisted of modest numbers of CD3[+] cells (Chertov et al. 1996). Interestingly, in contrast to in vitro results, in this animal model study, the infiltration of PMNs was observed. However, it is unclear whether PMN infiltration was caused by direct chemotaxis or indirect effects of the peptide treatment. Further studies demonstrated that these peptides specifically lead to chemotaxis of immature dendritic cells and naïve, but not memory, T cells (Yang et al. 2000a).

Collectively, these data indicate that neutrophil granules contain important chemotactic factors which promote the infiltration of cells of both the innate and adaptive immune responses.

The β-defensins HBD1 and HBD2 are chemoattractants for immature dendritic cells and memory T, cells with peak activities occurring at 1 μg/ml (Yang et al. 1999). These activities are mediated through the chemokine receptor CCR6, which also binds the chemokine LARC. HBD2, but not HBD1, has also been demonstrated to be a chemotactic agent for TNF-α treated human neutrophils (Niyonsaba et al. 2004), a response that is also mediated through CCR6.

LL-37 has been demonstrated to be chemotactic for rat mast cells (Niyonsaba et al. 2002b), mouse mononuclear cells and PMNs (Chertov et al. 1996), as well as human neutrophils, monocytes and T cells (De et al. 2000). As LL-37 has been demonstrated to induce a number of chemokines, there has been some debate as to whether it induces chemotaxis directly or indirectly by induction of classical chemokines. In the rat mast cell model, it appears as though this chemotaxis is a direct effect: as when mast cells are cultured with LL-37 and the supernatants are used for the chemotaxis assay, chemotaxis can be blocked by anti-LL-37 antiserum (Niyonsaba et al. 2002b).

Host defence peptides may also indirectly enhance chemotaxis by inducing the production of chemokines from a variety of different cell types, including epithelial cells and monocytes. The HNPs, for example, have been demonstrated to induce IL-8 from lung epithelial cells and cell lines (Van Wetering et al. 1997; Sakamoto et al. 2004) and to induce the production of IL-1β and IL-8 mRNA production (Sakamoto et al. 2004) from a lung epithelial cell line.

It is unclear whether the β-defensins have similar chemokine-inducing activities. HBD2, for example, does not induce IL-8 expression in bronchial epithelial cells (Sakamoto et al. 2004). However, in BAL from patients with diffuse panbronchiolitis, the HBD2 concentration correlated significantly with the numbers of cells recovered from the BAL fluid (total cells, neutrophils, and lymphocytes) (Hiratsuka et al. 2003), implying that there might be a link between this peptide and cellular infiltration to the site of infection.

LL-37 has been demonstrated to induce MCP-1 and IL-8 release in a mouse macrophage and a human bronchial epithelial cell line, respectively, and both chemokines were increased upon stimulation with LL-37 in whole human blood (Scott et al. 2002). LL-37 has also been demonstrated to induce chemokine transcription (IL-8, MCP-1, MCP-3) and release (IL-8) in a mitogen-activated protein kinase (MAPK)-dependent manner in human peripheral blood derived monocytes (Bowdish 2004). Both LL-37 and the HNPs are neutrophil-derived peptides which are released upon neutrophil degranulation. These peptides induce the transcription and release of chemokines, specifically IL-8, which preferentially attract neutrophils. Consequently the presence of these peptides correlates well with that of IL-8, a potent chemoat-tractant for neutrophils (Ashitani et al. 2002), as well as the presence of other neutrophil components such as elastase (Zhang et al. 2002).

It appears that host defence peptides induce chemotaxis in two ways: first through direct chemotactic activity of PMNs and mononuclear cells mediated through CCR6 and other as yet to be identified receptors and second through inducing chemokine production which would hypothetically increase the numbers of neutrophils and monocytes at sites of infection. This then would have the net effect of promoting or marshalling cells important in innate immunity to the sites of excessive production (through induction) or deposition (through neutrophil degranulation) of these host defence peptides. This then begs the question as to whether host defence peptides are overtly pro-inflammatory.

2.3
Anti-inflammatory (Anti-endotoxin) Roles of Host Defence Peptides

Early experiments determined that a number of host defence peptides from various sources bound to LPS from diverse Gram-negative bacteria and re-duced LPS-induced release of pro-inflammatory cytokines (e.g. TNF-α, IL-1, IL-6) and nitric oxide from monocyte or macrophages and protected mice from LPS lethality (Larrick et al. 1994, 1995; VanderMeer et al. 1995; Kirikae et al. 1998).

Initial studies focussed on the unprocessed form of cathelicidin, hCAP-18 (Kirikae et al. 1998); however, it was later found that the LPS-binding properties of the peptide were contained within the processed 37-amino acid C-terminal domain, LL-37 (Turner et al. 1998). It has been proposed that the anti-endotoxic properties of these peptides are the result of the inhibition of binding of LPS to CD14 (Nagaoka et al. 2001) and lipopolysaccharide-binding protein (LBP) (Scott et al. 2000), and/or indirect effects on cells (Scott et al. 2002). LL-37 has been shown to block a number of LPS-induced inflammatory responses, including contractility and (nitric oxide) NO release in aortic rings (Ciornei et al. 2003), pro-inflammatory cytokine production in a macrophage cell line and in animal models (Scott et al. 2002) (Ohgami et al. 2003), suppression of leukocyte infiltration in a model of endotoxin-induced uveitis (Ohgami et al. 2003) and lethality in animal models of sepsis (Scott et al. 2002). These effects occur at concentrations in the physiological range for LL-37 (1–5 µg/ml) and may reflect a natural role for LL-37 in the body (e.g. balancing of the potential stimulus by endotoxin from commensals). This anti-endotoxin activity appears to correlate with an ability to dampen the pro-inflammatory effects of the Gram-positive surface molecule lipoteichoic acid (Scott et al. 2002).

It appears that there may be marked differences in the ability of LL-37 and the defensins to inhibit the pro-inflammatory effects of endotoxin. For example, HNP1 and HBD2 are not potent inhibitors of LPS–LBP binding (Scott et al. 2000). In ex vivo whole blood experiments, HNP1 was approximately 1,000-fold less potent than BPI at reducing TNF-α in response to Gram-negative bacteria and is much less potent in blocking endotoxin activity, as assessed by a surrogate assay, the *Limulus* amoebocyte lysate assay, or in priming PMN for arachidonate release or stimulating leukocyte oxidase activity (Levy et al. 1995). Thus, the ability to bind to and neutralise endotoxin-induced activity in humans may be more evident for LL-37 and other proteins such as bacterial permeability-inducing protein (BPI) and LBP (Weiss 2003).

2.4
Interactions with Effector Cells of the Innate Immune Response

2.4.1
NK Cells

Natural Killer cells are $CD56^-CD3^-$ lymphocytes that are an important component of the innate immune response. They kill transformed and infected cells, but unlike T cells they are active against cells that have decreased or ablated expression of major histocompatibility complex class 1 (MHC1) molecules. The cytolytic properties of NK cells are increased in the presence of cytokines produced by cells of the innate immune response. NK cells themselves produce cytokines, such as IFN-γ, which are involved in the enhancement of both the innate and adaptive immune responses. NK cells contain a wide variety of cytotoxic peptides of which granulysin (NK-lysin) is considered to be the most important (Kumar et al. 2001). Recently NK cells have also been demonstrated to express the transcripts for LL-37 and HNP1–3, and these peptides were found in the supernatants of IL-2-treated cells consistent with an involvement in the cytotoxic properties of these cells (Agerberth et al. 2000), or alternatively an immunomodulatory role. Consistent with these observations, it has been shown that both TLR2 and TLR5 agonists induce the release of HNP1–3 from NK cells into the supernatant and that this release is increased synergistically in the presence of other cytokines found at the site of inflammation (Chalifour et al. 2004).

It is not entirely clear what the role of defensins may be in modulating NK-induced cytotoxicity. In one study it was found that NK mediated cytolysis of the transformed cell line KN62 is decreased in the presence of HNP1–3 in a dose-dependent manner. As well, NK cells treated with HNPs had a decreased expression of both CD16 and CD56 (Zhang et al. 2004). This study also demonstrated that there are high concentrations of HNPs due to the infiltration of neutrophils in colorectal tumours, but not in surrounding healthy tissue. Thus the authors hypothesised that the presence of HNPs might actually protect cancerous cells from NK cytolysis. A conflicting study demonstrated that PBMCs, treated with opsonin-coated zymosan particles, induced the release of substances that enhanced NK-mediated cytotoxicity. These substances were identified as neutral serine proteases and HNPs. Of the peptides tested, HNP1 was the most potent, increasing NK-mediated cytolysis optimally at a concentration of 1.25 µg/ml (Lala et al. 1992). Clearly, further studies are required to fully elucidate the role that host defence peptides have on NK mediated cytoxicity.

2.4.2
Monocytes and Macrophages

Monocytes and macrophages do not express high levels of defensins or cathelicidins unless stimulated by LPS or pro-inflammatory mediators (Agerberth et al. 2000; Duits et al. 2002). However, when thus stimulated, they secrete as yet unidentified factors that stimulate epithelial cells and keratinocytes to produce host defence peptides (Liu et al. 2003). Monocytes and macrophages are, however, quite responsive to stimulation with these peptides and both LL-37 and the defensins have been demonstrated to induce chemotaxis (Territo et al. 1989; De et al. 2000). It has been noted that host defence peptides are strong inducers of chemokine activity in monocytes (Chaly et al. 2000; Bowdish 2004). Interestingly it has been demonstrated that the HNPs are able to prevent HIV replication in monocytes and monocyte-derived macrophages and that this property may be due to their ability to induce chemokine production and/or receptor antagonism (Guo et al. 2004). HNP1 and HNP2 were both demonstrated to induce production of MIP-α and MIP-1β, the ligands for CCR5 in monocyte-derived macrophages and to prevent replication of a CCR5 tropic strain of the virus, presumably by blocking virus binding to CCR5 (Guo et al. 2004).

There has been some evidence that host defence peptides might work as opsonins (Fleischmann et al. 1985; Sawyer et al. 1988). Although this property would be predicted to generally enhance that antimicrobial activity associated with these peptides, one study demonstrated that an LL-37 derivative actually promoted infectivity of *Coxiella burnetii*, an intracellular pathogen of macrophages (Aragon et al. 1995).

Generally, host defence peptides are thought to possess anti-inflammatory properties, as described above. However, in some cases, they may actually enhance some aspects of a pro-inflammatory response. LL-37, for example, has been demonstrated to enhance IL-1β processing and release in LPS-primed primary human monocytes (Elssner et al. 2004). This property appears to be conserved across a range of host defence peptides from a number of different species (Perregaux et al. 2002).

2.4.3
Mast Cells

Mast cells are distributed throughout the body and are also found in low amounts in the blood. These cells rapidly accumulate at sites of infection, and upon encountering certain bacterial components or pro-inflammatory stimuli they promote the inflammatory response by releasing histamine, which causes vasodilation and thus assists in the recruitment of cells and substances from the blood. Two host defence peptides, LL-37 and HBD2 have been demonstrated to be chemotactic for rat mast cells, although they may work by different mechanisms (Niyonsaba et al. 2002b, 2003). Thus mast cells may accumulate at sites of high concentrations of host defence peptides such as at sites of neutrophil degranulation or at epithelial surfaces in vivo. HBD2 and LL-37 as well as the HNP1–3 and HNP homologues from rabbits and guinea pigs have also been demonstrated to induce histamine release (Befus et al. 1999; Niyonsaba et al. 2001). This property may be especially important in the development of host defence peptides as drugs, as mast cell degranulation is a potentially detrimental side effect.

2.4.4
Epithelial Cells

Host defence peptides have been demonstrated to interact with epithelial cells. Neutrophil peptides have been demonstrated to induce proliferation (M. Nishimura et al. 2004), induce chemokine production (Van Wetering et al. 1997) and stimulate cell signalling pathways (Bowdish 2004). LL-37 has been demonstrated to bind to a lung epithelial cell line in a manner which suggests that it may have more than one receptor (Lau et al. 2005). It has also been demonstrated that binding and subsequent internalisation is required in order to induce IL-8 production (Lau et al. 2005).

3
Host Defence Peptides in the Adaptive Immune Response

3.1
Adjuvant Activity

In addition to apparently having multiple roles in innate immunity, it is becoming clear that host defence peptides can modulate the adaptive immune
response, and several studies have now demonstrated adjuvant activities of
host defence peptides in vivo. The mechanisms involved remain unclear, although these activities could reflect the innate immunity modulating activity
of host defence peptides and the fact that there appears to be a strong inter-
connection between innate and adaptive immunity.

The relatively non-immunogenic model antigen ovalbumin (OVA) is
widely used to study adaptive immune responses. Intranasal co-administration of human α-defensins HNP1–3 with OVA was shown to enhance the
production of OVA-specific IgG antibodies and OVA-specific CD4$^+$ T cells,
which produced significantly more IFNγ, IL-5, IL-6, and IL-10 (Lillard et al.
1999). This indicated the capacity of α-defensins to alter the host response
to OVA, acting as adjuvants to promote a mixed T helper (Th) cell response.
In two other recent studies, HNP1, the human β-defensins HBD1 and HBD2
(Brogden et al. 2003), and a simple synthetic peptide KLKL$_5$KLK (Fritz et al.
2004) were also demonstrated to be effective adjuvants. The observation that
such effects are observed with the model peptide KLKL$_5$KLK suggests the
possibility of a relatively non-specific mechanism, and that such activities
may therefore be seen with a broad range of host defence peptides. However,
the nature of the enhanced responses may depend both on the antigen and
the peptide used. In contrast to the mixed Th-1/Th-2 response enhanced in
the HNP1–3-treated animals (Lillard et al. 1999), OVA-stimulated splenic
lymphoid cell cultures were found to produce significantly decreased levels
of IFN-γ, when taken from HBD2-treated mice (Brogden et al. 2003). On
the other hand, although KLKL$_5$KLK induced a strong Th-2 type response
when co-administered with OVA, it enhanced a mixed response when the
trivalent influenza split-vaccine FLUVIRIN was used as antigen, with the
production of both IgG$_1$ and IgG$_2$ antibodies (Fritz et al. 2004). Interestingly,
this report also demonstrated that a peptide could markedly enhance antigen
association with a monocytic cell line in vitro, and that co-administration
in vivo could result in the formation of a transient depot of antigen at
the site of injection. These observations indicate that antigen uptake by
antigen-presenting cells (APCs) might be enhanced in the presence of the
peptide, and thus influence responses in the presence of KLKL$_5$KLK in vivo.
Although these studies clearly showed altered humoral and Th responses to

antigens, the functional consequences of these alterations were not clearly demonstrated.

In another study, mice were given an intraperitoneal vaccination combining a B-cell lymphoma idiotype antigen and daily 1 μg injections of human α-defensins. This study also demonstrated adjuvant activity, whereby the defensins led to increased levels of antigen-specific IgG antibodies and enhanced IFN-γ production by splenic cells (Tani et al. 2000). Moreover, defensins showed mitogenic properties (with a significant increase in the number of splenic B cells) and led to an increase in resistance to tumour challenge. The latter observation raised the possibility that an antigen-specific cytotoxic T cell response was being generated in addition to a humoral response.

These studies collectively demonstrate that co-administration of host defence peptides with antigens can enhance and perhaps alter the nature of the host's specific adaptive immune responses in vivo. This raises the question of whether host defence peptides might naturally act as endogenous adjuvants to enhance normal immunological responses, since many peptides can be up-regulated or secreted at sites of infection and inflammation. It is unclear whether the doses used in such studies to assess in vivo immunological processes are within relevant physiological ranges. The physiological significance should be addressed by examining transgenic mice with defective production of host defence peptides, although the issue of possible functional redundancy amongst the many murine defensins must be considered when examining single gene knockouts. The published characterisations of such mice have concentrated on innate responses and have generally not described defects in adaptive immune responses (Nizet et al. 2001; Morrison et al. 2002; Moser et al. 2002). However, one mBD-1 knockout model was found to display a defect in generating antibodies to the carbohydrate capsule of pneumococci (C. Moser, personal communication). While this is consistent with an in vivo role for this constitutively expressed defensin in generating an effective humoral response, it is clearly an area requiring further study. Regardless of possible physiological significance, the adjuvant effects of host defence peptides are clearly of interest from an immunotherapeutic and vaccinology perspective.

In contrast to the studies that have co-administered host defence peptides and antigens, other groups have taken an alternative DNA-vaccine approach. This methodology involved immunizing mice with DNA plasmids encoding non-immunogenic lymphoma antigens fused to murine β-defensins (Biragyn et al. 2001). Successfully transfected cells of an undefined nature should then express the peptide/lymphoma antigen fusion proteins. This strategy represents an attempt to target antigen to immature dendritic cells (iDCs), by exploiting the affinity of the β-defensin portion of the fusion proteins for the chemokine receptor CCR6, expressed on iDCs. This approach also demonstrated an adjuvant capacity for host defence peptides; however, IgG responses were only observed when the plasmid encoded a fusion of the antigen and peptide, and not observed after simple co-administration of peptide and antigen. Interestingly, anti-tumour activity was also generated in these mice (most effectively with murine β-defensin 2), but did not correlate with the amplitude of the humoral response (superior with murine β-defensin 3). Furthermore, this anti-tumour activity could be transferred to other mice with the delivery of splenocytes, but not serum, from vaccinated animals, indicating the generation of cytotoxic T cells in response to non-immunogenic antigens when fused to peptides. In another recent study, using a similar approach, immunisation of mice with a plasmid fusing the human cathelicidin LL-37 to M-CSFR (acting as a tumour antigen in this model) also generated enhanced antigen-specific humoral and cytotoxic responses, and prolonged survival in a tumour model (An et al. 2004). LL-37 fusion plasmids were found to be significantly more effective than the M-CSFR plasmid alone, or co-administration of separately encoded M-CSFR and LL-37 plasmids.

These animal studies all demonstrate the adjuvant capacity of host defence peptides in vivo, but the mechanisms underlying these observations have not been fully elucidated. A variety of hypotheses can be proposed, including direct modulation of lymphocyte responses, mitogenic effects, chemotactic capacity, increased APC antigen uptake and consequently enhanced presentation, activity as endogenous danger signals, alterations to the APC cytokine environment, or direct modulation of APC function (Fig. 1).

The most obvious mechanisms might include altered antigen uptake (Fritz et al. 2004) and direct modulation of lymphocyte activity and proliferation (Tani et al. 2000), boosting APC presentation, cellular and humoral responses. This could be further enhanced by direct chemotactic effects of host defence peptides, resulting in the chemotaxis of monocytes, neutrophils, macrophages, iDCs, mast cells and T lymphocytes (Territo et al. 1989; Chertov et al. 1996; Yang et al. 1999, 2000a, 2000b; Niyonsaba et al. 2002a, 2002b), and the enhancement of chemokine receptor expression on these cells (Scott et al. 2002). In addition, host defence peptides could act indirectly to stimulate the release of potent traditional chemokines (such as IL-8) from epithelial cells (Van Wetering et al. 1997), and/or cause mast cell degranulation (Niyonsaba et al. 2001), enhancing vascular permeability. These direct and indirect chemotactic effects could amplify the inflammatory response and bring key cells of the adaptive immune response to the location of the antigen. While recruiting memory T cells to an infection site may induce a more rapid cellular response to previously encountered antigens, the recruitment of monocytes and iDC is likely to be critical to generating the initial response.

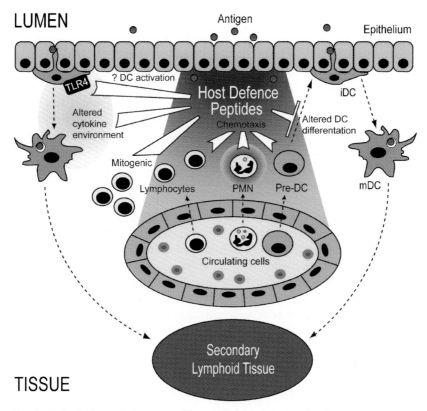

Fig. 1 Cationic host defence peptides modulate the adaptive immune response. Cationic host defence peptides, produced by epithelial cells, neutrophils (*PMN*), transfected cells, or added exogenously, might alter the adaptive immune response to antigen by inducing: (a) chemotaxis of immature dendritic cells (*iDC*), monocytes/Pre-DCs, PMN and T cells, (b) modulation of lymphocyte activity and/or proliferation, (c) alteration of the local cytokine environment, (d) direct iDC activation via TLR4, or (e) generation of primed iDCs with enhanced antigen uptake and presentation capacity. (Reproduced with kind permission from *Leukemia Research*)

Dendritic cells are sentinel leukocytes that capture antigen in the peripheral tissues and then initiate and orchestrate T cell helper (Th-1) responses, the nature of which determines the character of the adaptive immune response (Moser and Murphy 2000). This process is critical to generating a successful defence against harmful microbial non-self antigens while maintaining tolerance to self. It is dependent upon the antigen-capturing capabilities of iDCs, and antigen-presenting capabilities of mature dendritic cells (mDCs). iDCs are derived from circulating haematopoietic precursor cells and preDC populations (monocytes and plasmacytoid cells) under the influence of specific cytokines and growth factors (Liu 2001; Pulendran et al. 2001). In the tissues, these cells encounter and take up antigen. Stimulation of iDCs by conserved structures on certain microbial antigens, acting via the Toll-like receptors (TLRs) of the innate immune system (Medzhitov and Janeway 2000) or by signals from host cytokines, results in DC activation. These activated cells mature to become effective antigen-processing and presenting mDCs, migrate to the secondary lymphoid organs and interact with naïve T-lymphocytes (Banchereau et al. 2000). The characteristics of the mDCs determine the nature and consequences of this interaction, resulting in proliferation and differentiation, or deletion of T cells, and determine the polarisation of the Th response (Lanzavecchia and Sallusto 2001). Whereas steady-state trafficking of non-activated iDCs carrying self-antigen is thought to help maintain tolerance, it has been proposed that sustained trafficking of large numbers of highly stimulatory mDCs to the T cell areas is necessary for the generation of an effective T cell proliferative response (Lanzavecchia and Sallusto 2001). This would require extensive, repeated recruitment of circulating preDCs to the site of infection, with rapid differentiation to replace the "first-line" resident iDCs. Thus, at the simplest level, it is conceivable that the in vivo effects of host defence peptides on the adaptive immune response are the result of direct and indirect chemotaxis of iDCs and monocytes to the site of inflammation.

3.2
Mechanisms

Thus, chemotaxis, altered antigen uptake, and mitogenic effects on lymphocytes offer potential mechanisms by which host defence peptides may enhance responses to immunogenic antigens. However, these explanations do not account for the generation of humoral and cytotoxic T lymphocyte responses to non-immunogenic antigens observed in vivo. In these examples, an increase in the number of DCs encountered and the amount of antigen taken up should make no difference to the response in the absence of an activating signal. Indeed, theoretically, this might serve to increase host tolerance to these antigens. Despite this, host defence peptides clearly enhance an adaptive immune response to non-immunogenic antigens in vivo.

On the basis of current literature, we propose three hypotheses that might explain these observations. The first two theories propose that these peptides directly or indirectly provide an activating signal to differentiated iDCs concurrent with these cells encountering antigen, while the third proposes peptide modulation of DC differentiation from precursor cells.

In one intriguing report, it was demonstrated that murine β-defensin 2 fusion proteins were capable of activating iDCs directly in a TLR-4 dependent manner, to produce T helper (Th-1) polarised responses (Biragyn et al. 2002). In the context of these DNA plasmid vaccines, stimulation of the innate pattern recognition pathways through TLR would occur in close spatial and temporal conjunction to an otherwise non-immunogenic antigen. This suggests that host defence peptides might be capable of functioning as endogenous ligands of innate pattern recognition receptors. However, activation of iDCs was not observed with murine β-defensin 2 in the absence of fusion to lymphoma antigen. Although this appears to make it improbable that this mechanism is responsible for most of the above-described in vivo observations, it is possible that peptide and antigen concentrations are much higher in co-administration studies, than when relying on DNA plasmid expression. Thus, if the temporal coordination of TLR4 stimulation and antigen presentation are critical, this may be achieved by high concentration co-administration and depot formation at the site of delivery but, when utilizing a DNA vaccine approach, require peptide fusion. However, despite proving effective as an adjuvant for humoral responses (Biragyn et al. 2001), murine β-defensin 3 fusion proteins did not have TLR-4 dependent iDC-activating capabilities (Biragyn et al. 2002). Furthermore we have seen no evidence of an ability to directly mature human monocyte-derived DC in vitro when studying a range of peptides at or above the putative physiological concentrations, (DJ Davidson, AJ Currie,, REW Hancock, DP Speert, unpublished data). These data indicate that direct activation of iDCs may not be an inherent property of host defence peptides. Thus, although direct activation of iDCs is unlikely to be the basic mechanism underlying host defence peptide adjuvant activities, we cannot rule out a peptide-specific effect in which temporal coordination of TLR4 ligation and chemokine receptor-directed antigen uptake by the same cell are critical.

An alternative mechanism to explain altered iDC activation is to suggest that the effects of host defence peptides might be indirect, acting to alter the milieu in which these cells encounter antigen. Defensins have been shown to increase expression of various cytokines, including IL-8, IL-6, MCP-1 and GM-CSF (van Wetering et al. 2002), in different airway epithelial cells, while LL-37 can induce IL-8 and MCP-1 expression in epithelial and monocytic cells (Scott et al. 2002). Changes to the cytokine environment may induce a myriad of effects, from the chemotactic activities of MCP-1 and IL-8, and cellular differentiation effects of GM-CSF, to the enhancement of B cell proliferation and blockade of the suppressive effects of regulatory T cells by IL-6 (Pasare and Medzhitov 2003). Possibly other factors are induced that might activate iDCs even in the presence of non-immunogenic antigens. Following activation of monocytes with *S. aureus* or phorbol myristate, human α-defensins at concentrations as low as 1 nM, can increase the expression of TNF-α and IL-1β (Chaly et al. 2000). These cytokines have the potential to directly induce DC maturation, sharing components of activating pathways with TLR, and thus potentially enhancing the generation of highly stimulatory mDCs. However, while such mechanisms might therefore be proposed at sites of inflammation, similar activities in the presence of non-immunogenic antigens are only speculative.

The third hypothesis relates to our recent discovery that the human cathelicidin LL-37 can modulate the differentiation of iDCs from precursor cells, with consequent impact on Th cell polarisation (Davidson et al. 2004). The stimulatory nature of DCs is subject to dynamic temporal regulation (Langenkamp et al. 2000) and can be modified by precursor cell lineage, the specific antigen captured, the receptors engaged, and the microenvironment for both differentiation and maturation (Liu 2001; Pulendran et al. 2001; de Jong et al. 2002; Boonstra et al. 2003). We demonstrated that LL-37 has the potential to act as an endogenous environmental modifier of DC differentiation (Davidson et al. 2004). LL-37-primed DCs displayed significantly upregulated endocytic capacity, modified phagocytic receptor expression and function, up-regulated co-stimulatory molecule expression, enhanced secretion of Th-1-inducing cytokines, and promoted Th-1 responses in vitro. These results suggest the potential for host defence peptides to exert effects on the adaptive immune system by priming newly differentiating DCs to enhance their antigen uptake and presentation capabilities and influence the nature of the response they will subsequently generate. According to this hypothesis, host defence peptides would not simply affect "first-line" resident iDCs, but act upon the differentiating "second-line" DCs that can sustain highly stimulatory presentation of antigen to generate an effective T cell proliferative response. In the context of a physiological role, LL-37-primed iDCs might be generated at sites where LL-37 is up-regulated in response to infection or inflammation, be matured by immunogenic antigens, and promote a more robust adaptive immune response. However, LL-37-primed iDCs also have increased expression of the co-stimulatory molecule CD86 in vitro in the absence of activating stimuli and any other signs of maturation. If such cells were generated in vivo in the presence of a high concentration depot of host defence peptides at the site of vaccination, or in an area of peptide overexpression by host cells transfected with a DNA vaccine, they might be capable of presenting non-immunogenic antigen in a stimulatory context. Although enhanced humoral and cytotoxic T-lymphocyte (CTL) responses in DNA vaccinated mice are dependent on the fusion of LL-37 and the antigen (An et al. 2004), this might again relate to issues of local concentration and co-presentation to the same cells. Interestingly however, enhanced splenocyte IFN-γ responses were observed not only in mice vaccinated with the LL-37 fusion plasmid, but also in those given separately encoded LL-37 and antigen plasmids. This might reflect the enhanced IFN-γ responses of T cells stimulated with LL-37-primed DCs in vitro. However, further research is required to establish the in vivo significance of the LL-37-priming of DCs observed in vitro. Furthermore, the effects of other host defence peptides on DC differentiation have not been described, raising uncertainty about this hypothesis in the context of the in vivo studies using defensins, or synthetic peptides as adjuvants.

In conclusion, the potential for host defence peptides to modulate the adaptive immune response is evident, but remains largely undescribed. In addition to further exploration of the effects in vitro, innovative in vivo modelling is a priority to dissect the mechanisms underlying these observations. A clear understanding of the extent and mechanisms of the immunomodulatory effects of host defence peptides will be fundamental to their future development as novel therapeutic agents. However, these early in vivo studies demonstrate great potential for targeting tumours, recalcitrant, antibiotic-resistant pathogens, infections for which effective vaccines do not exist, and vaccines, which generate suboptimal responses of an inappropriate nature.

4
Conclusion

Mammalian host defence peptides were originally discovered as components of the non-oxidative killing mechanisms of neutrophils. In the granules of neutrophils, these peptides are found at sufficiently high concentrations to be antimicrobial. However, it is less clear that this is the case at mucosal surfaces or in other body fluids, especially at sites that already support a rich and diverse normal flora. Certain body fluids, including sinus fluid and gastric juices, have innate antimicrobial activity against certain bacteria, and the components that appear to contribute to this include a variety of antimicrobial proteins (e.g. lysozyme, secretory phospholipase A_2), as well as peptides (e.g. defensins) (Cole et al. 2002). However, the specific contributions of each of these components to overall antimicrobial activity has not been determined, and given the moderate levels of peptides often found in these fluids, synergy between individual agents working in combination may be important, as been demonstrated for some peptides, and combinations of lysozyme and peptides in vitro (Singh et al. 2000; Yan and Hancock 2001). This is still further complicated at sites such as the mucosa when considering the abilities of host defence peptides to modulate innate immunity, as discussed extensively in this review.

One approach to trying to resolve these mechanisms is to use genetic strategies using either knockout models in animals or specific genetic deficiencies in humans (Ganz et al. 1988; Nizet et al. 2001; Putsep et al. 2002; Salzman et al. 2003). Such studies have clearly demonstrated that defensins and cathelicidins are integral components of host defence in mammals, and that these peptides are required to reduce bacterial load and inhibit infection. However, they do not always permit the discrimination between the various potential mechanisms of host defence peptides, namely direct antimicrobial activity, synergistic activity with other antimicrobial components and/or the broad range of abilities to modulate immunity. Indeed, these distinctions may be unimportant, as they all have the same net result, namely the control of potentially dangerous microbes. We hypothesise that all of these mechanisms operate in the body of mammals, and that any given peptide may have different roles in anti-infective immunity according to the body site it is found, its local concentration, the prevailing physiological conditions, and the other antimicrobial and cellular components of immunity at that site. A clear illustration of this complexity is provided by a study of the use of human defensin 1 (HNP-1) to treat experimental peritoneal *K. pneumoniae* infections (Welling et al. 1998). In this model, HNP1 injection was shown to markedly reduce bacterial numbers, but the antibacterial effect was associated with an increased influx of leukocytes into the peritoneal cavity, and this was strongly related to the antibacterial effect, as no such activity was observed in leukocytopenic mice.

A further challenge to our thinking, and possibly the most profound question in innate immunity, is how mammals manage to support a complex normal flora while retaining the ability to respond to potentially dangerous pathogens. The Toll-like receptors (TLRs), which represent one of the major "triggers" of innate immunity, do not really distinguish between the conserved surface molecules from pathogens and commensal organisms. Thus it is of interest as to whether host defence peptides may play a role in this delicate dance between symbiosis and pathogenesis. As shown by E. Nishimura et al., many commensal bacteria from the oral cavity are quite susceptible to HBD2, while Chung and Dale indicated that both commensals and pathogenic bacteria can induce this defensin (Chung and Dale 2004; E. Nishimura et al. 2004). Conversely, Putsep et al. compared germ-free and normal mice to conclude that an intestinal microflora does not have a major influence on the production or processing of defensins (Putsep et al. 2000). However, we consider at least one activity of these peptides might play a role for host defence peptides in homeostasis, namely anti-endotoxic activity, which in our experience is expressed at lower concentrations of LL-37 than other immunomodulatory activities. It seems possible that this would provide a mechanism for balancing the potential stimulation of TLRs by surface molecules from commensal organisms. Innate immunity would then be triggered by local perturbations of peptide concentrations though mechanisms such as degranulation of phagocytes, or specific up-regulation by certain cytokines. In addition, the efficiency of these peptides might the increased by other local factors, for example phagocytes that enter the local site (Welling et al. 1998) or local cytokines such as GM-CSF that has been shown to enhance MAPK signalling by LL-37 (Bowdish 2004).

The full spectrum of the immunomodulatory properties of these peptides is not yet known and each new report demonstrates that the range and importance of these immunomodulatory effects is greater than initially suspected. Most likely both antimicrobial and immunomodulatory activities are to some degree involved, as this is consistent with the redundant and efficient nature of evolution, and with the concept of innate immunity as a network of overlapping mechanisms. Understanding the interplay between host defence peptides and innate and adaptive immunity will expand our knowledge of immunity in general and allow us to develop anti-infective therapies adapted from nature's design that will enhance the efficiency of immune defences.

Acknowledgements We would like to acknowledge support for the authors' research from the Applied Food and Materials Network, the Canadian Bacterial Diseases Network and the Functional Pathogenomics of Mucosal Immunity program funded by Genome Prairie and Genome BC, with additional funding from Inimex Pharmaceuticals. REWH holds a Canada Research Chair, DMEB is supported by a CIHR studentship, and DJD was funded by a Wellcome Trust UK, International Prize Travelling Research Fellowship (060168).

References

Aarbiou J, Ertmann M, van Wetering S, van Noort P, Rook D, Rabe KF, Litvinov SV, van Krieken JH, de Boer WI, Hiemstra PS (2002) Human neutrophil defensins induce lung epithelial cell proliferation in vitro. J Leukoc Biol 72:167–174

Aarbiou J, Verhoosel RM, Van Wetering S, De Boer WI, Van Krieken JH, Litvinov SV, Rabe KF, Hiemstra PS (2004) Neutrophil defensins enhance lung epithelial wound closure and mucin gene expression in vitro. Am J Respir Cell Mol Biol 30:193–201

Agerberth B, Charo J, Werr J, Olsson B, Idali F, Lindbom L, Kiessling R, Jornvall H, Wigzell H, Gudmundsson GH (2000) The human antimicrobial and chemotactic peptides LL-37 and alpha-defensins are expressed by specific lymphocyte and monocyte populations. Blood 96:3086–3093

An LL, Yang YH, Ma XT, Lin YM, Li G, Song YH, Wu KF (2004) LL-37 enhances adaptive immune response in murine model challenged with tumour cells. Leuk Res 29:535–543

Aragon AS, Pereira HA, Baca OG (1995) A cationic antimicrobial peptide enhances the infectivity of Coxiella burnetii. Acta Virol 39:223–226

Ashitani J, Mukae H, Hiratsuka T, Nakazato M, Kumamoto K, Matsukura S (2002) Elevated levels of alpha-defensins in plasma and BAL fluid of patients with active pulmonary tuberculosis. Chest 121:519–526

Baconnais S, Tirouvanziam R, Zahm JM, de Bentzmann S, Peault B, Balossier G, Puchelle E (1999) Ion composition and rheology of airway liquid from cystic fibrosis fetal tracheal xenografts. Am J Respir Cell Mol Biol 20:605–611

Bals R, Goldman MJ, Wilson JM (1998a) Mouse beta-defensin 1 is a salt-sensitive antimicrobial peptide present in epithelia of the lung and urogenital tract. Infect Immun 66:1225–1232

Bals R, Wang X, Wu Z, Freeman T, Bafna V, Zasloff M, Wilson JM (1998b) Human beta-defensin 2 is a salt-sensitive peptide antibiotic expressed in human lung. J Clin Invest 102:874–880

Bals R, Wang X, Zasloff M, Wilson JM (1998c) The peptide antibiotic LL-37/hCAP-18 is expressed in epithelia of the human lung where it has broad antimicrobial activity at the airway surface. Proc Natl Acad Sci U S A 95:9541–9546

Bals R, Weiner DJ, Moscioni AD, Meegalla RL, Wilson JM (1999) Augmentation of innate host defense by expression of a cathelicidin antimicrobial peptide. Infect Immun 67:6084–6089

Banchereau J, Briere F, Caux C, Davoust J, Lebecque S, Liu YJ, Pulendran B, Palucka K (2000) Immunobiology of dendritic cells. Annu Rev Immunol 18:767–811

Becker MN, Diamond G, Verghese MW, Randell SH (2000) CD14-dependent lipopolysaccharide-induced beta-defensin-2 expression in human tracheo-bronchial epithelium. J Biol Chem 275:29731–29736

Befus AD, Mowat C, Gilchrist M, Hu J, Solomon S, Bateman A (1999) Neutrophil defensins induce histamine secretion from mast cells: mechanisms of action. J Immunol 163:947–953

Bensch KW, Raida M, Magert HJ, Schulz-Knappe P, Forssmann WG (1995) hBD-1: a novel beta-defensin from human plasma. FEBS Lett 368:331–335

Biragyn A, Surenhu M, Yang D, Ruffini PA, Haines BA, Klyushnenkova E, Oppenheim JJ, Kwak LW (2001) Mediators of innate immunity that target immature, but not mature, dendritic cells induce antitumor immunity when genetically fused with nonimmunogenic tumor antigens. J Immunol 167:6644–6653

Biragyn A, Ruffini PA, Leifer CA, Klyushnenkova E, Shakhov A, Chertov O, Shirakawa AK, Farber JM, Segal DM, Oppenheim JJ, Kwak LW (2002) Toll-like receptor 4-dependent activation of dendritic cells by beta -defensin 2. Science 298:1025–1029

Boonstra A, Asselin-Paturel C, Gilliet M, Crain C, Trinchieri G, Liu YJ, O'Garra A (2003) Flexibility of mouse classical and plasmacytoid-derived dendritic cells in directing T helper type 1 and 2 cell development: dependency on antigen dose and differential toll-like receptor ligation. J Exp Med 197:101–109

Bowdish DME, Davidson DJ, Speert DP, Hancock REW (2004) The Human cationic peptide LL-37 induces activation of the extracellular signal-regulated kinase and p38 kinase pathways in primary human monocytes. J Immunol 172:3758–3765

Bowdish DM, Davidson DJ, Lau YE, Lee K, Scott MG, Hancock RE (2005) Impact of LL-37 on anti-infective immunity. J Leukoc Biol 77:451–459

Brogden KA, Heidari M, Sacco RE, Palmquist D, Guthmiller JM, Johnson GK, Jia HP, Tack BF, McCray PB (2003) Defensin-induced adaptive immunity in mice and its potential in preventing periodontal disease. Oral Microbiol Immunol 18:95–99

Buhimschi IA, Jabr M, Buhimschi CS, Petkova AP, Weiner CP, Saed GM (2004) The novel antimicrobial peptide beta3-defensin is produced by the amnion: a possible role of the fetal membranes in innate immunity of the amniotic cavity. Am J Obstet Gynecol 191:1678–1687

Butmarc J, Yufit T, Carson P, Falanga V (2004) Human beta-defensin-2 expression is increased in chronic wounds. Wound Repair Regen 12:439–443

Chalifour A, Jeannin P, Gauchat JF, Blaecke A, Malissard M, N'Guyen T, Thieblemont N, Delneste Y (2004) Direct bacterial protein PAMP recognition by human NK cells involves TLRs and triggers alpha-defensin production. Blood 104:1778–1783

Chaly YV, Paleolog EM, Kolesnikova TS, Tikhonov II, Petratchenko EV, Voitenok NN (2000) Neutrophil alpha-defensin human neutrophil peptide modulates cytokine production in human monocytes and adhesion molecule expression in endothelial cells. Eur Cytokine Netw 11:257–266

Chen CI, Schaller-Bals S, Paul KP, Wahn U, Bals R (2004) Beta-defensins and LL-37 in bronchoalveolar lavage fluid of patients with cystic fibrosis. J Cyst Fibros 3:45–50

Chertov O, Michiel DF, Xu L, Wang JM, Tani K, Murphy WJ, Longo DL, Taub DD, Oppenheim JJ (1996) Identification of defensin-1, defensin-2, and CAP37/azurocidin as T-cell chemoattractant proteins released from interleukin-8-stimulated neutrophils. J Biol Chem 271:2935–2940

Chung WO, Dale BA (2004) Innate immune response of oral and foreskin keratinocytes: utilization of different signaling pathways by various bacterial species. Infect Immun 72:352–358

Ciornei CD, Egesten A, Bodelsson M (2003) Effects of human cathelicidin antimicrobial peptide LL-37 on lipopolysaccharide-induced nitric oxide release from rat aorta in vitro. Acta Anaesthesiol Scand 47:213–220

Cole AM, Dewan P, Ganz T (1999) Innate antimicrobial activity of nasal secretions. Infect Immun 67:3267–3275

Cole AM, Tahk S, Oren A, Yoshioka D, Kim YH, Park A, Ganz T (2001) Determinants of Staphylococcus aureus nasal carriage. Clin Diagn Lab Immunol 8:1064–1069

Cole AM, Liao HI, Stuchlik O, Tilan J, Pohl J, Ganz T (2002) Cationic polypeptides are required for antibacterial activity of human airway fluid. J Immunol 169:6985–6991

Conejo-Garcia JR, Benencia F, Courreges MC, Kang E, Mohamed-Hadley A, Buckanovich RJ, Holtz DO, Jenkins A, Na H, Zhang L, Wagner DS, Katsaros D, Caroll R, Coukos G (2004) Tumor-infiltrating dendritic cell precursors recruited by a beta-defensin contribute to vasculogenesis under the influence of Vegf-A. Nat Med 10:950–958

Cunliffe RN, Rose FR, Keyte J, Abberley L, Chan WC, Mahida YR (2001) Human defensin 5 is stored in precursor form in normal Paneth cells and is expressed by some villous epithelial cells and by metaplastic Paneth cells in the colon in inflammatory bowel disease. Gut 48:176–185

Daher KA, Selsted ME, Lehrer RI (1986) Direct inactivation of viruses by human granulocyte defensins. J Virol 60:1068–1074

Davidson DJ, Currie AJ, Reid GS, Bowdish DM, MacDonald KL, Ma RC, Hancock RE, Speert DP (2004) The cationic antimicrobial peptide LL-37 modulates dendritic cell differentiation and dendritic cell-induced T cell polarization. J Immunol 172:1146–1156

De Jong EC, Vieira PL, Kalinski P, Schuitemaker JH, Tanaka Y, Wierenga EA, Yazdanbakhsh M, Kapsenberg ML (2002) Microbial compounds selectively induce Th1 cell-promoting or Th2 cell-promoting dendritic cells in vitro with diverse Th cell-polarizing signals. J Immunol 168:1704–1709

De Y, Chen Q, Schmidt AP, Anderson GM, Wang JM, Wooters J, Oppenheim JJ, Chertov O (2000) LL-37, the neutrophil granule- and epithelial cell-derived cathelicidin, utilizes formyl peptide receptor-like 1 (FPRL1) as a receptor to chemoattract human peripheral blood neutrophils, monocytes, and T cells. J Exp Med 192:1069–1074

Dhaliwal W, Bajaj-Elliott M, Kelly P (2003) Intestinal defensin gene expression in human populations. Mol Immunol 40:469–475

Dorschner RA, Pestonjamasp VK, Tamakuwala S, Ohtake T, Rudisill J, Nizet V, Agerberth B, Gudmundsson GH, Gallo RL (2001) Cutaneous injury induces the release of cathelicidin anti-microbial peptides active against group A Streptococcus. J Invest Dermatol 117:91–97

Duits LA, Nibbering PH, van Strijen E, Vos JB, Mannesse-Lazeroms SP, van Sterkenburg MA, Hiemstra PS (2003) Rhinovirus increases human beta-defensin-2 and -3 mRNA expression in cultured bronchial epithelial cells. FEMS Immunol Med Microbiol 38:59–64

Duits LA, Ravensbergen B, Rademaker M, Hiemstra PS, Nibbering PH (2002) Expression of beta-defensin 1 and 2 mRNA by human monocytes, macrophages and dendritic cells. Immunology 106:517–525

Durr M, Peschel A (2002) Chemokines meet defensins: the merging concepts of chemoattractants and antimicrobial peptides in host defense. Infect Immun 70:6515–6517

Elssner A, Duncan M, Gavrilin M, Wewers MD (2004) A novel P2X7 receptor activator, the human cathelicidin-derived peptide LL37, induces IL-1 beta processing and release. J Immunol 172:4987–4994

Emes RD, Goodstadt L, Winter EE, Ponting CP (2003) Comparison of the genomes of human and mouse lays the foundation of genome zoology. Hum Mol Genet 12:701–709

Erdag G, Morgan JR (2002) Interleukin-1alpha and interleukin-6 enhance the antibacterial properties of cultured composite keratinocyte grafts. Ann Surg 235:113–124

Fahlgren A, Hammarstrom S, Danielsson A, Hammarstrom ML (2003) Increased expression of antimicrobial peptides and lysozyme in colonic epithelial cells of patients with ulcerative colitis. Clin Exp Immunol 131:90–101

Fleischmann J, Selsted ME, Lehrer RI (1985) Opsonic activity of MCP-1 and MCP-2, cationic peptides from rabbit alveolar macrophages. Diagn Microbiol Infect Dis 3:233–242

Fritz JH, Brunner S, Birnstiel ML, Buschle M, Gabain A, Mattner F, Zauner W (2004) The artificial antimicrobial peptide KLKLLLLLKLK induces predominantly a TH2-type immune response to co-injected antigens. Vaccine 22:3274–3284

Frye M, Bargon J, Dauletbaev N, Weber A, Wagner TO, Gropp R (2000a) Expression of human alpha-defensin 5 (HD5) mRNA in nasal and bronchial epithelial cells. J Clin Pathol 53:770–773

Frye M, Bargon J, Lembcke B, Wagner TO, Gropp R (2000b) Differential expression of human alpha- and beta-defensins mRNA in gastrointestinal epithelia. Eur J Clin Invest 30:695–701

Ganz T, Selsted ME, Szklarek D, Harwig SS, Daher K, Bainton DF, Lehrer RI (1985) Defensins. Natural peptide antibiotics of human neutrophils. J Clin Invest 76:1427–1435

Ganz T, Metcalf JA, Gallin JI, Boxer LA, Lehrer RI (1988) Microbicidal/cytotoxic proteins of neutrophils are deficient in two disorders: Chediak-Higashi syndrome and "specific" granule deficiency. J Clin Invest 82:552–556

Garcia JR, Krause A, Schulz S, Rodriguez-Jimenez FJ, Kluver E, Adermann K, Forssmann U, Frimpong-Boateng A, Bals R, Forssmann WG (2001) Human beta-defensin 4: a novel inducible peptide with a specific salt-sensitive spectrum of antimicrobial activity. FASEB J 15:1819–1821

Giacometti A, Cirioni O, Ghiselli R, Mocchegiani F, D'Amato G, Circo R, Orlando F, Skerlavaj B, Silvestri C, Saba V, Zanetti M, Scalise G (2004) Cathelicidin peptide sheep myeloid antimicrobial peptide-29 prevents endotoxin-induced mortality in rat models of septic shock. Am J Respir Crit Care Med 169:187–194

Grutkoski PS, Graeber CT, Lim YP, Ayala A, Simms HH (2003) Alpha-defensin 1 (human neutrophil protein 1) as an antichemotactic agent for human polymorphonuclear leukocytes. Antimicrob Agents Chemother 47:2666–2668

Guo CJ, Tan N, Song L, Douglas SD, Ho WZ (2004) Alpha-defensins inhibit HIV infection of macrophages through upregulation of CC-chemokines. Aids 18:1217–1218

Halmerbauer G, Arri S, Schierl M, Strauch E, Koller DY (2000) The relationship of eosinophil granule proteins to ions in the sputum of patients with cystic fibrosis. Clin Exp Allergy 30:1771–1776

Harder J, Bartels J, Christophers E, Schroder JM (1997) A peptide antibiotic from human skin. Nature 387:861

Harder J, Bartels J, Christophers E, Schroder JM (2001) Isolation and characterization of human beta-defensin-3, a novel human inducible peptide antibiotic. J Biol Chem 276:5707–5713

Hase K, Murakami M, Iimura M, Cole SP, Horibe Y, Ohtake T, Obonyo M, Gallo RL, Eckmann L, Kagnoff MF (2003) Expression of LL-37 by human gastric epithelial cells as a potential host defense mechanism against Helicobacter pylori. Gastroenterology 125:1613–1625

Heilborn JD, Nilsson MF, Kratz G, Weber G, Sorensen O, Borregaard N, Stahle-Backdahl M (2003) The cathelicidin anti-microbial peptide LL-37 is involved in re-epithelialization of human skin wounds and is lacking in chronic ulcer epithelium. J Invest Dermatol 120:379–389

Hieshima K, Ohtani H, Shibano M, Izawa D, Nakayama T, Kawasaki Y, Shiba F, Shiota M, Katou F, Saito T, Yoshie O (2003) CCL28 has dual roles in mucosal immunity as a chemokine with broad-spectrum antimicrobial activity. J Immunol 170:1452–1461

Hiratsuka T, Mukae H, Iiboshi H, Ashitani J, Nabeshima K, Minematsu T, Chino N, Ihi T, Kohno S, Nakazato M (2003) Increased concentrations of human beta-defensins in plasma and bronchoalveolar lavage fluid of patients with diffuse panbronchiolitis. Thorax 58:425–430

Hornef MW, Frisan T, Vandewalle A, Normark S, Richter-Dahlfors A (2002) Toll-like receptor 4 resides in the Golgi apparatus and colocalizes with internalized lipopolysaccharide in intestinal epithelial cells. J Exp Med 195:559–570

Hoshino K, Ogawa K, Hishitani T, Kitazawa R (2003) Influence of heart surgery on magnesium concentrations in pediatric patients. Pediatr Int 45:39–44

Islam D, Bandholtz L, Nilsson J, Wigzell H, Christensson B, Agerberth B, Gudmunds-son G (2001) Downregulation of bactericidal peptides in enteric infections: a novel immune escape mechanism with bacterial DNA as a potential regulator. Nat Med 7:180–185

Jones DE, Bevins CL (1992) Paneth cells of the human small intestine express an antimicrobial peptide gene. J Biol Chem 267:23216–23225

Jones DE, Bevins CL (1993) Defensin-6 mRNA in human Paneth cells: implications for antimicrobial peptides in host defense of the human bowel. FEBS Lett 315:187–192

Kelly P, Feakins R, Domizio P, Murphy J, Bevins C, Wilson J, McPhail G, Poulsom R, Dhaliwal W (2004) Paneth cell granule depletion in the human small intestine under infective and nutritional stress. Clin Exp Immunol 135:303–309

Kirikae T, Hirata M, Yamasu H, Kirikae F, Tamura H, Kayama F, Nakatsuka K, Yokochi T, Nakano M (1998a) Protective effects of a human 18-kilodalton cationic antimicro-bial protein (CAP18)-derived peptide against murine endotoxemia. Infect Immun 66:1861–1868

Koczulla R, von Degenfeld G, Kupatt C, Krotz F, Zahler S, Gloe T, Issbrucker K, Unterberger P, Zaiou M, Lebherz C, Karl A, Raake P, Pfosser A, Boekstegers P, Welsch U, Hiemstra PS, Vogelmeier C, Gallo RL, Clauss M, Bals R (2003) An angiogenic role for the human peptide antibiotic LL-37/hCAP-18. J Clin Invest 111:1665–1672

Krisanaprakornkit S, Kimball JR, Weinberg A, Darveau RP, Bainbridge BW, Dale BA (2000) Inducible expression of human beta-defensin 2 by Fusobacterium nuclea-tum in oral epithelial cells: multiple signaling pathways and role of commensal bacteria in innate immunity and the epithelial barrier. Infect Immun 68:2907–2915

Krisanaprakornkit S, Kimball JR, Dale BA (2002) Regulation of human beta-defensin-2 in gingival epithelial cells: the involvement of mitogen-activated protein kinase pathways, but not the NF-kappaB transcription factor family. J Immunol 168:316–324

Kumar J, Okada S, Clayberger C, Krensky AM (2001) Granulysin: a novel antimicrobial. Expert Opin Investig Drugs 10:321–329

Lala A, Lindemann RA, Miyasaki KT (1992) The differential effects of polymorphonu-clear leukocyte secretion on human natural killer cell activity. Oral Microbiol Immunol 7:89–95

Langenkamp A, Messi M, Lanzavecchia A, Sallusto F (2000) Kinetics of dendritic cell activation: impact on priming of TH1, TH2 and nonpolarized T cells. Nat Immunol 1:311–316

Lanzavecchia A, Sallusto F (2001) Regulation of T cell immunity by dendritic cells. Cell 106:263–266

Larrick JW, Hirata M, Balint RF, Lee J, Zhong J, Wright SC (1995) Human CAP18: a novel antimicrobial lipopolysaccharide-binding protein. Infect Immun 63:1291–1297

Larrick JW, Hirata M, Zheng H, Zhong J, Bolin D, Cavaillon JM, Warren HS, Wright SC (1994) A novel granulocyte-derived peptide with lipopolysaccharide-neutralizing activity. J Immunol 152:231–240

Larrick JW, Lee J, Ma S, Li X, Francke U, Wright SC, Balint RF (1996) Structural, functional analysis and localization of the human CAP18 gene. FEBS Lett 398:74–80

Lau YE, Rozek A, Scott MG, Goosney DL, Davidson DJ, Hancock RE (2005) Interaction and cellular localization of the human host defense peptide LL-37 with lung epithelial cells. Infect Immun 73:583–591

Lehrer RI, Ganz T (2002) Defensins of vertebrate animals. Curr Opin Immunol 14:96–102

Lehrer RI, Ganz T, Szklarek D, Selsted ME (1988) Modulation of the in vitro candidacidal activity of human neutrophil defensins by target cell metabolism and divalent cations. J Clin Invest 81:1829–1835

Levy O, Ooi CE, Elsbach P, Doerfler ME, Lehrer RI, Weiss J (1995) Antibacterial proteins of granulocytes differ in interaction with endotoxin. Comparison of bactericidal/permeability-increasing protein, p15s, and defensins. J Immunol 154:5403–5410

Lillard JW Jr, Boyaka PN, Chertov O, Oppenheim JJ, McGhee JR (1999) Mechanisms for induction of acquired host immunity by neutrophil peptide defensins. Proc Natl Acad Sci U S A 96:651–656

Liu L, Roberts AA, Ganz T (2003) By IL-1 signaling, monocyte-derived cells dramatically enhance the epidermal antimicrobial response to lipopolysaccharide. J Immunol 170:575–580

Liu YJ (2001) Dendritic cell subsets and lineages, and their functions in innate and adaptive immunity. Cell 106:259–262

Maxwell AI, Morrison GM, Dorin JR (2003) Rapid sequence divergence in mammalian beta-defensins by adaptive evolution. Mol Immunol 40:413–421

Medzhitov R, Janeway C Jr (2000) Innate immunity. N Engl J Med 343:338–344

Morrison G, Kilanowski F, Davidson D, Dorin J (2002) Characterization of the mouse beta defensin 1, defb1, mutant mouse model. Infect Immun 70:3053–3060

Moser C, Weiner DJ, Lysenko E, Bals R, Weiser JN, Wilson JM (2002) Beta-defensin 1 contributes to pulmonary innate immunity in mice. Infect Immun 70:3068–3072

Moser M, Murphy KM (2000) Dendritic cell regulation of TH1-TH2 development. Nat Immunol 1:199–205

Muller CA, Markovic-Lipkovski J, Klatt T, Gamper J, Schwarz G, Beck H, Deeg M, Kalbacher H, Widmann S, Wessels JT, Becker V, Muller GA, Flad T (2002) Human alpha-defensins HNPs-1, -2, and -3 in renal cell carcinoma: influences on tumor cell proliferation. Am J Pathol 160:1311–1324

Murakami M, Ohtake T, Dorschner RA, Gallo RL (2002a) Cathelicidin antimicrobial peptides are expressed in salivary glands and saliva. J Dent Res 81:845–850

Murakami M, Ohtake T, Dorschner RA, Schittek B, Garbe C, Gallo RL (2002b) Cathelicidin anti-microbial peptide expression in sweat, an innate defense system for the skin. J Invest Dermatol 119:1090–1095

Murakami M, Lopez-Garcia B, Braff M, Dorschner RA, Gallo RL (2004) Postsecretory processing generates multiple cathelicidins for enhanced topical antimicrobial defense. J Immunol 172:3070–3077

Murakami M, Dorschner RA, Stern LJ, Lin KH, Gallo RL (2005) Expression and secretion of cathelicidin antimicrobial peptides in murine mammary glands and human milk. Pediatr Res 57:10–15

Murphy CJ, Foster BA, Mannis MJ, Selsted ME, Reid TW (1993) Defensins are mitogenic for epithelial cells and fibroblasts. J Cell Physiol 155:408–413

Nagaoka I, Hirota S, Niyonsaba F, Hirata M, Adachi Y, Tamura H, Heumann D (2001) Cathelicidin family of antibacterial peptides CAP18 and CAP11 inhibit the expression of TNF-alpha by blocking the binding of LPS to CD14(+) cells. J Immunol 167:3329–3338

Nagaoka I, Hirota S, Yomogida S, Ohwada A, Hirata M (2000) Synergistic actions of antibacterial neutrophil defensins and cathelicidins. Inflamm Res 49:73–79

Naik S, Kelly EJ, Meijer L, Pettersson S, Sanderson IR (2001) Absence of Toll-like receptor 4 explains endotoxin hyporesponsiveness in human intestinal epithelium. J Pediatr Gastroenterol Nutr 32:449–453

Nell MJ, Sandra Tjabringa G, Vonk MJ, Hiemstra PS, Grote JJ (2004) Bacterial products increase expression of the human cathelicidin hCAP-18/LL-37 in cultured human sinus epithelial cells. FEMS Immunol Med Microbiol 42:225–231

Nishimura E, Eto A, Kato M, Hashizume S, Imai S, Nisizawa T, Hanada N (2004) Oral streptococci exhibit diverse susceptibility to human beta-defensin-2: antimicrobial effects of hBD-2 on oral streptococci. Curr Microbiol 48:85–87

Nishimura M, Abiko Y, Kurashige Y, Takeshima M, Yamazaki M, Kusano K, Saitoh M, Nakashima K, Inoue T, Kaku T (2004) Effect of defensin peptides on eukaryotic cells: primary epithelial cells, fibroblasts and squamous cell carcinoma cell lines. J Dermatol Sci 36:87–95

Niyonsaba F, Someya A, Hirata M, Ogawa H, Nagaoka I (2001) Evaluation of the effects of peptide antibiotics human beta-defensins-1/-2 and LL-37 on histamine release and prostaglandin D(2) production from mast cells. Eur J Immunol 31:1066–1075

Niyonsaba F, Iwabuchi K, Matsuda H, Ogawa H, Nagaoka I (2002a) Epithelial cell-derived human beta-defensin-2 acts as a chemotaxin for mast cells through a pertussis toxin-sensitive and phospholipase C-dependent pathway. Int Immunol 14:421–426

Niyonsaba F, Iwabuchi K, Someya A, Hirata M, Matsuda H, Ogawa H, Nagaoka I (2002b) A cathelicidin family of human antibacterial peptide LL-37 induces mast cell chemotaxis. Immunology 106:20–26

Niyonsaba F, Hirata M, Ogawa H, Nagaoka I (2003) Epithelial cell-derived antibacterial peptides human beta-defensins and cathelicidin: multifunctional activities on mast cells. Curr Drug Targets Inflamm Allergy 2:224–231

Niyonsaba F, Ogawa H, Nagaoka I (2004) Human beta-defensin-2 functions as a chemotactic agent for tumour necrosis factor-alpha-treated human neutrophils. Immunology 111:273–281

Nizet V, Ohtake T, Lauth X, Trowbridge J, Rudisill J, Dorschner RA, Pestonjamasp V, Piraino J, Huttner K, Gallo RL (2001) Innate antimicrobial peptide protects the skin from invasive bacterial infection. Nature 414:454–457

Nomura I, Goleva E, Howell MD, Hamid QA, Ong PY, Hall CF, Darst MA, Gao B, Boguniewicz M, Travers JB, Leung DY (2003) Cytokine milieu of atopic dermatitis, as compared to psoriasis, skin prevents induction of innate immune response genes. J Immunol 171:3262–3269

Nyberg P, Rasmussen M, Bjorck L (2004) Alpha 2-macroglobulin-proteinase complexes protect Streptococcus pyogenes from killing by the antimicrobial peptide LL-37. J Biol Chem 279:52820–52823

Ogushi K, Wada A, Niidome T, Mori N, Oishi K, Nagatake T, Takahashi A, Asakura H, Makino S, Hojo H, Nakahara Y, Ohsaki M, Hatakeyama T, Aoyagi H, Kurazono H, Moss J, Hirayama T (2001) Salmonella enteritidis FliC (flagella filament protein) induces human beta-defensin-2 mRNA production by Caco-2 cells. J Biol Chem 276:30521–30526

Ohgami K, Ilieva IB, Shiratori K, Isogai E, Yoshida K, Kotake S, Nishida T, Mizuki N, Ohno S (2003) Effect of human cationic antimicrobial protein 18 peptide on endotoxin-induced uveitis in rats. Invest Ophthalmol Vis Sci 44:4412–4418

O'Neil DA, Porter EM, Elewaut D, Anderson GM, Eckmann L, Ganz T, Kagnoff MF (1999) Expression and regulation of the human beta-defensins hBD-1 and hBD-2 in intestinal epithelium. J Immunol 163:6718–6724

Ong PY, Ohtake T, Brandt C, Strickland I, Boguniewicz M, Ganz T, Gallo RL, Leung DY (2002) Endogenous antimicrobial peptides and skin infections in atopic dermatitis. N Engl J Med 347:1151–1160

Palfree RG, Sadro LC, Solomon S (1993) The gene encoding the human corticostatin HP-4 precursor contains a recent 86-base duplication and is located on chromosome 8. Mol Endocrinol 7:199–205

Panyutich AV, Panyutich EA, Krapivin VA, Baturevich EA, Ganz T (1993) Plasma defensin concentrations are elevated in patients with septicemia or bacterial meningitis. J Lab Clin Med 122:202–207

Parmley RT, Gilbert CS, Boxer LA (1989) Abnormal peroxidase-positive granules in "specific granule" deficiency. Blood 73:838–844

Pasare C, Medzhitov R (2003) Toll pathway-dependent blockade of CD4+CD25+ T cell-mediated suppression by dendritic cells. Science 299:1033–1036

Perregaux DG, Bhavsar K, Contillo L, Shi J, Gabel CA (2002) Antimicrobial peptides initiate IL-1 beta posttranslational processing: a novel role beyond innate immunity. J Immunol 168:3024–3032

Pulendran B, Banchereau J, Maraskovsky E, Maliszewski C (2001) Modulating the immune response with dendritic cells and their growth factors. Trends Immunol 22:41–47

Putsep K, Axelsson LG, Boman A, Midtvedt T, Normark S, Boman HG, Andersson M (2000) Germ-free and colonized mice generate the same products from enteric prodefensins. J Biol Chem 275:40478–40482

Putsep K, Carlsson G, Boman HG, Andersson M (2002) Deficiency of antibacterial peptides in patients with morbus Kostmann: an observation study. Lancet 360:1144–1149

Quayle AJ, Porter EM, Nussbaum AA, Wang YM, Brabec C, Yip KP, Mok SC (1998) Gene expression, immunolocalization, and secretion of human defensin-5 in human female reproductive tract. Am J Pathol 152:1247–1258

Ross DJ, Cole AM, Yoshioka D, Park AK, Belperio JA, Laks H, Strieter RM, Lynch JP, 3rd, Kubak B, Ardehali A, Ganz T (2004) Increased bronchoalveolar lavage human beta-defensin type 2 in bronchiolitis obliterans syndrome after lung transplantation. Transplantation 78:1222–1224

Sakamoto N, Mukae H, Fujii T, Ishii H, Yoshioka S, Kakugawa T, Sugiyama K, Mizuta Y, Kadota JI, Nakazato M, Kohno S (2005) Differential effects of alpha- and beta-defensin on cytokine production by cultured human bronchial epithelial cells. Am J Physiol Lung Cell Mol Physiol 288:L508–L513

Salzman NH, Ghosh D, Huttner KM, Paterson Y, Bevins CL (2003) Protection against enteric salmonellosis in transgenic mice expressing a human intestinal defensin. Nature 422:522–526

Sawa T, Kurahashi K, Ohara M, Gropper MA, Doshi V, Larrick JW, Wiener-Kronish JP (1998) Evaluation of antimicrobial and lipopolysaccharide-neutralizing effects of a synthetic CAP18 fragment against *Pseudomonas aeruginosa* in a mouse model. Antimicrob Agents Chemother 42:3269–3275

Sawyer JG, Martin NL, Hancock RE (1988) Interaction of macrophage cationic proteins with the outer membrane of *Pseudomonas aeruginosa*. Infect Immun 56:693–698

Schutte BC, Mitros JP, Bartlett JA, Walters JD, Jia HP, Welsh MJ, Casavant TL, McCray PB Jr (2002) Discovery of five conserved beta -defensin gene clusters using a computational search strategy. Proc Natl Acad Sci U S A 99:2129–2133

Scott MG, Davidson DJ, Gold MR, Bowdish D, Hancock RE (2002) The human antimicrobial peptide LL-37 is a multifunctional modulator of innate immune responses. J Immunol 169:3883–3891

Scott MG, Vreugdenhil AC, Buurman WA, Hancock RE, Gold MR (2000) Cutting edge: cationic antimicrobial peptides block the binding of lipopolysaccharide (LPS) to LPS binding protein. J Immunol 164:549–553

Selsted ME, Harwig SS, Ganz T, Schilling JW, Lehrer RI (1985) Primary structures of three human neutrophil defensins. J Clin Invest 76:1436–1439

Singh A, Bateman A, Zhu QZ, Shimasaki S, Esch F, Solomon S (1988) Structure of a novel human granulocyte peptide with anti-ACTH activity. Biochem Biophys Res Commun 155:524–529

Singh PK, Jia HP, Wiles K, Hesselberth J, Liu L, Conway BA, Greenberg EP, Valore EV, Welsh MJ, Ganz T, Tack BF, McCray PB Jr (1998) Production of beta-defensins by human airway epithelia. Proc Natl Acad Sci U S A 95:14961–14966

Singh PK, Tack BF, McCray PB Jr, Welsh MJ (2000) Synergistic and additive killing by antimicrobial factors found in human airway surface liquid. Am J Physiol Lung Cell Mol Physiol 279: L799–L805

Sorensen OE, Follin P, Johnsen AH, Calafat J, Tjabringa GS, Hiemstra PS, Borregaard N (2001) Human cathelicidin, hCAP-18, is processed to the antimicrobial peptide LL-37 by extracellular cleavage with proteinase 3. Blood 97:3951–3959

Sorensen OE, Gram L, Johnsen AH, Andersson E, Bangsboll S, Tjabringa GS, Hiemstra PS, Malm J, Egesten A, Borregaard N (2003) Processing of seminal plasma hCAP-18 to ALL-38 by gastricsin: a novel mechanism of generating antimicrobial peptides in vagina. J Biol Chem 278:28540–28546

Sparkes RS, Kronenberg M, Heinzmann C, Daher KA, Klisak I, Ganz T, Mohandas T (1989) Assignment of defensin gene(s) to human chromosome 8p23. Genomics 5:240–244

Spencer LT, Paone G, Krein PM, Rouhani FN, Rivera-Nieves J, Brantly ML (2003) The role of human neutrophil peptides in lung inflammation associated with α1-antitrypsin deficiency. Am J Physiol Lung Cell Mol Physiol 286:L514–520

Tamamura H, Imai M, Ishihara T, Masuda M, Funakoshi H, Oyake H, Murakami T, Arakaki R, Nakashima H, Otaka A, Ibuka T, Waki M, Matsumoto A, Yamamoto N, Fujii N (1998) Pharmacophore identification of a chemokine receptor (CXCR4) antagonist, T22 ([Tyr(5,12),Lys7]-polyphemusin II), which specifically blocks T cell-line-tropic HIV-1 infection. Bioorg Med Chem 6:1033–1041

Tanaka D, Miyasaki KT, Lehrer RI (2000) Sensitivity of *Actinobacillus actinomycetem-comitans* and *Capnocytophaga* spp. to the bactericidal action of LL-37: a cathelicidin found in human leukocytes and epithelium. Oral Microbiol Immunol 15:226–231

Tani K, Murphy WJ, Chertov O, Salcedo R, Koh CY, Utsunomiya I, Funakoshi S, Asai O, Herrmann SH, Wang JM, Kwak LW, Oppenheim JJ (2000) Defensins act as potent adjuvants that promote cellular and humoral immune responses in mice to a lymphoma idiotype and carrier antigens. Int Immunol 12:691–700

Territo MC, Ganz T, Selsted ME, Lehrer R (1989) Monocyte-chemotactic activity of defensins from human neutrophils. J Clin Invest 84:2017–2020

Turner J, Cho Y, Dinh NN, Waring AJ, Lehrer RI (1998) Activities of LL-37, a cathelin-associated antimicrobial peptide of human neutrophils. Antimicrob Agents Chemother 42:2206–2214

Van Wetering S, Mannesse-Lazeroms SP, Van Sterkenburg MA, Daha MR, Dijkman JH, Hiemstra PS (1997a) Effect of defensins on interleukin-8 synthesis in airway epithelial cells. Am J Physiol 272: L888–L896

Van Wetering S, Mannesse-Lazeroms SP, van Sterkenburg MA, Hiemstra PS (2002) Neutrophil defensins stimulate the release of cytokines by airway epithelial cells: modulation by dexamethasone. Inflamm Res 51:8–15

VanderMeer TJ, Menconi MJ, Zhuang J, Wang H, Murtaugh R, Bouza C, Stevens P, Fink MP (1995) Protective effects of a novel 32-amino acid C-terminal fragment of CAP18 in endotoxemic pigs. Surgery 117:656–662

Vora P, Youdim A, Thomas LS, Fukata M, Tesfay SY, Lukasek K, Michelsen KS, Wada A, Hirayama T, Arditi M, Abreu MT (2004) β-defensin-2 expression is regulated by TLR signaling in intestinal epithelial cells. J Immunol 173:5398–5405

Wang TT, Nestel FP, Bourdeau V, Nagai Y, Wang Q, Liao J, Tavera-Mendoza L, Lin R, Hanrahan JW, Mader S, White JH (2004) Cutting edge: 1,25-dihydroxyvitamin D3 is a direct inducer of antimicrobial peptide gene expression. J Immunol 173:2909–2912

Wang Y, Agerberth B, Lothgren A, Almstedt A, Johansson J (1998) Apolipoprotein A-I binds and inhibits the human antibacterial/cytotoxic peptide LL-37. J Biol Chem 273:33115–33118

Wehkamp J, Schmidt K, Herrlinger KR, Baxmann S, Behling S, Wohlschlager C, Feller AC, Stange EF, Fellermann K (2003) Defensin pattern in chronic gastritis: HBD-2 is differentially expressed with respect to *Helicobacter pylori* status. J Clin Pathol 56:352–357

Weiss J (2003) Bactericidal/permeability-increasing protein (BPI) and lipopolysaccharide-binding protein (LBP): structure, function and regulation in host defence against Gram-negative bacteria. Biochem Soc Trans 31:785–790

Welling MM, Hiemstra PS, van den Barselaar MT, Paulusma-Annema A, Nibbering PH, Pauwels EK, Calame W (1998) Antibacterial activity of human neutrophil defensins in experimental infections in mice is accompanied by increased leukocyte accumulation. J Clin Invest 102:1583–1590

White SH, Wimley WC, Selsted ME (1995) Structure, function, and membrane integration of defensins. Curr Opin Struct Biol 5:521–527

Wilde CG, Griffith JE, Marra MN, Snable JL, Scott RW (1989) Purification and characterization of human neutrophil peptide 4, a novel member of the defensin family. J Biol Chem 264:11200–11203

Wilson CL, Ouellette AJ, Satchell DP, Ayabe T, Lopez-Boado YS, Stratman JL, Hultgren SJ, Matrisian LM, Parks WC (1999) Regulation of intestinal alpha-defensin activation by the metalloproteinase matrilysin in innate host defense. Science 286:113–117.

Yan H, Hancock RE (2001) Synergistic interactions between mammalian antimicrobial defense peptides. Antimicrob Agents Chemother 45:1558–1560

Yang D, Chertov O, Bykovskaia SN, Chen Q, Buffo MJ, Shogan J, Anderson M, Schroder JM, Wang JM, Howard OM, Oppenheim JJ (1999) Beta-defensins: linking innate and adaptive immunity through dendritic and T cell CCR6. Science 286:525–528

Yang D, Chen Q, Chertov O, Oppenheim JJ (2000a) Human neutrophil defensins selectively chemoattract naive T and immature dendritic cells. J Leukoc Biol 68:9–14

Yang D, Chen Q, Schmidt AP, Anderson GM, Wang JM, Wooters J, Oppenheim JJ, Chertov O (2000b) LL-37, the neutrophil granule- and epithelial cell-derived cathelicidin, utilizes formyl peptide receptor-like 1 (FPRL1) as a receptor to chemoattract human peripheral blood neutrophils, monocytes, and T cells. J Exp Med 192:1069–1074

Yang D, Biragyn A, Kwak LW, Oppenheim JJ (2002) Mammalian defensins in immunity: more than just microbicidal. Trends Immunol 23:291–296

Yang D, Chen Q, Hoover DM, Staley P, Tucker KD, Lubkowski J, Oppenheim JJ (2003) Many chemokines including CCL20/MIP-3alpha display antimicrobial activity. J Leukoc Biol 74:448–455

Zaiou M, Nizet V, Gallo RL (2003) Antimicrobial and protease inhibitory functions of the human cathelicidin (hCAP18/LL-37) prosequence. J Invest Dermatol 120:810–816

Zanetti M (2004) Cathelicidins, multifunctional peptides of the innate immunity. J Leukoc Biol 75:39–48

Zhang H, Downey GP, Suter PM, Slutsky AS, Ranieri VM (2002) Conventional mechanical ventilation is associated with bronchoalveolar lavage-induced activation of polymorphonuclear leukocytes: a possible mechanism to explain the systemic consequences of ventilator-induced lung injury in patients with ARDS. Anesthesiology 97:1426–1433

Zhang K, Lu Q, Zhang Q, Hu X (2004) Regulation of activities of NK cells and CD4 expression in T cells by human HNP-1, -2, and -3. Biochem Biophys Res Commun 323:437–444

Zhao C, Wang I, Lehrer RI (1996) Widespread expression of beta-defensin hBD-1 in human secretory glands and epithelial cells. FEBS Lett 396:319–322

CTMI (2006) 306:67–90

Host Antimicrobial Defence Peptides in Human Disease

B. Agerberth[1] · G. H. Guðmundsson[2] (✉)

[1]Department of Medical Biochemistry and Biophysics, Karolinska Institutet,
SE 171 77 Stockholm, Sweden

[2]Biology Institute, University of Iceland, Sturlugata 7, 101 Reykjavik, Iceland
ghrafn@hi.is

Abstract Antimicrobial peptides or host defence peptides are endogenous peptide antibiotics, which have been confirmed as an essential part of the immune system. Apart from direct killing of bacteria, a role for the peptides in antiviral and immunomodulatory functions has recently been claimed. In this chapter we have focused on the host contact with microbes, where these host defence peptides are key players. The interplay with commensals and pathogens in relation to antimicrobial peptide expression is discussed, with specific emphasis on the respiratory and the alimentary systems. A possible novel difference in epithelial interactions between commensals and pathogens is considered in relation to disease.

1
Introduction

All eukaryotic organisms live in proximity with microbes, and this contact has been an important determinant for eukaryotic evolution. Battles have been fought and equilibriums have been established. The evolutionary outcome has resulted in different forms of symbiosis, with the mitochondria a fascinating example that can be traced to early eukaryotic evolution. Today our body continues to deal with microbes on a daily basis. Our natural flora or commensal bacteria are indeed an important part of us and are in equilibrium with our body surfaces without causing disease, but are instead beneficial to our health. Some microbes are professional pathogens and have inherent strategies for invading host cells and cause disease. However, human pathogens can be in equilibrium with other hosts. Also, microbes of the normal flora can behave like pathogens when our defence system is weakened. These microbes are opportunistic and utilize a ruptured equilibrium. Definition of the parameters of the equilibrium is a major challenge in host–microbe interaction, which is extremely complex and species-specific. In this setting, our immune system is of great importance and many effector molecules of innate immunity line the epithelia, which is the surface first in contact with the microbes. In this context, antimicrobial peptides, our endogenous antibiotics, are crucial effector molecules.

Antimicrobial peptides are gene-encoded and synthesized as prepro-proteins. The active peptides are cleaved off within vacuoles or after secretory release from its propart upon stimuli (Scocchi et al. 1992; Sorensen et al. 2001). The first antimicrobial peptides were identified in the moth *Hyalophora cecropia* and designated cecropins (Steiner et al. 1981). A few years later, the defensins were characterized in rabbit macrophages (Selsted et al. 1983). Today, roughly 1,000 peptides have been identified in different eukaryotic organisms from yeast to humans (see database http://www.bbcm.units.it/~tossi/pag1.htm). Nowadays, antimicrobial peptides are accepted as important effectors of innate immunity, not as relics or evolutionary leftovers. Recently, an increased interest in antimicrobial peptides has arisen, since they have been attributed additional activities such as chemotaxis (Yang et al. 2001) and stimulation of critical cytokines and chemokines that are important for the direction of the immune response (Bowdish et al. 2005). Thus, the emerging picture is that these host defence peptides are included in the innate–adaptive axis. In humans, the main families of antimicrobial peptides are the defensins and the cathelicidins. The defensins are identified by having six conserved cysteine residues that are folded in a characteristic disulphide pattern (Ganz, 2003). According to

the connection between the cysteines, the defensins are subdivided into α-
and β-defensins. A third group, the θ-defensins, has been characterized in
the neutrophils of the macaque monkey and are cyclic peptides, which are
synthesized through a novel pathway (Tang et al. 1999; Tran et al. 2002).
As in humans, the θ-defensins have been silenced in our closest relatives
among the primates with a mutation in an ancestral species (Cole et al. 2004).
The highest concentration of α-defensins are found in neutrophils (Ganz
and Lehrer 1997), but also mononuclear leukocytes express α-defensins
(Agerberth et al. 2000). The defensins of the β- type have mainly been
detected in epithelial surfaces and constitute an important function in
epithelial barriers. In humans, four α-defensins appear in neutrophils (Ganz
and Lehrer 1997) and two in the Paneth cells of the small intestine (Jones
and Bevins 1992, 1993). Until now, four β-defensin peptides have been
characterized (Bensch et al. 1995; Harder et al. 1997; Garcia et al. 2001; Harder
et al. 2001); however, 28 additional human genes have been found in a cluster,
constituting five syntenic chromosomal regions (Schutte et al. 2002).

The cathelicidins are recognized by having a conserved proregion, cathelin,
with a highly variable C-terminal antimicrobial domain (Zanetti et al. 1995).
Upon activation, the proprotein is cleaved by proteases, liberating a mature
active peptide (Scocchi et al. 1992; Sorensen et al. 2001). In contrast to the defensins, where a distinct expression pattern is associated with the subgroups,
cathelicidins are expressed both in leukocytes and epithelial cells (Zanetti
2004) The sole cathelicidin in humans is LL-37/hCAP-18, where hCAP-18
refers to the proregion and LL-37 to the mature active peptide, which consists
of 37 residues, forming an amphipathic α-helical structure in its active conformation (Gudmundsson and Agerberth 2004). Interestingly, it was recently
demonstrated that also the propart, cathelin, exhibits antimicrobial activity,
which is complementary to that of LL-37 (Zaiou et al. 2003).

In this chapter, we will focus on defensins and cathelicidins with respect
to human health and disease.

2
Mechanism of Action

Most antimicrobial peptides kill bacteria by attacking their membrane and
cause lysis. This is a very rapid procedure difficult to analyse in molecular
detail in terms of the mechanisms behind this process in vivo. Biophysical
studies have resulted in three different models based on the amphipathic character of the peptides. They are the barrel-stave pore, the thoroidal pore and the
carpet models (Brogden, 2005). In the barrel-stave pore model, the peptides

transverse the membrane with the hydrophobic side directed towards the lipid membrane and the hydrophilic side assembled to form a pore, allowing water and electrolyte leakage (Oren and Shai 1998). In the thoroidal pore model, the peptides are assumed to aggregate and induce the lipid monolayer to bend through the pore, causing the lipid head groups to be directed towards the water core (Brogden 2005). The carpet model describes a mechanism, where the peptides after electrostatic interaction cover the microbial membrane, disrupting it in a detergent-like manner (Shai 1995). The architecture of microbial membranes is delicate, with vital functions for the bacteria. Since the targets of antimicrobial peptides are membranes, traditional resistance is rarely observed (Hancock and Scott 2000). However, several bacteria are resistant because of production of proteases or channel systems (Shafer et al. 1998; Schmidtchen et al. 2002). Microbial membranes are not the only targets of antimicrobial peptides. Eukaryotic membranes are also affected, but to lyse eukaryotic membranes, the concentration of the peptides has to reach levels higher than those of the bactericidal concentration (Lehrer et al. 1993; Johansson et al. 1998). The extent of affinity between the peptides and prokaryotic and eukaryotic membranes is most likely the reason for this selectivity. Eukaryotic organisms contain cholesterol and zwitterionic phospholipids in the outer leaflet of their membranes, while prokaryotic membranes have acidic phospholipids. Antimicrobial peptides are mainly cationic, with a high ratio of basic residues, and accordingly the peptides have higher affinity to prokaryotic membranes (Zasloff 2002).

Additional mechanisms of action have been demonstrated for antimicrobial peptides, e.g. proline-rich peptides have been shown to inhibit intracellular processes such as protein and DNA syntheses (Boman et al. 1993; Gennaro et al. 2002). The antifungal mechanism of histatins has been suggested to target the mitochondria, causing efflux of ATP (Kavanagh and Dowd 2004).

In the complex elimination of microbial intruders, antimicrobial peptides are not solo players. Synergy has been demonstrated between peptides such as HBD2 and LL-37 (Ong et al. 2002), but also between the antimicrobial proteins lysozyme and lactoferrin (Singh et al. 2000). Since the peptides are synthesized as proproteins, proteases are involved in the activation process and are thus important factors in this setting. Other players are protease inhibitors that may balance the active concentration of the peptides. Interestingly, several protease inhibitors exhibit intrinsic antimicrobial activity such as the secretory leukocyte protease inhibitor (SLPI) (Hiemstra et al. 1996). It is intriguing to consider protease inhibitors as important sensors for the presence of bacteria, since they could become occupied when bacteria are present and released from the protease, which in turn will then process proeffectors of innate immunity.

3
Immune Modulation

The multifunctional character of antimicrobial peptides was already emphasized in 1999 (Gudmundsson and Agerberth 1999). However, it must be stressed that these activities have been detected in vitro. In vivo experiments with confirmatory results have only been obtained for the angiogenetic effect of LL-37 (Koczulla et al. 2003). LPS-binding properties and mitogenic effects in addition to chemotaxis were reported a decade ago (Larrick et al. 1991; Murphy et al. 1993; Chertov et al. 1996). The ability of α-defensins 1 and 2 to attract T cells was first demonstrated by Oppenheim's group (Chertov et al. 1996). These studies have been extended, showing that β-defensins and LL-37 also work as chemoattractants for specific immune cells (Yang et al. 1999; Agerberth et al. 2000; Yang et al. 2000a, b). The chemotactic activity is mediated by receptors on the target cells, where β-defensin acts on CCR6 to attract immature dendritic cells and memory T cells (Yang et al. 1999). On the other hand, several receptors have been shown to bind LL-37, the first being formyl peptide receptor-like 1 (FPRL1), which can mediate its effect on monocytes, neutrophils and T cells (Yang et al. 2000). More recently, the purinergic receptor R2X7 has been identified to interact with LL-37, inducing the processing of IL-1β followed by release of this central cytokine in immunity (Elssner et al. 2004). If this signalling occurs in vivo, LL-37 is in the early front of immune stimulation and may partially exert its effect through the action of IL-1β. Furthermore, the peptides have been shown to interact with lung epithelial cells, where LL-37 transactivates the epidermal growth factor receptor (EGFR). Downstream events are activation of mitogen-activated protein kinase (MAPK)/extracellular signal-regulated kinase (ERK) with subsequent release of IL-8 (Tjabringa et al. 2003), signalling pathways of great importance in lung epithelial immunity. The MAPK/ERK pathway is also activated by LL-37 in monocytes (Bowdish et al. 2004).

The emerging picture connected with different activities of antimicrobial peptides includes immune modulation and might thus affect the direction of the immune responses. The total picture is not yet fully clear and it is important to confirm the relevance in vivo. Nevertheless, LL-37 and its truncated versions have been shown to affect gene expression of chemokines and cytokines and to promote maturation and differentiation of dendritic cells (Davidson et al. 2004; Braff et al. 2005). Interestingly, the effect on IL-8 release is mediated by both D and L isomers of LL-37, indicating that a specific ligand–receptor interaction is not involved, rather a membrane interaction (Braff et al. 2005), which needs further research.

4
Relation to Disease

4.1
Respiratory System

A major milestone in the field of antimicrobial peptides relates to studies on the autosomal recessive disease cystic fibrosis (CF) (Smith et al. 1996). The disease is caused by mutations in the gene encoding the cystic fibrosis transmembrane conductance regulator (CFTR), a protein that functions as a chloride channel. The formation of thick, sticky mucus in the lungs with recurrent bacterial infections is characteristic of the disease.

The original claim in relation to antimicrobial peptides was that in CF patients, the peptides in the lung mucosa were inactivated because of an increased salt concentration in the liquid of the airway surface (Smith et al. 1996). This was the first report connecting a disease to experimental data of antimicrobial peptides, thus linking susceptibility to infections to breached activity of the peptides. Thereby, pulmonary medicine was the first medical discipline to emphasize the importance of these effectors in innate immunity. However, a relation to CF was prophetically suggested already in 1987 (Zasloff 1987). Further research has shown that the role of the peptides in relation to CF is much more complex. The main innate defence polypeptides are synthesized at the same level in normal lung as in the lung of CF patients (Bals et al. 2001). Some, such as the defensins, are inactivated at high salt concentrations, but measurements on the composition of airway surface fluid are difficult to obtain.

In mild CF, the levels of the β-defensins hBD1 and hBD2 as well as of the cathelicidin LL-37 have been associated with the severity of the disease. The level of LL-37 was elevated and suggested to be a marker for bronchoalveolar inflammation, while the levels of the β-defensins were reduced in more advanced disease (Chen et al. 2004). Today, it is clear that the CFTR does not only function as a chloride channel, but also regulates other ion channels, affects vesicular trafficking, stimulates ATP release, and is linked to the expression of inflammatory mediators in epithelial cells (Donaldson and Boucher 2003). To investigate the molecular mechanism in CF is thus complicated. Malfunctions of innate defences have been generally accepted as the underlying cause of CF, where multiple functions contribute to the severity of the disease. This includes the action of antimicrobial peptides, mucociliary movements and glanular secretion, resulting in a dysfunctional barrier. However, it is clear that removal and efficient killing of bacteria are defected and can stimulate the formation of a bacterial biofilm.

The role of antimicrobial peptides in the lung of CF patients is far from known in detail. However, the relation to CF has had a great impact for the interest in peptides in lung disease, and in innate defences of the lung. The innate defence of the lung involves many peptides/proteins, which directly interact with inhaled bacteria. Killing of bacteria is mediated by α- and β-defensins as well as by cathelicidin LL-37, and the innate effector proteins lysozyme and lactoferrrin (Bals and Hiemstra 2004). Several opsonizing factors, such as immunoglobulins, complement factors, fibronectin, and LPS binding proteins, are also crucial in the defence system of the lung (Ng et al. 2004). Furthermore, the microenvironment of the airway surface is tightly regulated with respect to pH associated with peptide activity, since ion composition and pH are crucial for the action of antimicrobial peptides. These parameters affect the peptide folding and hence the activity, as demonstrated for LL-37 (Johansson et al. 1998).

In another lung disease, antimicrobial peptides have been analyzed in diffuse panbronchiolitis (Mukae et al. 1995). This disease was defined in a Japanese population and exhibits similar symptoms to CF. Infiltration of immune cells such as monocytes and plasma cells together with neutrophils and T cells are also characteristic features of this disease. Infections with *Pseudomonas aueriginosa* and *Haemophilus influenzae* occur frequently despite the increased concentration of α- and β-defensins (Ashitani et al. 1998; Hiratsuka et al. 2003). The higher level of α-defensins most likely reflects neutrophil influx, a characteristic of this disease. In addition, hBD2 has been detected with enhanced levels in the plasma of these patients. Elevated IL-8 has also been found to be a potent chemoattractant for neutrophils. The susceptibility to lung infections, a potential relationship to antimicrobial peptides, and additional innate effectors are unclear but might be similar to CF (Hiratsuka et al. 2003). The understanding of the immune failure in these diseases could be of great importance and would most likely have a therapeutic value.

Moreover, the molecular mechanism behind the innate defence of the healthy lung is incomplete. Pattern recognition receptors that sense the presence of inhaled bacteria are expressed on epithelial cells of the lung. Interestingly, the toll-like receptor 2 (TLR2) has recently been linked to the induction of β-defensin in tracheobronchial cells (Hertz et al. 2003). In general, activation of TLRs by microbial exposure is thought to be associated with the expression of antimicrobial peptides, cytokines and chemokines (Bals and Hiemstra 2004). At the same time, it is important to gain information about the constitutive pathways likely to be linked to cellular differentiation (Schauber et al. 2003). Gene-deficient mice have been valuable tools and informative results have been obtained. With respect to antimicrobial peptides, studies in knock-out mice have recently started. Deletion of the CRAMP gene encoding the sole cathelicidin in mouse has been shown to be detrimental for surface defences in the skin (Nizet et al. 2001). In relation to lung defences, mice deficient in one β-defensin gene (mBD1) have delayed clearance of *Haemophilus influenzae* and *Staphylococcus aureus*, supporting the importance of mBD1 in these infections (Morrison et al. 2002; Moser et al. 2002). In contrast, enhancement of this defence system was obtained by over-expression of the human cathelicidin (hCAP18/LL-37), using adenoviral gene transfer in a bronchial xenograft model (Bals et al. 1999). With this information at hand, one might expect that genetic polymorphism in antimicrobial peptide genes is associated with susceptibility to lung infections. Interestingly, reports have recently appeared indicating the presence of certain genotypes of human β-defensins that may be linked to high-risk subgroups of chronic obstructive pulmonary disease (COPD) (Matsushita et al. 2002; Hu et al. 2004). We have observed enhancement of the total antimicrobial activity in sarcoidosis as analysed in bronchoalveolar lavage (BAL) fluid from sarcoidosis patients. This observation might be connected to the low frequency of respiratory infections in this disease (Agerberth et al. 1999).

With the current data available, an emerging picture may be summarized. In the normal situation, a number of bacteria are inhaled that encounter constitutively expressed innate effectors, where lysozyme is present in the highest concentration (Travis et al. 2001). In addition, pattern recognition receptors sense the presence of microbes, which may result in synthesis of chemokines, cytokines, and antimicrobial peptides/proteins (Bals and Hiemstra 2004), thus strengthening the defence barrier of airway epithelia. Proteases are also important actors in this defence system. Obviously, this is successful during healthy conditions, otherwise chronic lung infections would be the rule.

The level of peptide induction in relation to microbe clearance may be linked to the killing capacity of antimicrobial peptides. With inefficient clearance of pathogenic strains, a stage of colonization will be reached, leading to further induction and recruitment of inflammatory cells. Approaching this point, where pathogens have invaded the body, the innate response together with the recognition of the intruder will trigger the adaptive immunity. With respect to the total scenario, antimicrobial peptides might serve as double edge daggers, i.e. in addition to killing bacteria the peptides are immunomodulators (Bowdish et al. 2005). Both LL-37 and the defensins have been shown to stimulate the secretion of IL-8, a potent chemoattractant, from lung epithelial cells and thus lead to further enhancement of the immune response (Van Wetering et al. 1997; Tjabringa et al. 2003). Furthermore, LL-37 is able to induce gene expression of markers for dendritic cell function, manipulating the outcome of T cell responses (Davidson et al. 2004). Thus, the concentration of peptides such as LL-37 may represent one of the central sensors of innate immunity, affecting the amplitude and character of the total immune response. In light of all these activities of antimicrobial peptides demonstrated in vitro, questions have been raised concerning the main activities in vivo. Some have indeed suggested immune modulation as the main function of antimicrobial peptides (Bowdish et al. 2005). However, the relevance of immune modulation has not been established in proper model systems. Ultimately, this will be achieved in model systems, utilizing gene-deficient mice. Furthermore, concerns of the therapeutic value of the peptides have been raised, because of these multiple activities (Elsbach 2003). In the in vivo situation, the peptides are likely under stringent control at transcriptional, translational and processing levels in order to be in balance with the natural microflora. Furthermore, antimicrobial peptides are also cytotoxic at elevated concentrations (Lehrer et al. 1993; Johansson et al. 1998) and thus harmful to host cells, causing tissue damage, although their contribution to tissue damage in inflammation has not been determined.

4.2
Digestive System

Just as the microbes are inhaled and encounter innate defences of the respiratory system, microbes ingested will face the mucosal immunity of the gastrointestinal (GI) tract. However, there are fundamental differences between these two systems. To keep a sterile environment applies to the respiratory system, while microbes are an important part of the digestive system. In the alimentary canal, microbes are welcomed with their beneficial properties, contributing to our well-being. In the GI tract, microbes fulfill essential functions of vital importance. They are involved in nutrient exchange, epithelial development and instruction of innate immunity (Eckburg et al. 2005). This vital relationship applies mainly to the colon, where the host microbe interaction is mutually beneficial. The term "commensals" will be used here according to tradition (Backhed et al. 2005).

The digestive system may be divided into separate microbial niches, where the stomach represents a hostile environment for micro-organisms. In the small intestine, microbes are relatively few compared to the colon, which harbours commensals outnumbering our own cells. Hence, the colon can be considered as a special ecosystem or a microbial organ. Defining the composition and parameters of this niche is an active research field (Eckburg et al. 2005). One of these parameters is linked to immediate mucosal defences, where the antimicrobial peptides have a central role. How much antimicrobial peptides contribute to defining each microbial niche with respect to bacterial strain and species is indeed one major question in our research field. The variation of the peptides has often been viewed as a result of co-evolution of the host with the quickly evolving microbes. Some of the primate genes encoding antimicrobial peptides show positive selection (Crovella et al. 2005), indicating that co-evolution with rapidly changing microbes could apply. The species variation in peptide armament might be the key factor for the microflora variance among species.

A second central question in our research field relates to differences between commensal and pathogenic bacteria vs antimicrobial peptides. Upon contact with the epithelia, the bacteria are exposed to antimicrobial peptides, which are synthesized constitutively. Commensal and pathogenic bacteria exhibit similar structures, which are recognized by pattern recognition receptors; and both types of bacteria are sensitive to the action of antimicrobial peptides. It may be argued that the commensal bacteria are in equilibrium without harming the mucosal cells, while the pathogens affect peptide concentration by degrading the peptides or down-regulating their expression (Islam et al. 2001; Schmidtchen et al. 2002). Such immediate immune escape will lead to a shifted equilibrium of the bacteria towards proximity to epithelial cells, with increased risk of infections. A similar situation might arise with commensals upon mechanical injury. The critical border of proximity could be outlined by the glycocalyx with the commensals located mainly outside this border and unable to trigger inflammatory responses. An additional question concerning health and infectious disease of humans is host–microbe crosstalk, which is important for livestock in general, i.e. honeybees and cows. Antimicrobial peptides are crucial factors in this setting, since they are potent killers of both pathogens and commensals. In addition, they cover all mucosal linings but are not solo players in the initial contact with bacteria. Recent data indicate that the peptides are decisive for downstream regulation, induction of cytokines, dendritic cell (DC) presentation, T cell response and amplification of the immunoglobin response. In association with the alimentary canal many pathogens originate from water and food, also of the commensals.

4.2.1
Oral Cavity

Upon ingestion, the fight starts already in the mouth: saliva contains active peptides and proteins (Murakami et al. 2002; Kavanagh and Dowd 2004), most likely working in synergy. Furthermore, the epithelial cells in the oral cavity are also active producers of antimicrobial peptides (Weinberg et al. 1998). A pioneer study has been conducted in which oral epithelial cells were used as responder cells for a well-defined commensal *Fusobacterium nucleatum* and a pathogen *Porphyromonas gingivalis*. The commensal bacteria were shown to induce hBD2 production in the cells, while the pathogen did not (Krisanaprakornkit et al. 2000). Further, the induction of the commensals was independent of NF-κB; instead mitogen-activated protein (MAP) kinase was involved (Chung and Dale, 2004). Some of the key elements in mucosal immunity might be hidden in these results, since the induction of the commensals is activated without the involvement of the transcription factor NF-κB, which has a key role in inflammatory signalling. Since the commensals and the pathogens have general structures in common, pattern recognition will be similar. Thus, proximity of live bacteria might be the main concept in this defence system. Bacterial proteases often linked to pathogens might mediate this proximity, reaching receptors that commensals rarely touch. In this context, it is interesting that in the colon, with its enormous bacterial community, some pattern recognition receptors are hidden in intracellular compartments (Girardin et al. 2003; Hornef et al. 2003). A microbe binding to these intracellular receptors has definitely come too close, resulting in an additional alarm signal in the host, including secretion of cytokines, chemokines and antimicrobial peptides that can trigger the adaptive immune system. If this scenario is true it is not implausible to understand the beneficial effects of probiotics, which might partially enhance the effect of innate actors such as antimicrobial peptides, without inflammatory activation.

Continuing with the oral compartment, the human cathelicidin LL-37 is present in normal saliva (Murakami et al. 2002). In the rare recessive neutrophenia Morbus Kostmann, LL-37 is absent in saliva (Putsep et al. 2002). The genetic component of this disease is unknown but compromised immunity and frequent infections are hallmarks for the symptoms. Morbus Kostmann is a fatal disease that since 1990 has been treated with granulocyte colony-stimulating factor (G-CSF) and, when necessary, antibiotics. With the G-CSF treatment, the level of neutrophils is restored and the quality of life for patients is improved. Still, infections are common and the patients suffer from periodontal disease. Furthermore, the level of neutrophil defensins is only 30% of that in healthy individuals and LL-37 is completely abolished. Only upon bone marrow transplantation the levels of these antimicrobial peptides are restored (Putsep et al. 2002). A correlation between dental health and the levels of the peptides has been suggested, emphasizing the importance of the peptides. Mutations in regulatory factors involved in gene expression of antimicrobial peptides have been suggested as the underlying mechanism in Morbus Kostmann (Putsep et al. 2002; Gudmundsson and Agerberth 2004). Details of the regulation for genes encoding antimicrobial peptides are still unresolved. Susceptibility of Kostmann patients to infections is in line with results from experimental infection models, where gene-deficient mice show increased susceptibility or delayed clearance. In other granulocytic disorders, such as specific granule deficiency, infections are frequent with decreased levels of α-defensins (Ganz et al. 1988).

In the field of antimicrobial peptides, the main emphasis has been on interactions with bacteria. However, other important microbes in our environment include viruses, fungi and eukaryotic parasites. Only few studies have indicated a possible role for peptides in anti-viral defences. A breakthrough, regarding antimicrobial peptides and viral infections, was obtained from studies on mucosal transmission of HIV-1, a strain that dominates in acute HIV infections (Quinones-Mateu et al. 2003). Clearly, there are differences in viral passage dependent on mucosal surfaces. Passage through oral mucosa is rare and infectious viruses have not been isolated from saliva of infected individuals. β-Defensins that are abundant in the oral cavity have been characterized as inhibiting viral progression and shown to interact directly with the viron by modulation of CXCR4, a co-receptor utilized by the virus upon cellular entry. Furthermore, viral contact with epithelial cells of the oral cavity induces expression of β-defensins (Quinones-Mateu et al. 2003). A single nucleotide polymorphism in the human β-defensin 1 gene has been implicated as a risk factor in HIV-1 infections in Italian children (Braida et al. 2004). The molecular details are unknown but deserve full attention as there might be important approaches in therapy based on these results. There are additional studies related to HIV, indicating that the peptides deserve research attention. α-Defensins have been found to block HIV-1 infection at nuclear import and transcription (Chang et al. 2005). Furthermore, the recently discovered cyclic peptide, θ-defensin, isolated from the neutrophils of macaque monkey, protect mononuclear cells from infection by HIV-1 (Selsted 2004). Notably, the human counterpart of the θ-defensin gene contains a premature stop codon within the signal-encoding sequence. Further analysis showed that this silencing mutation occurred around the divergence of the orangutan lineage some 7.5–10 million years ago (Cole et al. 2004). Interestingly, the human gene is actively transcribed and a synthetic variant of the gene product has been made, according to the human coding sequence, where the stop codon has been corrected (Cole et al. 2004). This ancient peptide has been designated retrocylin and exhibits modest antibacterial activity. Surprisingly, retrocyclin is able to protect CD4 T cells from being infected by several strains of HIV. The bases for this inhibition have been associated with lectin activity of retrocyclin, binding carbohydrate structures that are necessary for viral entry to cells (Cole et al. 2004). In terms of molecular evolution, with increasing sequencing data, we are becoming aware that genes encoding antimicrobial peptides are divergent in order to cope with the quickly evolving microbes. Very likely there are microbes that exert a stringent selection on all eukaryotes. Concerning the θ-defensins, the primate lineage has gone through an evolutionary bottleneck, resulting in a fixation of this mutation. However, the circumstances allowing this fixation can only remain speculative at the moment.

4.2.2
Stomach

After the alimentary canal, the next niche that has received a great deal of attention with respect to antimicrobial peptides is the stomach. This is a hostile environment for bacteria and many are eliminated because of low pH and proteolytic enzymes. However, β-defensins and lysozyme are synthesized by gastric epithelial cells (O'Neil et al. 2000), hBD1 constitutively and hBD2 induced in response to proinflammatory cytokines and microbial infections (O'Neil et al. 2000). Recently, human cathelicidin LL-37 was also found to be produced by gastric epithelial cells with active LL-37 secreted into the gastric juice (Hase et al. 2003). The expression of LL-37 was analysed in cells infected with *Helicobacter pylorii*. This bacterium remains one of the most prevalent bacterial pathogens in the world, involved in gastritis and peptic ulcer disease (Blanchard et al. 2004). *H. pylorii* adheres to the mucus layer of the stomach and can chronically colonize the gastric mucosa. Furthermore, *H. pylorii* induces transcription of the gene encoding LL-37 and this induction is dependent on the *cag* pathogenicity island of the bacterium (Hase et al. 2003). The products of this pathogenicity island are known to activate NF-κB. Whether this is connected to the induced expression of LL-37 is not yet known. Several strains of *H. pylorii* were sensitive to LL-37 and enhanced antibacterial action was detected when hBD1 was included in the assay (Hase et al. 2003). Indeed, these two peptides are expressed by the gastric mucosa. Thus, the peptides constitute a parameter to count with in our understanding of *H. pylorii* infections and host interactions. It is possible that the expression levels of these two peptides are associated with susceptibility to *H. pylorii* infection. Interestingly, high expression levels of LL-37 have been demonstrated to correlate with a low level of inflammation (Hase et al. 2003).

4.2.3
Small Intestine

In the small intestine, with relatively low microbial density, infectious diseases are infrequent. To some extent this is attributed to the bacterial clearance in the stomach and release of bile in the upper part of the small intestine. Furthermore, a complex defence system, where antimicrobial peptides are included, operates from specialized epithelial cells. The necessity of a powerful microbial clearance system may be connected to the function of the small intestine, where adsorption of nutrients takes place and the luminal content may serve as a rich medium for bacterial growth. In the small intestine, antimicrobial peptides are expressed in the specialized Paneth cells, which have been in focus regarding innate immunity. The Paneth cells are located at the base of the crypts of Lieberkühn and are secretory granulocyte-like cells derived from stem cells located close to the base of the crypts. The secreted effector molecules of Paneth cells involved in innate defences include peptides/proteins, such as α-defensins HD5 and HD6, lysozyme, type II secretory phospholipase A_2, ($sPLA_2$), DNAse and angiogenin-4 (Ouellette 2004). The total concentration of the α-defensins in the crypt has been estimated at levels far above documented minimal inhibitory concentrations (Ouellette 2004). HD5 and HD6 are secreted from Paneth cells and are implicated in the defence against pathogens and in the protection of stem cells. In addition, they have been suggested to limit the number of commensals also influencing their composition (Bevins 2004). The defensins of the Paneth cells are secreted upon contact with micro-organisms and most likely work in synergy with other components in the mucosa of the small intestine. There is a great deal of evidence supporting vital functions of defensins in the gut of mice. Matrilysin gene-deficient mice, lacking the main processing enzyme for mouse α-defensins (originally named cryptins) in the gut, are sensitive to orally ingested bacteria, more specifically to a virulent strain of *Salmonella typhimurium* (Wilson et al. 1999). In humans, the processing enzyme for Paneth cell defensins is trypsin (Ghosh et al. 2002), showing that variations in this defence system do not only apply to the primary structure of the peptides. An additional important functional finding in gut immunity relates to the Paneth cell defensins. A transgene mouse, containing the human HD5 gene with Paneth cell expression, exhibits a dramatic increase in resistance to intestinal infections with *S. typhimurium* (Salzman et al. 2003a). This strain of *Salmonella* gives rise to severe infection in wild type litter mates and is sensitive to HD5 activity, confirming the importance of a single antimicrobial gene in relation to susceptibility to a pathogen. Thus, the peptide armament of each species might be critical in distinguishing a pathogen from a commensal bacterium.

Stretching the concept further, one could expect gene polymorphism in antimicrobial peptide genes associated with susceptibility, i.e. the genotype is connected to different expression levels of the peptides.

Antimicrobial peptides and diseases of the gut also raise interest in Crohn's disease. It must be underlined that this disease has been under intensive research and scrutiny for decades. Crohn's disease is multifactorial, and a role for luminal bacteria has been suggested for disease development. A major advance in understanding the disease was the finding that subset of Crohn's patients have a mutation in the NOD2 gene (Hugot et al. 2001; Ogura et al. 2001). NOD2 encodes a receptor for muramyl dipeptide, a degradation fragment of bacterial cell walls. The NOD2 protein has a cytoplasmic location in Paneth cells and the mutant genotype has been related to diminished HD5 and HD6 expression (Wehkamp et al. 2004). However, the signalling pathways linking NOD2 and HD5 and HD6 expression in Paneth cells are not known. Further studies have demonstrated that NOD2 mutation potentiates the expression of NF-κB (Maeda et al. 2005) and NOD2 gene-deficient mice have low expression of α-defensin genes (Kobayashi et al. 2005). Again the proximity of bacteria to epithelia in the gut might be the key. Lower levels of peptides increase proximity of the microbial flora, followed by strong induction of inflammatory pathways.

4.2.4
Colon

When entering the colon, a world of microbes will be encountered. The colon is indeed a microbial organ, where only fragmentary knowledge has been obtained concerning the composition of this complex bacterial niche. With the quantities of bacteria present in colon, co-evolution of surface defences and the commensals must have been critical. The current picture of a natural flora in balance with the host includes important host factors as gate-keepers, but also inter-microbial competition. In this setting, antimicrobial peptides are important host factors. Colon epithelial cells are known to produce hBD1 constitutively both in surface and crypt epithelial cells, while hBD2 is induced upon inflammation (O'Neil et al. 1999) and is a target gene for NF-κB. Furthermore, human cathelicidin LL-37 is, like hBD1, also constitutively expressed, mainly in surface epithelia lining the gut lumen (Hase et al. 2002; Schauber et al. 2003). In an experimental model utilizing the colonic cell-line HT-29, we have shown that expression of LL-37 is associated with pathways including MEK/ERK kinases and p38 (Schauber et al. 2003). Numerous pathogens are known to infect the colon, such as *Shigella* and *Vibrio*, which both give rise to severe diarrheal infections. *Shigella* is sensitive to the action of antimicrobial peptides such as LL-37. Thus, this pathogen encounters endogenous peptide antibiotics at the epithelial surface, most likely at concentrations that can eliminate the bacteria. Interestingly, we have demonstrated that these pathogens, i.e. *Shigella* and *Vibrio*, down-regulate the expression of both LL-37 and hBD-1 (Islam et al. 2001). This down-regulation might facilitate contact and invasion through the M cell in the colon epithelia. Furthermore, the suppression of these peptides in *Shigella* infections is long-lasting, while the duration is short in *Vibrio* (Islam et al. 2001). Notably, recovery from

Vibrio infections is also fast. Reduction of epithelial defences is therefore one strategy that pathogens have evolved as an immune escape mechanism. However, there are also examples of pathogens that secrete proteases that degrade antimicrobial peptides (Schmidtchen et al. 2002), in addition to certain pump systems that are protective for the microbe (Shafer et al. 1998). In both cases, it may be argued that this decreased concentration of effective defence molecules will bring the pathogens into closer contact with the epithelial cells. It is conceivable that the down-regulation of innate effectors is a general mechanism for professional pathogens, since a similar reaction in relation to pathogenic *Salmonella* has been observed (Salzman et al. 2003b). Furthermore, we have detected transcriptional down-regulation in a cervical epithelial cell line infected with *Neisseria gonorrhoeae*, showing that this strategy is used by pathogens at other mucosal surfaces (Bergman et al. 2005). Understanding the molecular mechanism behind this immune escape and the interference between pathogens and mucosal surfaces may lead to targets in preventive medicine and even therapy.

4.3
Skin

Cutaneous immunity has been important for studies of antimicrobial peptides in health and disease. Recent studies indicate an essential role of antimicrobial peptides in skin immunity. One major milestone in this research field, confirming the importance of the expression of a single peptide, was obtained in a gene knockout mouse, infected with the skin pathogen group A Streptococcus. The gene-deficient mice developed severe wounds and bacterial clearance was delayed (Nizet et al. 2001). A direct connection was established between the absence of peptide expression and disease stage. Induced antimicrobial peptide expression in relation to disease was first observed in skin diseases. We observed enhanced expression of the sole human cathelicidin LL-37 in skin biopsies derived from affected psoriatic lesions (Frohm et al. 1997). This observation seemed to be associated with the rarity of bacterial infections in psoriatic lesions, although the skin barrier is disrupted. We detected an immense total antibacterial activity in protein extracts of psoriatic scales. Today, many innate peptides and proteins have been identified in skin such as BPI, lysozyme, lactoferrin and RNase 7 (Harder and Schroder 2005). Furthermore, psoriatic scales have served as an important source for isolation of novel peptides such as human β-defensisns hBD2 and -3 (Harder et al. 1997, 2001). Very recently, the same group has characterized psoriasin, a very potent defence molecule against Gram-positive bacteria (Glaser et al. 2005). Psoriasin has also been implicated in host defence mechanism of vernix caseosa

(Yoshio et al. 2003). Antimicrobial peptides in skin have today been analysed in relation to several diseases and have been associated with susceptibility in atopic allergy (Ong et al. 2002) and antiviral defences (Howell et al. 2004). Most interestingly, the peptides generated from cathelin in humans have been characterized in multiple processing forms with additional activities (Murakami et al. 2004). In short, the peptides have been confirmed as an essential part of cutaneous innate immunity, but this interesting research field is covered in detail in another chapter of this book.

5
Future Perspective and Therapy

Working with antimicrobial peptides, which kill bacteria and are able to eliminate viruses, it is not far-fetched to see a possibility in using the peptides in medical therapy. However, classical antibiotics still dominate every treatment of bacterial infection. Excessive use and/or failure to follow recommended protocols of antibiotics have caused severe problems, because of resistant bacterial strains. The problems have not been microbial in character, but rather economical.

Alternative ways to treat or stop the spread of antibiotic resistance strains continue to be sought. Antimicrobial peptides have been a target for such research but still none is available in treatment. However, this growing field has enhanced our understanding of the host–microbe interaction and identified new targets that might be used in an efficient way to treat infections. In our opinion, the most fruitful approach would include in situ induction of antimicrobial host defence peptides. Thus, antimicrobial peptides are no longer outdated relics, but instead active immune substances, an important outcome of adaptive evolution.

Acknowledgements We wish to thank Prof. Hans Jörnvall for critical reading of the manuscript. The authors are supported by grants from The Swedish Foundation for International Cooperation in Research and Higher Education (STINT), The Swedish Research Council, The Icelandic Centre for Research, the Ruth and Richard Julin Foundation, the Prof. Nanna Svartz Foundation and Karolinska Institutet.

References

Agerberth B, Charo J, Werr J, Olsson B, Idali F, Lindbom L, Kiessling R, Jornvall H, Wigzell H, Gudmundsson GH (2000) The human antimicrobial and chemotactic peptides LL-37 and alpha-defensins are expressed by specific lymphocyte and monocyte populations. Blood 96:3086–3093

Agerberth B, Grunewald J, Castanos-Velez E, Olsson B, Jornvall H, Wigzell H, Eklund A, Gudmundsson GH (1999) Antibacterial components in bronchoalveolar lavage fluid from healthy individuals and sarcoidosis patients. Am J Respir Crit Care Med 160:283–290

Ashitani J, Mukae H, Nakazato M, Ihi T, Mashimoto H, Kadota J, Kohno S, Matsukura S (1998) Elevated concentrations of defensins in bronchoalveolar lavage fluid in diffuse panbronchiolitis. Eur Respir J 11:104–111

Backhed F, Ley RE, Sonnenburg JL, Peterson DA, Gordon JI (2005) Host-bacterial mutualism in the human intestine. Science 307:1915–1920

Bals R, Hiemstra PS (2004) Innate immunity in the lung: how epithelial cells fight against respiratory pathogens. Eur Respir J 23:327–333

Bals R, Weiner DJ, Meegalla RL, Accurso F, Wilson JM (2001) Salt-independent abnormality of antimicrobial activity in cystic fibrosis airway surface fluid. Am J Respir Cell Mol Biol 25:21–25

Bals R, Weiner DJ, Meegalla RL, Wilson JM (1999) Transfer of a cathelicidin peptide antibiotic gene restores bacterial killing in a cystic fibrosis xenograft model. J Clin Invest 103:1113–1117

Bensch KW, Raida M, Magert HJ, Schulz-Knappe P, Forssmann WG (1995) hBD-1: a novel beta-defensin from human plasma. FEBS Lett 368:331–335

Bergman P, Johansson L, Asp V, Plant L, Gudmundsson GH, Jonsson A-B, Agerberth B (2005) Neisseria gonorrhoeae down-regulates expression of the human antimicrobial peptide LL-37. Cellular Microbiology 7:1009–1017

Bevins CL (2004) The Paneth cell and the innate immune response. Curr Opin Gastroenterol 20:572–580

Blanchard TG, Drakes ML, Czinn SJ (2004) Helicobacter infection: pathogenesis. Curr Opin Gastroenterol 20:10–15

Boman HG, Agerberth B, Boman A (1993) Mechanisms of action on Escherichia coli of cecropin P1 and PR-39, two antibacterial peptides from pig intestine. Infect Immun 61:2978–2984

Bowdish DM, Davidson DJ, Speert DP, Hancock RE (2004) The human cationic peptide LL-37 induces activation of the extracellular signal-regulated kinase and p38 kinase pathways in primary human monocytes. J Immunol 172:3758–3765

Bowdish DM, Davidson DJ, Scott MG, Hancock RE (2005) Immunomodulatory activities of small host defense peptides. Antimicrob Agents Chemother 49:1727–1732

Braff MH, Hawkins MA, Di Nardo A, Lopez-Garcia B, Howell MD, Wong C, Lin K, Streib JE, Dorschner R, Leung DY, Gallo RL (2005) Structure-function relationships among human cathelicidin peptides: dissociation of antimicrobial properties from host immunostimulatory activities. J Immunol 174:4271–4278

Braida L, Boniotto M, Pontillo A, Tovo PA, Amoroso A, Crovella S (2004) A single-nucleotide polymorphism in the human beta-defensin 1 gene is associated with HIV-1 infection in Italian children. Aids 18:1598–1600

Brogden KA (2005) Antimicrobial peptides: pore formers or metabolic inhibitors in bacteria? Nat Rev Microbiol 3:238–250

Chang TL, Vargas J Jr, DelPortillo A, Klotman ME (2005) Dual role of alpha-defensin-1 in anti-HIV-1 innate immunity. J Clin Invest 115:765–773

Chen CI, Schaller-Bals S, Paul KP, Wahn U, Bals R (2004) Beta-defensins and LL-37 in bronchoalveolar lavage fluid of patients with cystic fibrosis. J Cyst Fibros 3:45–50

Chertov O, Michiel DF, Xu L, Wang JM, Tani K, Murphy WJ, Longo DL, Taub DD, Oppenheim JJ (1996) Identification of defensin-1, defensin-2, and CAP37/azurocidin as T-cell chemoattractant proteins released from interleukin-8-stimulated neutrophils. J Biol Chem 271:2935–2940

Chung WO, Dale BA (2004) Innate immune response of oral and foreskin keratinocytes: utilization of different signaling pathways by various bacterial species. Infect Immun 72:352–358

Cole AM, Wang W, Waring AJ, Lehrer RI (2004) Retrocyclins: using past as prologue. Curr Protein Pept Sci 5:373–381

Crovella S, Antcheva N, Zelezetsky I, Boniotto M, Pacor S, Verga Falzacappa MV, Tossi A (2005) Primate beta-defensins—structure, function and evolution. Curr Protein Pept Sci 6:7–21

Davidson DJ, Currie AJ, Reid GS, Bowdish DM, MacDonald KL, Ma RC, Hancock RE, Speert DP (2004) The cationic antimicrobial peptide LL-37 modulates dendritic cell differentiation and dendritic cell-induced T cell polarization. J Immunol 172:1146–1156

Donaldson SH, Boucher RC (2003) Update on pathogenesis of cystic fibrosis lung disease. Curr Opin Pulm Med 9:486–491

Eckburg PB, Bik EM, Bernstein CN, Purdom E, Dethlefsen L, Sargent M, Gill SR, Nelson KE, Relman DA (2005) Diversity of the human intestinal microbial flora. Science 308:1635–1638

Elsbach P (2003) What is the real role of antimicrobial polypeptides that can mediate several other inflammatory responses? J Clin Invest 111:1643–1645

Elssner A, Duncan M, Gavrilin M, Wewers MD (2004) A novel P2X7 receptor activator, the human cathelicidin-derived peptide LL37, induces IL-1 beta processing and release. J Immunol 172:4987–4994

Frohm M, Agerberth B, Ahangari G, Stahle-Backdahl M, Liden S, Wigzell H, Gudmundsson GH (1997) The expression of the gene coding for the antibacterial peptide LL-37 is induced in human keratinocytes during inflammatory disorders. J Biol Chem 272:15258–15263

Ganz T (2003) Defensins: antimicrobial peptides of innate immunity. Nat Rev Immunol 3:710–720

Ganz T, Lehrer RI (1997) Antimicrobial peptides of leukocytes. Curr Opin Hematol 4:53–58

Ganz T, Metcalf JA, Gallin JI, Boxer LA, Lehrer RI (1988) Microbicidal/cytotoxic proteins of neutrophils are deficient in two disorders: Chediak-Higashi syndrome and "specific" granule deficiency. J Clin Invest 82:552–556

Garcia JR, Krause A, Schulz S, Rodriguez-Jimenez FJ, Kluver E, Adermann K, Forssmann U, Frimpong-Boateng A, Bals R, Forssmann WG (2001) Human beta-defensin 4: a novel inducible peptide with a specific salt-sensitive spectrum of antimicrobial activity. FASEB J 15:1819–1821

Gennaro R, Zanetti M, Benincasa M, Podda E, Miani M (2002) Pro-rich antimicrobial peptides from animals: structure, biological functions and mechanism of action. Curr Pharm Des 8:763–778

Ghosh D, Porter E, Shen B, Lee SK, Wilk D, Drazba J, Yadav SP, Crabb JW, Ganz T, Bevins CL (2002) Paneth cell trypsin is the processing enzyme for human defensin-5. Nat Immunol 3:583–590

Girardin SE, Hugot JP, Sansonetti PJ (2003) Lessons from Nod2 studies: towards a link between Crohn's disease and bacterial sensing. Trends Immunol 24:652–658

Glaser R, Harder J, Lange H, Bartels J, Christophers E, Schroder JM (2005) Antimicrobial psoriasin (S100A7) protects human skin from Escherichia coli infection. Nat Immunol 6:57–64

Gudmundsson GH, Agerberth B (1999) Neutrophil antibacterial peptides, multifunctional effector molecules in the mammalian immune system. J Immunol Methods 232:45–54

Gudmundsson GH, Agerberth B (2004) Biology and expression of the human cathelicidin LL-37. In: Devine DA, Hancock REW (eds.) Mammalian host defence peptides. Cambridge University Press, Cambridge, pp 139–160

Hancock RE, Scott MG (2000) The role of antimicrobial peptides in animal defenses. Proc Natl Acad Sci U S A 97:8856–8861

Harder J, Schroder JM (2005) Psoriatic scales: a promising source for the isolation of human skin-derived antimicrobial proteins. J Leukoc Biol 77:476–486

Harder J, Bartels J, Christophers E, Schroder JM (1997) A peptide antibiotic from human skin. Nature 387:861

Harder J, Bartels J, Christophers E, Schroder JM (2001) Isolation and characterization of human beta -defensin-3, a novel human inducible peptide antibiotic. J Biol Chem 276:5707–5713

Hase K, Eckmann L, Leopard JD, Varki N, Kagnoff MF (2002) Cell differentiation is a key determinant of cathelicidin LL-37/human cationic antimicrobial protein 18 expression by human colon epithelium. Infect Immun 70:953–963

Hase K, Murakami M, Iimura M, Cole SP, Horibe Y, Ohtake T, Obonyo M, Gallo RL, Eckmann L, Kagnoff MF (2003) Expression of LL-37 by human gastric epithelial cells as a potential host defense mechanism against Helicobacter pylori. Gastroenterology 125:1613–1625

Hertz CJ, Wu Q, Porter EM, Zhang YJ, Weismuller KH, Godowski PJ, Ganz T, Randell SH, Modlin RL (2003) Activation of Toll-like receptor 2 on human tracheobronchial epithelial cells induces the antimicrobial peptide human beta defensin-2. J Immunol 171:6820–6826

Hiemstra PS, Maassen RJ, Stolk J, Heinzel-Wieland R, Steffens GJ, Dijkman JH (1996) Antibacterial activity of antileukoprotease. Infect Immun 64:4520–4524

Hiratsuka T, Mukae H, Iiboshi H, Ashitani J, Nabeshima K, Minematsu T, Chino N, Ihi T, Kohno S, Nakazato M (2003) Increased concentrations of human beta-defensins in plasma and bronchoalveolar lavage fluid of patients with diffuse panbronchiolitis. Thorax 58:425–430

Hornef MW, Normark BH, Vandewalle A, Normark S (2003) Intracellular recognition of lipopolysaccharide by toll-like receptor 4 in intestinal epithelial cells. J Exp Med 198:1225–1235

Howell MD, Jones JF, Kisich KO, Streib JE, Gallo RL, Leung DY (2004) Selective killing of vaccinia virus by LL-37: implications for eczema vaccinatum. J Immunol 172:1763–1767

Hu RC, Xu YJ, Zhang ZX, Ni W, Chen SX (2004) Correlation of HDEFB1 polymorphism and susceptibility to chronic obstructive pulmonary disease in Chinese Han population. Chin Med J (Engl) 117:1637–1641

Hugot JP, Chamaillard M, Zouali H, Lesage S, Cezard JP, Belaiche J, Almer S, Tysk C, O'Morain CA, Gassull M, Binder V, Finkel Y, Cortot A, Modigliani R, Laurent-Puig P, Gower-Rousseau C, Macry J, Colombel JF, Sahbatou M, Thomas G (2001) Association of NOD2 leucine-rich repeat variants with susceptibility to Crohn's disease. Nature 411:599–603

Islam D, Bandholtz L, Nilsson J, Wigzell H, Christensson B, Agerberth B, Gudmundsson G (2001) Downregulation of bactericidal peptides in enteric infections: a novel immune escape mechanism with bacterial DNA as a potential regulator. Nat Med 7:180–185

Johansson J, Gudmundsson GH, Rottenberg ME, Berndt KD, Agerberth B (1998) Conformation-dependent antibacterial activity of the naturally occurring human peptide LL-37. J Biol Chem 273:3718–3724

Jones DE, Bevins CL (1992) Paneth cells of the human small intestine express an antimicrobial peptide gene. J Biol Chem 267:23216–23225

Jones DE, Bevins CL (1993) Defensin-6 mRNA in human Paneth cells: implications for antimicrobial peptides in host defense of the human bowel. FEBS Lett 315:187–192

Kavanagh K, Dowd S (2004) Histatins: antimicrobial peptides with therapeutic potential. J Pharm Pharmacol 56:285–289

Kobayashi KS, Chamaillard M, Ogura Y, Henegariu O, Inohara N, Nunez G, Flavell RA (2005) Nod2-dependent regulation of innate and adaptive immunity in the intestinal tract. Science 307:731–734

Koczulla R, von Degenfeld G, Kupatt C, Krotz F, Zahler S, Gloe T, Issbrucker K, Unterberger P, Zaiou M, Lebherz C, Karl A, Raake P, Pfosser A, Boekstegers P, Welsch U, Hiemstra PS, Vogelmeier C, Gallo RL, Clauss M, Bals R (2003) An angiogenic role for the human peptide antibiotic LL-37/hCAP-18. J Clin Invest 111:1665–1672

Krisanaprakornkit S, Kimball JR, Weinberg A, Darveau RP, Bainbridge BW, Dale BA (2000) Inducible expression of human beta-defensin 2 by Fusobacterium nucleatum in oral epithelial cells: multiple signaling pathways and role of commensal bacteria in innate immunity and the epithelial barrier. Infect Immun 68:2907–2915

Larrick JW, Morgan JG, Palings I, Hirata M, Yen MH (1991) Complementary DNA sequence of rabbit CAP18—a unique lipopolysaccharide binding protein. Biochem Biophys Res Commun 179:170–175

Lehrer RI, Lichtenstein AK, Ganz T (1993) Defensins: antimicrobial and cytotoxic peptides of mammalian cells. Annu Rev Immunol 11:105–128

Maeda S, Hsu LC, Liu H, Bankston LA, Iimura M, Kagnoff MF, Eckmann L, Karin M (2005) Nod2 mutation in Crohn's disease potentiates NF-kappaB activity and IL-1beta processing. Science 307:734–738

Matsushita I, Hasegawa K, Nakata K, Yasuda K, Tokunaga K, Keicho N (2002) Genetic variants of human beta-defensin-1 and chronic obstructive pulmonary disease. Biochem Biophys Res Commun 291:17–22

Morrison G, Kilanowski F, Davidson D, Dorin J (2002) Characterization of the mouse beta defensin 1, Defb1, mutant mouse model. Infect Immun 70:3053–3060

Moser C, Weiner DJ, Lysenko E, Bals R, Weiser JN, Wilson JM (2002) Beta-defensin 1 contributes to pulmonary innate immunity in mice. Infect Immun 70:3068–3072

Mukae H, Ashitani J, Nakazato M, Taniguchi H, Date Y, Ihi T, Shiomi K, Mashimoto H, Kadota J, Kohno S et al (1995) [A study of defensins in bronchoalveolar lavage fluid in patients with diffuse panbronchiolitis]. Kansenshogaku Zasshi 69:975–981

Murakami M, Ohtake T, Dorschner RA, Gallo RL (2002) Cathelicidin antimicrobial peptides are expressed in salivary glands and saliva. J Dent Res 81:845–850

Murakami M, Lopez-Garcia B, Braff M, Dorschner RA, Gallo RL (2004) Postsecretory processing generates multiple cathelicidins for enhanced topical antimicrobial defense. J Immunol 172:3070–3077

Murphy CJ, Foster BA, Mannis MJ, Selsted ME, Reid TW (1993) Defensins are mitogenic for epithelial cells and fibroblasts. J Cell Physiol 155:408–413

Ng AW, Bidani A, Heming TA (2004) Innate host defense of the lung: effects of lung-lining fluid pH. Lung 182:297–317

Nizet V, Ohtake T, Lauth X, Trowbridge J, Rudisill J, Dorschner RA, Pestonjamasp V, Piraino J, Huttner K, Gallo RL (2001) Innate antimicrobial peptide protects the skin from invasive bacterial infection. Nature 414:454–457

O'Neil DA, Porter EM, Elewaut D, Anderson GM, Eckmann L, Ganz T, Kagnoff MF (1999) Expression and regulation of the human beta-defensins hBD-1 and hBD-2 in intestinal epithelium. J Immunol 163:6718–6724

O'Neil DA, Cole SP, Martin-Porter E, Housley MP, Liu L, Ganz T, Kagnoff MF (2000) Regulation of human beta-defensins by gastric epithelial cells in response to infection with Helicobacter pylori or stimulation with interleukin-1. Infect Immun 68:5412–5415

Ogura Y, Bonen DK, Inohara N, Nicolae DL, Chen FF, Ramos R, Britton H, Moran T, Karaliuskas R, Duerr RH, Achkar JP, Brant SR, Bayless TM, Kirschner BS, Hanauer SB, Nunez G, Cho JH (2001) A frameshift mutation in NOD2 associated with susceptibility to Crohn's disease. Nature 411:603–606

Ong PY, Ohtake T, Brandt C, Strickland I, Boguniewicz M, Ganz T, Gallo RL, Leung DY (2002) Endogenous antimicrobial peptides and skin infections in atopic dermatitis. N Engl J Med 347:1151–1160

Oren Z, Shai Y (1998) Mode of action of linear amphipathic alpha-helical antimicrobial peptides. Biopolymers 47:451–463

Ouellette AJ (2004) Defensin-mediated innate immunity in the small intestine. Best Pract Res Clin Gastroenterol 18:405–419

Putsep K, Carlsson G, Boman HG, Andersson M (2002) Deficiency of antibacterial peptides in patients with morbus Kostmann: an observation study. Lancet 360:1144–1149

Quinones-Mateu ME, Lederman MM, Feng Z, Chakraborty B, Weber J, Rangel HR, Marotta ML, Mirza M, Jiang B, Kiser P, Medvik K, Sieg SF, Weinberg A (2003) Human epithelial beta-defensins 2 and 3 inhibit HIV-1 replication. Aids 17:F39–F48

Salzman NH, Chou MM, de Jong H, Liu L, Porter EM and Paterson Y (2003b) Enteric salmonella infection inhibits Paneth cell antimicrobial peptide expression. Infect Immun 71:1109–1115

Salzman NH, Ghosh D, Huttner KM, Paterson Y, Bevins CL (2003a) Protection against enteric salmonellosis in transgenic mice expressing a human intestinal defensin. Nature 422:522–526

Schauber J, Svanholm C, Termen S, Iffland K, Menzel T, Scheppach W, Melcher R, Agerberth B, Luhrs H, Gudmundsson GH (2003) Expression of the cathelicidin LL-37 is modulated by short chain fatty acids in colonocytes: relevance of signalling pathways. Gut 52:735–741

Schmidtchen A, Frick IM, Andersson E, Tapper H, Bjorck L (2002) Proteinases of common pathogenic bacteria degrade and inactivate the antibacterial peptide LL-37. Mol Microbiol 46:157–168

Schutte BC, Mitros JP, Bartlett JA, Walters JD, Jia HP, Welsh MJ, Casavant TL, McCray PB Jr (2002) Discovery of five conserved beta -defensin gene clusters using a computational search strategy. Proc Natl Acad Sci U S A 99:2129–2133

Scocchi M, Skerlavaj B, Romeo D, Gennaro R (1992) Proteolytic cleavage by neutrophil elastase converts inactive storage proforms to antibacterial bactenecins. Eur J Biochem 209:589–595

Selsted ME (2004) Theta-defensins: cyclic antimicrobial peptides produced by binary ligation of truncated alpha-defensins. Curr Protein Pept Sci 5:365–371

Selsted ME, Brown DM, DeLange RJ, Lehrer RI (1983) Primary structures of MCP-1 and MCP-2, natural peptide antibiotics of rabbit lung macrophages. J Biol Chem 258:14485–14489

Shafer WM, Qu X, Waring AJ, Lehrer RI (1998) Modulation of Neisseria gonorrhoeae susceptibility to vertebrate antibacterial peptides due to a member of the resistance/nodulation/division efflux pump family. Proc Natl Acad Sci U S A 95:1829–1833

Shai Y (1995) Molecular recognition between membrane-spanning polypeptides. Trends Biochem Sci 20:460–464

Singh PK, Tack BF, McCray PB Jr, Welsh MJ (2000) Synergistic and additive killing by antimicrobial factors found in human airway surface liquid. Am J Physiol Lung Cell Mol Physiol 279:L799–L805

Smith JJ, Travis SM, Greenberg EP, Welsh MJ (1996) Cystic fibrosis airway epithelia fail to kill bacteria because of abnormal airway surface fluid. Cell 85:229–236

Sorensen OE, Follin P, Johnsen AH, Calafat J, Tjabringa GS, Hiemstra PS, Borregaard N (2001) Human cathelicidin, hCAP-18, is processed to the antimicrobial peptide LL-37 by extracellular cleavage with proteinase 3. Blood 97:3951–3959

Steiner H, Hultmark D, Engstrom A, Bennich H, Boman HG (1981) Sequence and specificity of two antibacterial proteins involved in insect immunity. Nature 292:246–248

Tang YQ, Yuan J, Osapay G, Osapay K, Tran D, Miller CJ, Ouellette AJ, Selsted ME (1999) A cyclic antimicrobial peptide produced in primate leukocytes by the ligation of two truncated alpha-defensins. Science 286:498–502

Tjabringa GS, Aarbiou J, Ninaber DK, Drijfhout JW, Sorensen OE, Borregaard N, Rabe KF, Hiemstra PS (2003) The antimicrobial peptide LL-37 activates innate immunity at the airway epithelial surface by transactivation of the epidermal growth factor receptor. J Immunol 171:6690–6696

Tran D, Tran PA, Tang YQ, Yuan J, Cole T, Selsted ME (2002) Homodimeric theta-defensins from rhesus macaque leukocytes: isolation, synthesis, antimicrobial activities, and bacterial binding properties of the cyclic peptides. J Biol Chem 277:3079–3084

Travis SM, Singh PK, Welsh MJ (2001) Antimicrobial peptides and proteins in the innate defense of the airway surface. Curr Opin Immunol 13:89–95

Van Wetering S, Mannesse-Lazeroms SP, Van Sterkenburg MA, Daha MR, Dijkman JH, Hiemstra PS (1997) Effect of defensins on interleukin-8 synthesis in airway epithelial cells. Am J Physiol 272:L888–L896

Wehkamp J, Harder J, Weichenthal M, Schwab M, Schaffeler E, Schlee M, Herrlinger KR, Stallmach A, Noack F, Fritz P, Schroder JM, Bevins CL, Fellermann K, Stange EF (2004) NOD2 (CARD15) mutations in Crohn's disease are associated with diminished mucosal alpha-defensin expression. Gut 53:1658–1664

Weinberg A, Krisanaprakornkit S, Dale BA (1998) Epithelial antimicrobial peptides: review and significance for oral applications. Crit Rev Oral Biol Med 9:399–414

Wilson CL, Ouellette AJ, Satchell DP, Ayabe T, Lopez-Boado YS, Stratman JL, Hultgren SJ, Matrisian LM, Parks WC (1999) Regulation of intestinal alpha-defensin activation by the metalloproteinase matrilysin in innate host defense. Science 286:113–117

Yang D, Chertov O, Bykovskaia SN, Chen Q, Buffo MJ, Shogan J, Anderson M, Schroder JM, Wang JM, Howard OM, Oppenheim JJ (1999) Beta-defensins: linking innate and adaptive immunity through dendritic and T cell CCR6. Science 286:525–528

Yang D, Chen Q, Chertov O, Oppenheim JJ (2000a) Human neutrophil defensins selectively chemoattract naive T and immature dendritic cells. J Leukoc Biol 68:9–14

Yang D, Chen Q, Schmidt AP, Anderson GM, Wang JM, Wooters J, Oppenheim JJ, Chertov O (2000b) LL-37, the neutrophil granule- and epithelial cell-derived cathelicidin, utilizes formyl peptide receptor-like 1 (FPRL1) as a receptor to chemoattract human peripheral blood neutrophils, monocytes, and T cells. J Exp Med 192:1069–1074

Yang D, Chertov O, Oppenheim JJ (2001) The role of mammalian antimicrobial peptides and proteins in awakening of innate host defenses and adaptive immunity. Cell Mol Life Sci 58:978–989

Yoshio H, Tollin M, Gudmundsson GH, Lagercrantz H, Jornvall H, Marchini G, Agerberth B (2003) Antimicrobial polypeptides of human vernix caseosa and amniotic fluid: implications for newborn innate defense. Pediatr Res 53:211–216

Zaiou M, Nizet V, Gallo RL (2003) Antimicrobial and protease inhibitory functions of the human cathelicidin (hCAP18/LL-37) prosequence. J Invest Dermatol 120:810–816

Zanetti M (2004) Cathelicidins, multifunctional peptides of the innate immunity. J Leukoc Biol 75:39–48

Zanetti M, Gennaro R, Romeo D (1995) Cathelicidins: a novel protein family with a common proregion and a variable C-terminal antimicrobial domain. FEBS Lett 374:1–5

Zasloff M (1987) Magainins, a class of antimicrobial peptides from Xenopus skin: isolation, characterization of two active forms, and partial cDNA sequence of a precursor. Proc Natl Acad Sci U S A 84:5449–5453

Zasloff M (2002) Antimicrobial peptides of multicellular organisms. Nature 415:389–395

CTMI (2006) 306:91–110

Antimicrobial Peptides:
An Essential Component of the Skin Defensive Barrier

M. H. Braff · R. L. Gallo (✉)

University of California, San Diego, Mail Code 111B, San Diego, CA 92161, USA
rgallo@ucsd.edu

Abstract The skin is positioned at the interface between an organism's internal milieu and an external environment characterized by constant assault with potential microbial pathogens. While the skin was formerly considered an inactive physical protective barrier that participates in host immune defense merely by blocking entry of microbial pathogens, it is now apparent that a major role of the skin is to defend the body by rapidly mounting an innate immune response to injury and microbial insult. In the skin, both resident and infiltrating cells synthesize and secrete small peptides that demonstrate broad-spectrum antimicrobial activity against bacteria, fungi, and enveloped viruses. Antimicrobial peptides also act as multifunctional immune effectors by stimulating cytokine and chemokine production, angiogenesis, and wound healing. Cathelicidins and defensins comprise two major families of skin-derived antimicrobial peptides, although numerous others have been described. Many such immune defense molecules are currently being developed therapeutically in an attempt to combat growing bacterial resistance to conventional antibiotics.

1
Skin Host Defense

The skin innate immune system acts as a first line of defense against infection by responding to pathogen-associated molecular patterns such as lipopolysaccharide and peptidoglycan. Not surprisingly, circulating leukocytes such as neutrophils, macrophages, and natural killer cells are major contributors to the cellular innate immune system of the skin. More recently, it has become apparent that the epithelium also plays an active role in skin innate immunity. Keratinocytes, the most prominent epithelial cells of the skin, were previously regarded merely as forming a mechanical barrier to pathogen entry by presenting a covalently cross-linked stratum corneum at the interface. Today however, multiple observations suggest keratinocytes actively resist infection. As a dynamic and sensitive immune organ, the epithelium of the skin secretes various cytokines, chemokines, and small molecules that trigger recruitment of leukocytes or directly inhibit microbial growth and invasion. Prominent among these molecules are the gene-encoded antimicrobial peptides that are capable of killing a wide range of bacterial, fungal, and viral pathogens through membrane disruption (Brogden et al. 2003; Henzler Wildman et al. 2003). These immune defense peptides offer a practical alternative to the increasing dilemma of bacterial resistance to conventional antibiotics (Zasloff 2002; Ulvatne 2003).

2
Antimicrobial Peptides in Skin

There are many proteins and peptides present in the skin with antimicrobial activity. Table 1 provides an abbreviated list of skin antimicrobial peptides and their originally described functions. Most of these molecules were discovered due to other biological actions and later found to also have direct antimicrobial activity. Of the peptides that were first identified based on their antimicrobial activity, two major families serve as excellent examples of the potential function of antimicrobial peptides in skin defense: cathelicidins and defensins. Both infiltrating immune cells and resident epithelial cells synthesize these host defense peptides. These peptides display distinct, but overlapping, broad-spectrum antimicrobial activity against both Gram-positive and Gram-negative bacteria, with minimum inhibitory concentrations (MIC) ranging from 0.1 to 100 μg/ml (Bals 2000). Cathelicidin has been shown to act synergistically with defensins and other antimicrobial molecules involved in host immunity, such as lysozyme (Bals 2000; Singh et al. 2000). Interestingly,

Table 1 Diverse peptides and proteins present in skin may provide antimicrobial defense

Primary function	Examples
Antimicrobial activity	Cathelicidins, defensins
Protease inhibitor/enzyme activity	Cathelin-like prodomain, SKALP/elafin, SLPI/antileukoprotease, lysozyme
Chemokine activity	MIG/CXCL9, IP-10/CXCL10, I-TAC/CXCL11, psoriasin
Neuropeptide activity	α-MSH, substance P, neuropeptide Y, chromogranin A and B, adrenomedullin

it was recently shown that cathelicidins, but not defensins, inhibit replication of orthopox virus (vaccinia) (Howell et al. 2004). In addition to antimicrobial activity, it is now appreciated that these peptides modify host inflammatory responses and exert many other effects on the host, including the ability to promote chemotaxis, angiogenesis, and wound healing. Together, these multifunctional peptides contribute to skin innate immune defense through both antimicrobial and immunomodulatory activities. However, for discussion it is most appropriate to consider these peptides separately, as there is little similarity between cathelicidins and defensins other than their small size and cationic nature.

2.1
Cathelicidins

Humans, mice, pigs, sheep, and cows each express at least one cathelicidin (Lehrer and Ganz 2002). The human and mouse cathelicidins, LL-37 and CRAMP, respectively, are α-helical antimicrobial peptides that bear a net positive charge (+6) and are highly resistant to proteolytic degradation (Hornef et al. 2002). Structurally, cathelicidins are organized into an N-terminal signal sequence, followed by a conserved cathelin-like prodomain and variable antimicrobial domain (Ramanathan et al. 2002; Zaiou and Gallo 2002). The cathelin-like prodomain is named for its similarity to the cystatin family of cysteine protease inhibitors and shows approximately 60% sequence identity among species (Lehrer and Ganz 2002). The antimicrobial domain is released from the precursor peptide by proteolytic cleavage; this processing step is required for antimicrobial activity (Zanetti et al. 2000).

First observed in mammalian skin during analysis of wound healing (Gallo et al. 1994), it is now clear that several cell types in the skin produce cathelicidin, including keratinocytes (Frohm et al. 1997; Sorensen et al. 2003), neutrophils (Turner et al. 1998), mast cells (Di Nardo et al. 2003), and eccrine glands (Murakami et al. 2002). LL-37 levels in wounded skin can reach 2,000 ng/mg total protein, but the relative contribution of each cell type to skin immunity has not yet been elucidated (Dorschner et al. 2001; Zaiou and Gallo 2002; Heilborn et al. 2003). Secretions such as sweat and seminal fluid also contain high levels of cathelicidin, including novel processed cathelicidin peptides with enhanced antimicrobial activity (Murakami et al. 2004). The concentration of LL-37 has been shown to reach levels up to 360 µg/ml in sweat and 140 µg/ml in seminal fluid (Murakami et al. 2002).

Neutrophil secretory granules store a 17-kDa cathelicidin peptide, while a mature 5-kDa form is detected upon secretion (Gudmundsson et al. 1996; Sorensen et al. 2001). Proteinase 3 has been implicated in cathelicidin activation in human neutrophils (Sorensen et al. 2001). Cathelicidin processing also occurs in keratinocytes (Braff et al. 2005), although the protease responsible has not yet been identified. In sweat, LL-37 is further processed to smaller peptide fragments, namely KS-30 and RK-31 (Murakami et al. 2004). These peptide derivatives demonstrate more potent antimicrobial activity than LL-37 against a number of skin pathogens, but have reduced host stimulatory capacity. Therefore, the extent of cathelicidin processing may regulate the balance between antimicrobial activity and immunomodulatory functions (Fig. 1).

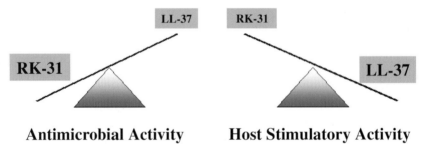

Antimicrobial Activity   **Host Stimulatory Activity**

Fig. 1 Cathelicidin processing regulates the balance between antimicrobial activity and immunomodulatory functions. In human skin, cathelicidin can be present in several peptide forms, including the 37-amino acid LL-37 peptide found in neutrophils and shorter forms such as the 31-amino acid RK-31 peptide found in sweat. In comparison with LL-37, the shorter peptide derivatives demonstrate more potent antimicrobial activity against a number of skin pathogens, but have reduced host stimulatory capacity. Alteration in the abundance of cathelicidin in the LL-37 form relative to the shorter RK-31 form predicts a change in the balance of antimicrobial to host stimulatory activity

2.2
Defensins

Like cathelicidins, defensins are small, cationic antimicrobial peptides with broad-spectrum antimicrobial activity against bacteria, fungi, and enveloped viruses (Ganz et al. 1985; Lehrer et al. 1985; Lehrer and Ganz 1999). Structurally, defensins form β-sheet structures and contain six cysteine residues that generate characteristic intermolecular disulfide bridges. Different disulfide linkage patterns separate these antimicrobial peptides into α-, β-, and θ-defensins, although humans do not make θ-defensins (Lehrer and Ganz 1999). Recent genomic analysis using computational gene discovery strategies suggests that many defensins have yet to be discovered (Schutte et al. 2002; Kao et al. 2003).

Characteristically, α-defensins contain three disulfide bridges in a 1–6, 2–4, 3–5 alignment. The α-defensin family contains four human neutrophil peptides, referred to as HNP1–4 (Harwig et al. 1994). HNPs are stored as processed, mature peptides in the azurophil granules of neutrophils. Two additional human α-defensins, named HD5 and HD6, are abundantly expressed in female urogenital tract epithelia (Harwig et al. 1994; Quayle et al. 1998) and as propeptides in Paneth cells, which are located in the crypts of the small intestine (Selsted et al. 1992; Jones and Bevins 1993). Resident skin cells such as keratinocytes have not been shown to produce α-defensins.

In contrast to α-defensins, the β-defensin disulfide bridges are spaced in a 1–5, 2–4, 3–6 pattern. The four well-studied human β-defensins, hBD1–4, are synthesized by various cell types, including epithelial cells and peripheral blood mononuclear cells (Harder et al. 1997, 2001; Duits et al. 2002; Fang et al. 2003). Keratinocytes produce hBD1, -2, and -3 in addition to LL-37 (Midorikawa et al. 2003; Sorensen et al. 2003). hBD1 is constitutively expressed in epithelia and has been detected at high levels in the kidney (Zhao et al. 1996; McCray and Bentley 1997; Valore et al. 1998). Alternatively, hBD2 is expressed at low levels in epithelia, but is highly up-regulated in inflamed skin (Harder et al. 1997; Bals et al. 1998) and in keratinocytes stimulated with cytokines or bacteria (Liu et al. 2002, 2003). hBD3 was purified from human psoriatic scales and calluses (Harder et al. 2001), and like hBD2, demonstrates inducible expression.

2.3
Other Antimicrobial Peptides

In addition to cathelicidins and defensins, many other skin-derived peptides possess innate antimicrobial activity (Table 1). For example, the cathelin-like prodomain, which has traditionally been considered merely as the conserved prosequence of cathelicidin peptides, possesses intrinsic protease inhibitory function against cathepsin L as well as antimicrobial activity against *E. coli* and methicillin-resistant *S. aureus* (MRSA) (Zaiou et al. 2003). LL-37 is not active against MRSA, although it shows broad-spectrum activity against various other pathogens. The negatively charged cathelin-like prodomain may also play a role in neutralizing the antimicrobial peptide domain prior to injury in order to prevent damage to storage granules and to prevent activation until proteases are released under inflammatory conditions (Sorensen et al. 1997). Secretory leukocyte proteinase inhibitor (SLPI), or antileuko-protease, demonstrates similar dual functionality. In addition to acting as an antimicrobial agent, SLPI demonstrates serine protease inhibitor activity toward neutrophil elastase, cathepsin G, and mast cell chymase (Franzke et al. 1996; Wiedow et al. 1998; Goetz et al. 2002). Further, SLPI-deficient mice demonstrate impaired cutaneous wound healing and increased inflammation (Ashcroft et al. 2000).

3
Antimicrobial Peptide Regulation in Keratinocytes

Multiple layers of keratinocytes at different stages of differentiation comprise the epidermis. Basal keratinocytes, which are attached to the basement membrane, are undifferentiated proliferative cells. As keratinocytes differentiate upwards to form the permeability barrier of skin, both their morphology and gene expression profile change. The spinous and granular layers lie directly above the basal keratinocytes and are covered by the stratum corneum, which consists of nonviable, terminally differentiated anucleated cells (Haake et al. 2003). Secretory granules such as lamellar bodies form in these differentiated epidermal layers. Lamellar bodies transport lipids and hydrolytic enzymes involved in stratum corneum formation and have been shown to store cytokines, including TNF-α and IL-1 (Nickoloff 1999; Oren et al. 2003), as well as antimicrobial peptides such as β-defensins and cathelicidins (Oren et al. 2003; Braff et al. 2005).

Inflammatory mediators and cell differentiation up-regulate the expression of cathelicidin and β-defensin antimicrobial peptides in keratinocytes (Frohm et al. 1997). In contrast, keratinocyte expression of the antimicrobial peptide dermcidin, which is constitutively expressed in eccrine sweat glands, is not induced under inflammatory conditions (Rieg et al. 2004). LL-37 expression by keratinocytes is induced by insulin-like growth factor-1 (IGF-1) and TGF-α, and upon contact with *S. aureus* (Midorikawa et al. 2003; Sorensen et al. 2003). Cytokines such as IL-1α and IL-6 up-regulate antimicrobial peptide transcript levels in skin constructs and enhance their antimicrobial activity against particular pathogens (Erdag and Morgan 2002). It is interesting to note that keratinocyte differentiation is a key requirement for stimulating LL-37 expression in human colon epithelium, while proinflammatory cytokines do not affect antimicrobial peptide expression (Hase et al. 2002), suggesting that the regulation of antimicrobial peptide expression may be tissue-specific. LPS, TNF-α, and UV-B irradiation induce the expression of β-defensins in keratinocyte cell lines (Seo et al. 2001). In separate studies, hBD2 expression was up-regulated in differentiated keratinocytes in response to IL-1 or in the presence of particular bacterial pathogens (Liu et al. 2002; Dinulos et al. 2003; Uehara et al. 2003). Treatment of keratinocytes with *M. furfur*, which is part of the normal cutaneous flora, up-regulates hBD2 expression in a protein kinase C (PKC)-dependent manner (Donnarumma et al. 2004). Interestingly, skin pathogenic and commensal bacteria, including Group A *Streptococcus* and *S. epidermidis*, respectively, utilize different signaling pathways to up-regulate keratinocyte hBD2 (Chung and Dale 2004). Recently, vitamin D was also found to induce keratinocyte LL-37 and hBD2 expression and antimicrobial activity against *P. aeruginosa* through consensus vitamin D response elements (Wang et al. 2004).

Down-regulation of expression or proteolytic inactivation provide further mechanisms of antimicrobial peptide regulation. While TNF-α induces hBD2, heat shock reduces hBD2 expression in keratinocytes (Bick et al. 2004). *Shigella* plasmid DNA has also been identified as a potential mediator for down-regulation of both LL-37 and hBD1 (Islam et al. 2001). Both host and bacterial proteases have recently been shown to inactivate antimicrobial peptides. Cathepsins B, L, and S, which are present in host respiratory epithelia, can degrade and inactivate hBD2 and hBD3 (Taggart et al. 2003). Proteinases of common pathogenic bacteria, including *P. aeruginosa, E. faecalis, P. mirabilis*, and Group A *Streptococcus*, degrade and inactivate LL-37 (Schmidtchen et al. 2002). Independent studies have since shown that the SIC protein secreted by Group A *Streptococcus* is responsible for LL-37 and HNP1 inactivation (Frick et al. 2003), while the ZapA metalloprotease made by *P. mirabilis* cleaves LL-37 and hBD1 (Belas et al. 2004). These studies confirmed that degradation and inactivation of antimicrobial peptides leads to increased bacterial survival.

4
Antimicrobial Activity

Many studies have shown that antimicrobial peptides can be inactivated at physiological salt concentrations or in the presence of serum (Goldman et al.

1997). However, most of these antimicrobial activity assays are done with bacterial growth media as opposed to physiologic mammalian cell culture media that more closely resembles the skin microenvironment. In this environment, and in the animal models discussed below, many antimicrobial peptides appear to maintain their effectiveness despite the presence of salt and serum. The mechanism behind these discordant results is not yet clear, and remains a controversial point in antimicrobial peptide biology.

In vivo studies have provided the best evidence to support the role of antimicrobial peptides as an important component of innate immunity. In neonatal mice and humans, in which the adaptive immune system has not yet developed, the skin expresses high levels of antimicrobial peptides that act together to kill the important pathogen Group B *Streptococcus* (Dorschner et al. 2003). Cathelicidin-deficient mice demonstrate increased susceptibility to skin infection caused by the invasive human pathogen Group A *Streptococcus* (Nizet et al. 2001). In this study, infected mice showed normal neutrophil recruitment, suggesting that the lack of cathelicidin in granulocytes and keratinocytes prevented efficient bacterial clearing. Using an alternative bacterial genetics approach, mice infected with cathelicidin-resistant Group A *Streptococcus* developed persistent skin lesions, as opposed to the nearly complete clearing of cathelicidin-sensitive bacteria (Nizet et al. 2001). The essential role of antimicrobial peptides in epithelial defense has been further demonstrated in matrilysin-deficient mice, which show reduced intestinal antimicrobial activity due to their inability to activate cryptdins (Wilson et al. 1999), and in pig wounds inhibited from processing porcine cathelicidins to their active peptide forms (Cole et al. 2001). In contrast to cathelicidin, which is encoded by a single gene in humans and mice, multiple defensins participate in host immune defense. Interestingly, mice deficient in murine β-defensin-1 (mBD1) do not display obvious innate immune defects, which may indicate the importance of redundancy in host defense. These mice show delayed, but not absent, clearance of *H. influenzae* from the lung (Moser et al. 2002). In addition, they harbor more *Staphylococcus* in their bladders, but are not impaired in their ability to clear *S. aureus* from the respiratory tract (Morrison et al. 2002). Overexpression of antimicrobial peptides in mice provides another means to study their in vivo function. For example, overexpression of hCAP18/LL-37 in a murine pulmonary infection model resulted in reduced bacterial load and inflammatory response following *P. aeruginosa* infection (Bals et al. 1999). In addition, mice engineered to overexpress HD5 were highly resistant to oral challenge with virulent *S. typhimurium* (Salzman et al. 2003). These studies provide further evidence that antimicrobial peptides are essential to epithelial immune defense.

Correlations between human disease and antimicrobial peptide expression have also begun to emerge. Clinically, patients with inflammatory skin diseases such as psoriasis or cholesteatoma demonstrate enhanced antimicrobial peptide expression, whereas those with acute or chronic atopic dermatitis show a relative deficiency in these host defense molecules (Ong et al. 2002; Jung et al. 2003). Not surprisingly, this difference is most pronounced in the granular keratinocytes and stratum corneum. The lack of LL-37 and hBD2 up-regulation in skin among atopic dermatitis patients may offer one explanation for their increased susceptibility to bacterial and viral infections (Gallo et al. 2002). Deficiencies in antimicrobial peptide expression may also account for the enhanced susceptibility to infection seen in Morbus Kostmann patients (Putsep et al. 2002). The relevance of altered antimicrobial peptide expression to skin infection is discussed in greater detail below.

5
Immunomodulatory Activity

In addition to their antimicrobial activity, cathelicidins and defensins modulate cytokine and chemokine production and otherwise act as multifunctional immune effectors through antitumor, mitogenic, signaling, and angiogenic activities (Gudmundsson and Agerberth 1999; Risso 2000; Bals and Wilson 2003; Kamysz and Okroj 2003; Zanetti 2004).

LL-37 acts on the G protein-coupled formyl peptide receptor-like 1 (FPRL1) to chemoattract peripheral blood neutrophils, monocytes, and T cells (Agerberth et al. 2000; De et al. 2000; Yang et al. 2001). LL-37-induced mast cell chemotaxis, however, appears to occur via a separate G_i protein-phospholipase C pathway (Niyonsaba et al. 2002, 2003). Recently, cathelicidin has also been assigned a role in angiogenesis through the direct activation of endothelial cell FPRL1 (Koczulla et al. 2003). LL-37 also regulates inflammation by binding to and neutralizing bacterial surface antigens such as lipopolysaccharide (LPS) and lipotechoic acid (LTA) (Larrick et al. 1995; Scott et al. 1999; Nagaoka et al. 2002).

hBD2 exerts chemotactic activity for immature dendritic cells and memory T cells by binding to CCR6 (Yang et al. 2001). hBD2 may also play an immunotherapeutic role as a vaccine adjuvant to enhance antibody production by promoting histamine release and prostaglandin D2 production in mast cells (Befus et al. 1999; Risso 2000). In addition, HNP1, -2, and -3 up-regulate the production of cytokines such as TNF-α and IL-1 in bacteria-activated monocytes and reduce the expression of the adhesion molecule VCAM-1 in TNF-α-activated endothelial cells (Chaly et al. 2000).

6
Bacterial Resistance to Skin Antimicrobial Peptides

Group A *Streptococcus* is a common skin pathogen that binds to injured keratinocytes to initiate infections such as impetigo, and Group B *Streptococcus* constitutes a major threat to neonates (Darmstadt et al. 1999; Dorschner et al. 2003). *S. aureus* is another important human pathogen that both causes disease directly and worsens inflammatory skin conditions such as psoriasis and atopic dermatitis (Travers et al. 2001). Recently, mutant bacterial strains that demonstrate increased sensitivity to antimicrobial peptides have been engineered. *S. aureus mprF* mutants lack the ability to modify membrane phosphatidylglycerol with L-lysine, leaving bacteria with negative surface charge and promoting increased binding of cationic peptides such as cathelicidins and defensins (Peschel et al. 2001; Kristian et al. 2003). Similarly, *S. aureus*, *L. monocytogenes*, and Group B *Streptococcus dlt* mutants lack the ability to modify teichoic acids with D-alanine, thereby rendering these bacteria more susceptible to killing by antimicrobial peptides and less virulent in murine infection models (Peschel et al. 1999; Abachin et al. 2002; Collins et al. 2002; Poyart et al. 2003).

Various other bacterial pathogens are also targeted by antimicrobial peptides in the skin. Mice that are unable to process cathelicidin due to disruption of the neutrophil elastase gene show increased susceptibility to *K. pneumoniae* and *E. coli* infections (Belaaouaj et al. 1998). Porcine neutrophil cathelicidins also show elastase-mediated antimicrobial activity against *L. monocytogenes* (Shi and Ganz 1998). *S. epidermidis*, *P. aeruginosa*, and *S. typhimurium* may serve as additional targets of cathelicidin and defensin antimicrobial activity (Turner et al. 1998; Liu et al. 2002; Rosenberger et al. 2004).

7
Role of Antimicrobial Peptides in Skin Infections

Psoriasis and atopic dermatitis provide examples of inflammatory skin diseases in which differential antimicrobial peptide expression plays in important role in susceptibility to infection (Christophers and Henseler 1987). In keratinocytes of inflamed psoriatic lesions, LL-37, hBD2, and hBD3 are up-regulated and secondary infection is rare (Henseler and Christophers 1995; Frohm et al. 1997; Harder et al. 1997, 2001; Nomura et al. 2003). Atopic dermatitis patients, on the other hand, do not up-regulate LL-37 and hBD2 expression and demonstrate increased susceptibility to bacterial and viral skin infections, particularly by *S. aureus* (Ong et al. 2002; Leung 2003; Howell et al. 2004; Leung et al. 2004). The resistance to infection in psoriasis, but not in atopic dermatitis, is likely to be due to differences in the regulation and expression of skin antimicrobial peptides.

The pathogenesis of acne vulgaris and superficial folliculitis may also be linked to antimicrobial peptide expression, although this correlation is less straightforward. In lesions of acne vulgaris, hBD1 and hBD2 are up-regulated in the suprabasal layers and in the permanent compartments of the hair follicle (Chronnell et al. 2001; Philpott 2003). Similarly, hBD2 and the HNPs are abundant in lesions of superficial folliculitis, a skin disease characterized by *S. aureus* infection and inflammation of the hair follicle (Oono et al. 2003). It is not clear whether antimicrobial peptide expression is up-regulated as a result of skin infection or whether failure to activate these peptides promotes infection.

Burns offer further evidence for the role of defensins in skin innate immune defense against infection. In acute burn wounds and burn blister fluid, hBD2 expression is dramatically decreased (Ortega et al. 2000; Milner et al. 2003). This relative absence may lead to infection by the common pathogen *P. aeruginosa*. Recently, a cutaneous gene therapy approach was designed to address the increased bacterial loads and loss of epidermal antimicrobial peptides in burn wound beds. In this study, keratinocytes overexpressing LL-37, hBD2, or hBD3 alone or in combination demonstrated increased capacity to inhibit bacterial growth (Carretero et al. 2004).

Dysregulation of antimicrobial peptide expression has been implicated in several other skin conditions. For instance, LL-37, but not hBD1, expression is up-regulated in lesions of erythema toxicum neonatorum, an inflammatory skin reaction common in healthy newborns (Marchini et al. 2002). Epidermal expression of LL-37 is also induced during the development of verruca vulgaris and condyloma acuminata, suggesting that it participates in the skin innate immune response to papillomavirus infection (Conner et al. 2002). Further, LL-37 is involved in re-epithelialization of skin wounds and its deficiency in chronic ulcer epithelium may partially explain the failure of these wounds to heal properly (Heilborn et al. 2003). Conversely, chronic wounds show constitutively high expression of hBD2, possibly due to prolonged tissue injury and microbial colonization (Butmarc et al. 2004). While these studies indicate important potential roles for antimicrobial peptides in host immune defense against skin infection, they also highlight the complexity of antimicrobial peptide regulation and function.

8
Conclusions

Antimicrobial peptides contribute importantly to skin defense, particularly when physical barriers of innate immunity have been breached (Fig. 2). In response to cutaneous injury or infection, antimicrobial peptides are synthesized by various cells, including keratinocytes and circulating neutrophils. Upon activation, antimicrobial peptides contribute to host defense against infection by inhibiting the growth of potentially harmful microbes and by stimulating additional immune functions. The necessity of these peptides is clearly shown by human diseases in which low or absent expression of antimicrobial peptides correlates with increased susceptibility to infection.

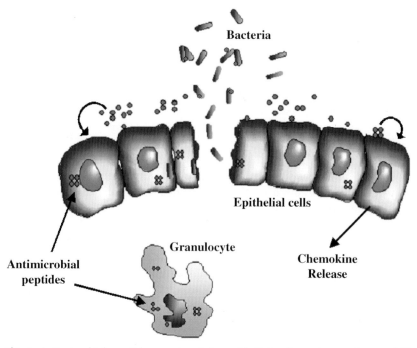

Fig. 2 Antimicrobial peptides produced by epithelial cells and granulocytes kill pathogens and stimulate additional host defense mechanisms. Antimicrobial peptides contribute importantly to skin defense, particularly when physical barriers of innate immunity have been breached. In response to cutaneous injury or infection, antimicrobial peptides are synthesized or deposited by various cells, including keratinocytes and neutrophils. Upon activation, antimicrobial peptides contribute to host defense against infection by inhibiting the growth of potentially harmful microbes and by stimulating additional immune functions such as chemotaxis

Therefore, the supplementation of antimicrobial peptides to immunocompromised skin will likely generate therapeutic effects through direct microbial killing, recruitment of immune cells, and enhancement of wound healing.

References

Abachin E, Poyart C, Pellegrini E, Milohanic E, Fiedler F, Berche P, Trieu-Cuot P (2002) Formation of D-alanyl-lipoteichoic acid is required for adhesion and virulence of Listeria monocytogenes. Mol Microbiol 43:1–14

Agerberth B, Charo J, Werr J, Olsson B, Idali F, Lindbom L, Kiessling R, Jornvall H, Wigzell H, Gudmundsson GH (2000) The human antimicrobial and chemotactic peptides LL-37 and alpha-defensins are expressed by specific lymphocyte and monocyte populations. Blood 96:3086–3093

Ashcroft GS, Lei K, Jin W, Longenecker G, Kulkarni AB, Greenwell-Wild T, Hale-Donze H, McGrady G, Song XY, Wahl SM (2000) Secretory leukocyte protease inhibitor mediates non-redundant functions necessary for normal wound healing. Nat Med 6:1147–1153

Bals R (2000) Epithelial antimicrobial peptides in host defense against infection. Respir Res 1:141–150

Bals R, Wilson JM (2003) Cathelicidins—a family of multifunctional antimicrobial peptides. Cell Mol Life Sci 60:711–720

Bals R, Wang X, Wu Z, Freeman T, Bafna V, Zasloff M, Wilson JM (1998) Human beta-defensin 2 is a salt-sensitive peptide antibiotic expressed in human lung. J Clin Invest 102:874–880

Bals R, Weiner DJ, Moscioni AD, Meegalla RL, Wilson JM (1999) Augmentation of innate host defense by expression of a cathelicidin antimicrobial peptide. Infect Immun 67:6084–6089

Befus AD, Mowat C, Gilchrist M, Hu J, Solomon S, Bateman A (1999) Neutrophil defensins induce histamine secretion from mast cells: mechanisms of action. J Immunol 163:947–953

Belaaouaj A, McCarthy R, Baumann M, Gao Z, Ley TJ, Abraham SN, Shapiro SD (1998) Mice lacking neutrophil elastase reveal impaired host defense against Gram-negative bacterial sepsis. Nat Med 4:615–618

Belas R, Manos J, Suvanasuthi R (2004) Proteus mirabilis ZapA metalloprotease degrades a broad spectrum of substrates, including antimicrobial peptides. Infect Immun 72:5159–5167

Bick RJ, Poindexter BJ, Bhat S, Gulati S, Buja M, Milner SM (2004) Effects of cytokines and heat shock on defensin levels of cultured keratinocytes. Burns 30:329–333

Braff MH, Di Nardo A, Gallo RL (2005) Keratinocytes store the antimicrobial peptide cathelicidin in lamellar bodies. J Invest Dermatol 124:394–400

Brogden KA, Ackermann M, McCray PB Jr, Tack BF (2003) Antimicrobial peptides in animals and their role in host defences. Int J Antimicrob Agents 22:465–478

Butmarc J, Yufit T, Carson P, Falanga V (2004) Human beta-defensin-2 expression is increased in chronic wounds. Wound Repair Regen 12:439–443

Carretero M, Del Rio M, Garcia M, Escamez MJ, Mirones I, Rivas L, Balague C, Jor-
    canoJL, Larcher F (2004) A cutaneous gene therapy approach to treat infection
    through keratinocyte-targeted overexpression of antimicrobial peptides. FASEB
    J 18:1931–1933
Chaly YV, Paleolog EM, Kolesnikova TS, Tikhonov II, Petratchenko EV, Voitenok NN
    (2000) Neutrophil alpha-defensin human neutrophil peptide modulates cytokine
    production in human monocytes and adhesion molecule expression in endothe-
    lial cells. Eur Cytokine Netw 11:257–266
Christophers E, Henseler T (1987) Contrasting disease patterns in psoriasis and atopic
    dermatitis. Arch Dermatol Res 279 [Suppl]:S48–S51
Chronnell CM, Ghali LR, Ali RS, Quinn AG, Holland DB, Bull JJ, Cunliffe WJ, McKay IA,
    Philpott MP, Muller-Rover S (2001) Human beta defensin-1 and -2 expression
    in human pilosebaceous units: upregulation in acne vulgaris lesions. J Invest
    Dermatol 117:1120–1125
Chung WO, Dale BA (2004) Innate immune response of oral and foreskin keratinocytes:
    utilization of different signaling pathways by various bacterial species. Infect
    Immun 72:352–358
Cole AM, Shi J, Ceccarelli A, Kim YH, Park A, Ganz T (2001) Inhibition of neutrophil
    elastase prevents cathelicidin activation and impairs clearance of bacteria from
    wounds. Blood 97:297–304
Collins LV, Kristian SA, Weidenmaier C, Faigle M, Van Kessel KP, Van Strijp JA, Gotz F,
    Neumeister B, Peschel A (2002) Staphylococcus aureus strains lacking D-alanine
    modifications of teichoic acids are highly susceptible to human neutrophil killing
    and are virulence attenuated in mice. J Infect Dis 186:214–219
Conner K, Nern K, Rudisill J, O'Grady T, Gallo RL (2002) The antimicrobial peptide LL-
    37 is expressed by keratinocytes in condyloma acuminatum and verruca vulgaris.
    J Am Acad Dermatol 47:347–350
Darmstadt GL, Mentele L, Fleckman P, Rubens CE (1999) Role of keratinocyte injury
    in adherence of Streptococcus pyogenes. Infect Immun 67:6707–6709
De Y, Chen Q, Schmidt AP, Anderson GM, Wang JM, Wooters J, Oppenheim JJ, Cher-
    tov O (2000) LL-37, the neutrophil granule- and epithelial cell-derived catheli-
    cidin, utilizes formyl peptide receptor-like 1 (FPRL1) as a receptor to chemoat-
    tract human peripheral blood neutrophils, monocytes, and T cells. J Exp Med
    192:1069–1074
Di Nardo A, Vitiello A, Gallo RL (2003) Cutting edge: mast cell antimicrobial activ-
    ity is mediated by expression of cathelicidin antimicrobial peptide. J Immunol
    170:2274–2278
Dinulos JG, Mentele L, Fredericks LP, Dale BA, Darmstadt GL (2003) Keratinocyte ex-
    pression of human beta defensin 2 following bacterial infection: role in cutaneous
    host defense. Clin Diagn Lab Immunol 10:161–166
Donnarumma G, Paoletti I, Buommino E, Orlando M, Tufano MA, Baroni A
    (2004) Malassezia furfur induces the expression of beta-defensin-2 in human
    keratinocytes in a protein kinase C-dependent manner. Arch Dermatol Res
    295:474–481
Dorschner RA, Lin KH, Murakami M, Gallo RL (2003) Neonatal skin in mice and
    humans expresses increased levels of antimicrobial peptides: innate immunity
    during development of the adaptive response. Pediatr Res 53:566–572

Dorschner RA, Pestonjamasp VK, Tamakuwala S, Ohtake T, Rudisill J, Nizet V, Ager-
    berth B, Gudmundsson GH, Gallo RL (2001) Cutaneous injury induces the release
    of cathelicidin anti-microbial peptides active against group A Streptococcus. J In-
    vest Dermatol 117:91–97
Duits LA, Ravensbergen B, Rademaker M, Hiemstra PS, Nibbering PH (2002) Expres-
    sion of beta-defensin 1 and 2 mRNA by human monocytes, macrophages and
    dendritic cells. Immunology 106:517–525
Erdag G, Morgan JR (2002) Interleukin-1alpha and interleukin-6 enhance the antibac-
    terial properties of cultured composite keratinocyte grafts. Ann Surg 235:113–124
Fang XM, Shu Q, Chen QX, Book M, Sahl HG, Hoeft A, Stuber F (2003) Differential
    expression of alpha- and beta-defensins in human peripheral blood. Eur J Clin
    Invest 33:82–87
Franzke CW, Baici A, Bartels J, Christophers E, Wiedow O (1996) Antileukoprotease in-
    hibits stratum corneum chymotryptic enzyme. Evidence for a regulative function
    in desquamation. J Biol Chem 271:21886–21890
Frick IM, Akesson P, Rasmussen M, Schmidtchen A, Bjorck L (2003) SIC, a secreted
    protein of Streptococcus pyogenes that inactivates antibacterial peptides. J Biol
    Chem 278:16561–16566
Frohm M, Agerberth B, Ahangari G, Stahle-Backdahl M, Liden S, Wigzell H, Gud-
    mundsson GH (1997) The expression of the gene coding for the antibacterial
    peptide LL-37 is induced in human keratinocytes during inflammatory disorders.
    J Biol Chem 272:15258–15263
Gallo RL, Murakami M, Ohtake T, Zaiou M (2002) Biology and clinical relevance of
    naturally occurring antimicrobial peptides. J Allergy Clin Immunol 110:823–831
Ganz T, Selsted ME, Szklarek D, Harwig SS, Daher K, Bainton DF, Lehrer RI (1985) De-
    fensins. Natural peptide antibiotics of human neutrophils. J Clin Invest 76:1427–
    1435
Goetz DH, Holmes MA, Borregaard N, Bluhm ME, Raymond KN, Strong RK (2002)
    The neutrophil lipocalin NGAL is a bacteriostatic agent that interferes with
    siderophore-mediated iron acquisition. Mol Cell 10:1033–1043
Goldman MJ, Anderson GM, Stolzenberg ED, Kari UP, Zasloff M, Wilson JM (1997)
    Human beta-defensin-1 is a salt-sensitive antibiotic in lung that is inactivated in
    cystic fibrosis. Cell 88:553–560
Gudmundsson GH, Agerberth B (1999) Neutrophil antibacterial peptides, multifunc-
    tional effector molecules in the mammalian immune system. J Immunol Methods
    232:45–54
Gudmundsson GH, Agerberth B, Odeberg J, Bergman T, Olsson B, Salcedo R (1996) The
    human gene FALL39 and processing of the cathelin precursor to the antibacterial
    peptide LL-37 in granulocytes. Eur J Biochem 238:325–332
Haake A, Scott GA, Holbrook KA (2003) Structure and function of the skin: overview
    of the epidermis and dermis. In: Freinkel RK, Woodley DT (eds) The biology of
    the skin. Parthenon, New York, pp 19–45
Harder J, Bartels J, Christophers E, Schroder JM (1997) A peptide antibiotic from
    human skin. Nature 387:861
Harder J, Bartels J, Christophers E, Schroder JM (2001) Isolation and characterization
    of human beta-defensin-3, a novel human inducible peptide antibiotic. J Biol
    Chem 276:5707–5713

Harwig SS, Ganz T, Lehrer RI (1994) Neutrophil defensins: purification, characterization, and antimicrobial testing. Methods Enzymol 236:160–172

Hase K, Eckmann L, Leopard JD, Varki N, Kagnoff MF (2002) Cell differentiation is a key determinant of cathelicidin LL-37/human cationic antimicrobial protein 18 expression by human colon epithelium. Infect Immun 70:953–963

Heilborn JD, Nilsson MF, Kratz G, Weber G, Sorensen O, Borregaard N, Stahle-Backdahl M (2003) The cathelicidin anti-microbial peptide LL-37 is involved in re-epithelialization of human skin wounds and is lacking in chronic ulcer epithelium. J Invest Dermatol 120:379–389

Henseler T, Christophers E (1995) Disease concomitance in psoriasis. J Am Acad Dermatol 32:982–986

Henzler Wildman KA, Lee DK, Ramamoorthy A (2003) Mechanism of lipid bilayer disruption by the human antimicrobial peptide, LL-37. Biochemistry 42:6545–6558

Hornef MW, Wick MJ, Rhen M, Normark S (2002) Bacterial strategies for overcoming host innate and adaptive immune responses. Nat Immunol 3:1033–1040

Howell MD, Jones JF, Kisich KO, Streib JE, Gallo RL, Leung DY (2004) Selective killing of vaccinia virus by LL-37: implications for eczema vaccinatum. J Immunol 172:1763–1767

Islam D, Bandholtz L, Nilsson J, Wigzell H, Christensson B, Agerberth B, Gudmundsson G (2001) Downregulation of bactericidal peptides in enteric infections: a novel immune escape mechanism with bacterial DNA as a potential regulator. Nat Med 7:180–185

Jones DE, Bevins CL (1993) Defensin-6 mRNA in human Paneth cells: implications for antimicrobial peptides in host defense of the human bowel. FEBS Lett 315:187–192

Jung HH, Chae SW, Jung SK, Kim ST, Lee HM, Hwang SJ (2003) Expression of a cathelicidin antimicrobial peptide is augmented in cholesteatoma. Laryngoscope 113:432–435

Kamysz W, Okroj M (2003) Novel properties of antimicrobial peptides. Acta Biochim Pol 50:461–469

Kao CY, Chen Y, Zhao YH, Wu R (2003) ORFeome-based search of airway epithelial cell-specific novel human [beta]-defensin genes. Am J Respir Cell Mol Biol 29:71–80

Koczulla R, von Degenfeld G, Kupatt C, Krotz F, Zahler S, Gloe T, Issbrucker K, Unterberger P, Zaiou M, Lebherz C, Karl A, Raake P, Pfosser A, Boekstegers P, Welsch U, Hiemstra PS, Vogelmeier C, Gallo RL, Clauss M, Bals R (2003) An angiogenic role for the human peptide antibiotic LL-37/hCAP-18. J Clin Invest 111:1665–1672

Kristian SA, Durr M, Van Strijp JA, Neumeister B, Peschel A (2003) MprF-mediated lysinylation of phospholipids in Staphylococcus aureus leads to protection against oxygen-independent neutrophil killing. Infect Immun 71:546–549

Larrick JW, Hirata M, Balint RF, Lee J, Zhong J, Wright SC (1995) Human CAP18: a novel antimicrobial lipopolysaccharide-binding protein. Infect Immun 63:1291–1297

Lehrer RI, Ganz T (1999) Antimicrobial peptides in mammalian and insect host defence. Curr Opin Immunol 11:23–27

Lehrer RI, Ganz T (2002) Cathelicidins: a family of endogenous antimicrobial peptides. Curr Opin Hematol 9:18–22

Lehrer RI, Daher K, Ganz T, Selsted ME (1985) Direct inactivation of viruses by MCP-1 and MCP-2, natural peptide antibiotics from rabbit leukocytes. J Virol 54:467–472

Leung DY (2003) Infection in atopic dermatitis. Curr Opin Pediatr 15:399–404

Leung DY, Boguniewicz M, Howell MD, Nomura I, Hamid QA (2004) New insights into atopic dermatitis. J Clin Invest 113:651–657

Liu AY, Destoumieux D, Wong AV, Park CH, Valore EV, Liu L, Ganz T (2002) Human beta-defensin-2 production in keratinocytes is regulated by interleukin-1, bacteria, and the state of differentiation. J Invest Dermatol 118:275–281

Liu L, Roberts AA, Ganz T (2003) By IL-1 signaling, monocyte-derived cells dramatically enhance the epidermal antimicrobial response to lipopolysaccharide. J Immunol 170:575–580

Marchini G, Lindow S, Brismar H, Stabi B, Berggren V, Ulfgren AK, Lonne-Rahm S, Agerberth B, Gudmundsson GH (2002) The newborn infant is protected by an innate antimicrobial barrier: peptide antibiotics are present in the skin and vernix caseosa. Br J Dermatol 147:1127–1134

McCray PB Jr, Bentley L (1997) Human airway epithelia express a beta-defensin. Am J Respir Cell Mol Biol 16:343–349

Midorikawa K, Ouhara K, Komatsuzawa H, Kawai T, Yamada S, Fujiwara T, Yamazaki K, Sayama K, Taubman MA, Kurihara H, Hashimoto K, Sugai M (2003) Staphylococcus aureus susceptibility to innate antimicrobial peptides, beta-defensins and CAP18, expressed by human keratinocytes. Infect Immun 71:3730–3739

Milner SM, Cole A, Ortega MR, Bakir MH, Gulati S, Bhat S, Ganz T (2003) Inducibility of HBD-2 in acute burns and chronic conditions of the lung. Burns 29:553–555

Morrison G, Kilanowski F, Davidson D, Dorin J (2002) Characterization of the mouse beta defensin 1, Defb1, mutant mouse model. Infect Immun 70:3053–3060

Moser C, Weiner DJ, Lysenko E, Bals R, Weiser JN, Wilson JM (2002) Beta-defensin 1 contributes to pulmonary innate immunity in mice. Infect Immun 70:3068–3072

Murakami M, Ohtake T, Dorschner RA, Schittek B, Garbe C, Gallo RL (2002) Cathelicidin anti-microbial peptide expression in sweat, an innate defense system for the skin. J Invest Dermatol 119:1090–1095

Murakami M, Lopez-Garcia B, Braff M, Dorschner RA, Gallo RL (2004) Postsecretory processing generates multiple cathelicidins for enhanced topical antimicrobial defense. J Immunol 172:3070–3077

Nagaoka I, Hirota S, Niyonsaba F, Hirata M, Adachi Y, Tamura H, Tanaka S, Heumann D (2002) Augmentation of the lipopolysaccharide-neutralizing activities of human cathelicidin CAP18/LL-37-derived antimicrobial peptides by replacement with hydrophobic and cationic amino acid residues. Clin Diagn Lab Immunol 9:972–982

Nickoloff BJ (1999) The immunologic and genetic basis of psoriasis. Arch Dermatol 135:1104–1110

Niyonsaba F, Iwabuchi K, Someya A, Hirata M, Matsuda H, Ogawa H, Nagaoka I (2002) A cathelicidin family of human antibacterial peptide LL-37 induces mast cell chemotaxis. Immunology 106:20–26

Niyonsaba F, Hirata M, Ogawa H, Nagaoka I (2003) Epithelial cell-derived antibacterial peptides human beta-defensins and cathelicidin: multifunctional activities on mast cells. Curr Drug Targets Inflamm Allergy 2:224–231

Nizet V, Ohtake T, Lauth X, Trowbridge J, Rudisill J, Dorschner RA, Pestonjamasp V, Piraino J, Huttner K, Gallo RL (2001) Innate antimicrobial peptide protects the skin from invasive bacterial infection. Nature 414:454–457

Nomura I, Goleva E, Howell MD, Hamid QA, Ong PY, Hall CF, Darst MA, Gao B, Boguniewicz M, Travers JB, Leung DY (2003) Cytokine milieu of atopic dermatitis, as compared to psoriasis, skin prevents induction of innate immune response genes. J Immunol 171:3262–3269

Ong PY, Ohtake T, Brandt C, Strickland I, Boguniewicz M, Ganz T, Gallo RL, Leung DY (2002) Endogenous antimicrobial peptides and skin infections in atopic dermatitis. N Engl J Med 347:1151–1160

Oono T, Huh WK, Shirafuji Y, Akiyama H, Iwatsuki K (2003) Localization of human beta-defensin-2 and human neutrophil peptides in superficial folliculitis. Br J Dermatol 148:188–191

Oren A, Ganz T, Liu L, Meerloo T (2003) In human epidermis, beta-defensin 2 is packaged in lamellar bodies. Exp Mol Pathol 74:180–182

Ortega MR, Ganz T, Milner SM (2000) Human beta defensin is absent in burn blister fluid. Burns 26:724–726

Peschel A, Otto M, Jack RW, Kalbacher H, Jung G, Gotz F (1999) Inactivation of the dlt operon in Staphylococcus aureus confers sensitivity to defensins, protegrins, and other antimicrobial peptides. J Biol Chem 274:8405–8410

Peschel A, Jack RW, Otto M, Collins LV, Staubitz P, Nicholson G, Kalbacher H, Nieuwenhuizen WF, Jung G, Tarkowski A, van Kessel KP, van Strijp JA (2001) Staphylococcus aureus resistance to human defensins and evasion of neutrophil killing via the novel virulence factor MprF is based on modification of membrane lipids with l-lysine. J Exp Med 193:1067–1076

Philpott MP (2003) Defensins and acne. Mol Immunol 40:457–462

Poyart C, Pellegrini E, Marceau M, Baptista M, Jaubert F, Lamy MC, Trieu-Cuot P (2003) Attenuated virulence of Streptococcus agalactiae deficient in D-alanyl-lipoteichoic acid is due to an increased susceptibility to defensins and phagocytic cells. Mol Microbiol 49:1615–1625

Putsep K, Carlsson G, Boman HG, Andersson M (2002) Deficiency of antibacterial peptides in patients with morbus Kostmann: an observation study. Lancet 360:1144–1149

Quayle AJ, Porter EM, Nussbaum AA, Wang YM, Brabec C, Yip KP, Mok SC (1998) Gene expression, immunolocalization, and secretion of human defensin-5 in human female reproductive tract. Am J Pathol 152:1247–1258

Ramanathan B, Davis EG, Ross CR, Blecha F (2002) Cathelicidins: microbicidal activity, mechanisms of action, and roles in innate immunity. Microbes Infect 4:361–372

Rieg S, Garbe C, Sauer B, Kalbacher H, Schittek B (2004) Dermcidin is constitutively produced by eccrine sweat glands and is not induced in epidermal cells under inflammatory skin conditions. Br J Dermatol 151:534–539

Risso A (2000) Leukocyte antimicrobial peptides: multifunctional effector molecules of innate immunity. J Leukoc Biol 68:785–792

Rosenberger CM, Gallo RL, Finlay BB (2004) Interplay between antibacterial effectors: a macrophage antimicrobial peptide impairs intracellular Salmonella replication. Proc Natl Acad Sci U S A 101:2422–2427

Salzman NH, Ghosh D, Huttner KM, Paterson Y, Bevins CL (2003) Protection against enteric salmonellosis in transgenic mice expressing a human intestinal defensin. Nature 422:522–526

Schmidtchen A, Frick IM, Andersson E, Tapper H, Bjorck L (2002) Proteinases of common pathogenic bacteria degrade and inactivate the antibacterial peptide LL-37. Mol Microbiol 46:157–168

Schutte BC, Mitros JP, Bartlett JA, Walters JD, Jia HP, Welsh MJ, Casavant TL, McCray PB Jr (2002) Discovery of five conserved beta -defensin gene clusters using a computational search strategy. Proc Natl Acad Sci U S A 99:2129–2133

Scott MG, Gold MR, Hancock RE (1999) Interaction of cationic peptides with lipoteichoic acid and Gram-positive bacteria. Infect Immun 67:6445–6453

Selsted ME, Miller SI, Henschen AH, Ouellette AJ (1992) Enteric defensins: antibiotic peptide components of intestinal host defense. J Cell Biol 118:929–936

Seo SJ, Ahn SW, Hong CK, Ro BI (2001) Expressions of beta-defensins in human keratinocyte cell lines. J Dermatol Sci 27:183–191

Shi J, Ganz T (1998) The role of protegrins and other elastase-activated polypeptides in the bactericidal properties of porcine inflammatory fluids. Infect Immun 66:3611–3617

Singh PK, Tack BF, McCray PB Jr, Welsh MJ (2000) Synergistic and additive killing by antimicrobial factors found in human airway surface liquid. Am J Physiol Lung Cell Mol Physiol 279:L799–L805

Sorensen OE, Follin P, Johnsen AH, Calafat J, Tjabringa GS, Hiemstra PS, Borregaard N (2001) Human cathelicidin, hCAP-18, is processed to the antimicrobial peptide LL-37 by extracellular cleavage with proteinase 3. Blood 97:3951–3959

Sorensen O, Arnljots K, Cowland JB, Bainton DF, Borregaard N (1997) The human antibacterial cathelicidin, hCAP-18, is synthesized in myelocytes and metamyelocytes and localized to specific granules in neutrophils. Blood 90:2796–2803

Sorensen OE, Cowland JB, Theilgaard-Monch K, Liu L, Ganz T, Borregaard N (2003) Wound healing and expression of antimicrobial peptides/polypeptides in human keratinocytes, a consequence of common growth factors. J Immunol 170:5583–5589

Taggart CC, Greene CM, Smith SG, Levine RL, McCray PB Jr, O'Neill S, McElvaney NG (2003) Inactivation of human beta-defensins 2 and 3 by elastolytic cathepsins. J Immunol 171:931–937

Travers JB, Norris DA, Leung DY (2001) The keratinocyte as a target for staphylococcal bacterial toxins. J Investig Dermatol Symp Proc 6:225–230

Turner J, Cho Y, Dinh NN, Waring AJ, Lehrer RI (1998) Activities of LL-37, a cathelin-associated antimicrobial peptide of human neutrophils. Antimicrob Agents Chemother 42:2206–2214

Uehara N, Yagihashi A, Kondoh K, Tsuji N, Fujita T, Hamada H, Watanabe N (2003) Human beta-defensin-2 induction in Helicobacter pylori-infected gastric mucosal tissues: antimicrobial effect of overexpression. J Med Microbiol 52:41–45

Ulvatne H (2003) Antimicrobial peptides: potential use in skin infections. Am J Clin Dermatol 4:591–595

Valore EV, Park CH, Quayle AJ, Wiles KR, McCray PB Jr, Ganz T (1998) Human beta-defensin-1: an antimicrobial peptide of urogenital tissues. J Clin Invest 101:1633–1642

Wang TT, Nestel FP, Bourdeau V, Nagai Y, Wang Q, Liao J, Tavera-Mendoza L, Lin R, Hanrahan JH, Mader S, White JH (2004) Cutting edge: 1,25-dihydroxyvitamin D3 is a direct inducer of antimicrobial peptide gene expression. J Immunol 173:2909–2912

Wiedow O, Harder J, Bartels J, Streit V, Christophers E (1998) Antileukoprotease in human skin: an antibiotic peptide constitutively produced by keratinocytes. Biochem Biophys Res Commun 248:904–909

Wilson CL, Ouellette AJ, Satchell DP, Ayabe T, Lopez-Boado YS, Stratman JL, Hultgren SJ, Matrisian LM, Parks WC (1999) Regulation of intestinal alpha-defensin activation by the metalloproteinase matrilysin in innate host defense. Science 286:113–117

Yang D, Chen Q, Le Y, Wang JM, Oppenheim JJ (2001) Differential regulation of formyl peptide receptor-like 1 expression during the differentiation of monocytes to dendritic cells and macrophages. J Immunol 166:4092–4098

Zaiou M, Gallo RL (2002) Cathelicidins, essential gene-encoded mammalian antibiotics. J Mol Med 80:549–561

Zaiou M, Nizet V, Gallo RL (2003) Antimicrobial and protease inhibitory functions of the human cathelicidin (hCAP18/LL-37) prosequence. J Invest Dermatol 120:810–816

Zanetti M (2004) Cathelicidins, multifunctional peptides of the innate immunity. J Leukoc Biol 75:39–48

Zanetti M, Gennaro R, Scocchi M, Skerlavaj B (2000) Structure and biology of cathelicidins. Adv Exp Med Biol 479:203–218

Zasloff M (2002) Antimicrobial peptides in health and disease. N Engl J Med 347:1199–1200

Zhao C, Wang I, Lehrer RI (1996) Widespread expression of beta-defensin hBD-1 in human secretory glands and epithelial cells. FEBS Lett 396:319–322

CTMI (2006) 306:111–152

Antimicrobial Peptides Versus Invasive Infections

M. R. Yeaman · A. S. Bayer (✉)

Division of Infectious Diseases, David Geffen School of Medicine at UCLA,
LAC-Harbor UCLA Medical Center 1124 West Carson Street/RB-2,
Torrance, CA 90502, USA
bayer@humc.edu, MRYeaman@ucla.edu

Abstract It has been estimated that there are more microorganisms within and upon the human body than there are human cells. By necessity, every accessible niche must be defended by innate mechanisms to prevent invasive infection, and ideally that precludes the need for robust inflammatory responses. Yet the potential for pathogens to transcend the integument actively or passively and access the bloodstream emphasizes the need for rapid and potent antimicrobial defense mechanisms within the vascular compartment. Antimicrobial peptides from leukocytes have long been contemplated as being integral to defense against these infections. Recently, platelets are increasingly recognized for their likely multiple roles in antimicrobial host defense. Platelets and leukocytes share many structural and functional archetypes. Once activated, both cell types respond in specific ways that emphasize key roles for their antimicrobial peptides in host defense efficacy: (a) targeted accumulation at sites of tissue injury or infection; (b) direct interaction with pathogens; and (c) deployment of intracellular (leukocyte phagosomes) or extracellular (platelet secretion) antimicrobial peptides. Antimicrobial peptides from these cells exert rapid, potent, and direct antimicrobial effects against organisms that commonly access the bloodstream. Experimental models in vitro and in vivo show that antimicrobial peptides from these cells significantly contribute to prevent or limit infection. Moreover, certain platelet antimicrobial proteins are multifunctional kinocidins (microbicidal chemokines) that recruit leukocytes to sites of infection, and potentiate the antimicrobial mechanisms of these cells. In turn, pathogens pre-decorated by kinocidins may be more efficiently phagocytosed and killed by leukocytes and their antimicrobial peptide arsenal. Hence, multiple and relevant interactions between platelets and leukocytes have immunologic functions yet to be fully understood. A clearer definition of these interactions, and the antimicrobial peptide effectors contributing to these functions, will significantly advance our understanding of antimicrobial host defense against invasive infection. In addition, this knowledge may accelerate development of novel anti-infective agents and strategies against pathogens that have become refractory to conventional antimicrobials.

1
Overview

Nature affords pathogens across the phylogenetic spectrum relentless opportunity to interact with and challenge potential hosts. In turn, natural selection has prioritized rapid mechanisms of host defense to thwart unfavorable microbial infection. Moreover, effective immune responses that do not evoke inflammation are advantageous to the host. Antimicrobial peptides represent a critical result of these co-evolutionary relationships.

Antimicrobial peptides have been isolated from essentially every source in which they have been sought. Numerous examples have now been identified in organisms from every phylogenetic kingdom, from ancient prokaryotes to human beings, and in species that occupy all niches of the biosphere. As remarkable is the wide range in diversity that is observed among the sequences and structures of these molecular effectors of immunity. Yet most antimicrobial peptides have broad antimicrobial spectra, including viruses and prokaryotes, fungi and even protozoal pathogens. Hence, a key question has emerged from these observations: what advantages justify the universal investment in expression of the broad repertoire of antimicrobial peptides in different organisms, or in different tissues of a single organism?

Recently, we proposed one answer to this question: immunorelativity (Yeaman and Yount 2005). The concept of antimicrobial peptide immunorelativity derives from the hypothesis that each of these molecules has evolved for optimal function in a specific anatomic or physiologic context, in response to corresponding signals associated with infection and to target cognate pathogens that challenge specific host contexts. Thus, evolutionary forces have likely fostered antimicrobial peptide immunorelativity in regard to innate immunity:

1. Favorable interactions among distinct antimicrobial peptides extend the functional repertoire of host defenses. Most antimicrobial peptides execute rapid and potent microbicidal effects against a wide range of microbial target cells. However, the mechanisms by which they do so, or the conditions under which they are most potent, often differ considerably. Thus, additivity or synergy among antimicrobial peptides with distinct mechanisms of action or conditional optima (perhaps originating from distinct physiologic or anatomic contexts) may increase the efficacy of antimicrobial host defense.

2. Certain host contexts likely benefit from dual molecular and cellular host defenses. Antimicrobial peptides are often considered front-line host defense molecules. Ostensibly, effective antimicrobial peptide response will preclude infection, without a need for inflammatory responses. However, certain host contexts or microbial pathogens are best defended through a combination of innate and adaptive immune mechanisms. For example, peptides that exert direct, potent microbicidal activity, and potentiate antimicrobial mechanisms of leukocytes, are likely advantageous in such contexts. Thus, relevant to this current review, peptides that confer multiple antimicrobial functions in a coordinated manner may provide optimal defenses against invasive infection in particularly vulnerable host contexts such as the vascular compartment.

3. Some antimicrobial peptides are limited to discrete contexts concordant with their propensity for host cytotoxicity. Beyond distinctions in source and sequence, antimicrobial peptides vary considerably in selective toxicity. Some are rather indiscriminate membrane perturbants that target and disrupt microbial and mammalian cell membranes with equal inten-

sity. Others are clearly much more selective, able to distinguish between host and pathogen with an accuracy that rivals that of the safest antibiotic. While much remains to be understood regarding the structural and mechanistic basis for their microbial vs host toxicity, the context in which a given antimicrobial peptide functions likely reflects its degree of selective toxicity. Hence, peptides that are restricted in location to exterior surfaces or cell interiors tend to be more toxic than those released to function in the bloodstream or other endogenous tissues per se. By way of example, many antimicrobial peptides have been isolated from relatively inert mucosal surfaces. However, these peptides can be very cytotoxic to human tissues, or inactivated in human blood and serum. Likewise, defensins normally restricted to the hostile confines of leukocyte phagolysosomes are often cytotoxic, and typically quickly inactivated in the bloodstream. In contrast, antimicrobial peptides from mammalian platelets have evolved to exert potent antimicrobial efficacy when liberated into the bloodstream. Illustrating this point, synthetic congeners of platelet peptides exert significantly greater antimicrobial efficacy in human blood and plasma ex vivo than in artificial medium, even after 2 h incubation in the biomatrices prior to inoculation of the target organisms (Yeaman et al. 2002).

From these perspectives, antimicrobial peptide immunorelativity represents a functional concatenation of molecular regulation and physicochemical properties of a given peptide, with its unique anatomic and physiologic host context, optimized to defend against the threat of cognate pathogens and microbiologic profile of that specific context. Moreover, evidence is mounting in support of the hypothesis that certain antimicrobial peptides act in synergy, deploy autonomous modules with complementary functions (e.g., direct microbicidal and leukocyte potentiating activities) to protect a given host context, and may immunologically condition specific contexts in favor of non-pathogens over pathogens (Yeaman and Yount 2005). Mindful of these latest themes in immunobiology, this review focuses on antimicrobial peptides that defend against invasive infections, emphasizing roles of antimicrobial peptides from platelets in defense of the bloodstream.

2
Strategies in Antimicrobial Peptide Defense Against Invasive Infection

Trophism of many pathogenic microorganisms for the bloodstream via invasion or trauma corresponds to the need for rapid and potent antimicrobial host defense systems that afford protection against successive steps in pathogenesis: (a) prevent pathogens from disrupting the integrity of skin or epithelial mucosa to access the vascular compartment; and (b) eliminate pathogens that reach the bloodstream. Ideally, these first-line defense mechanisms succeed without evoking a fulminant inflammatory response, but have the potential to potentiate and coordinate inflammatory responses if needed. The following examples highlight recent advances in light of these concepts.

2.1
Skin

Recent findings indicate that antimicrobial peptides are critical to defense of human skin against invasive infection. Remarkable studies by Ong and colleagues (2002) offer new insights into significant roles of the antimicrobial peptides LL-37 and β-defensins in innate immunity of the skin against potentially invasive infection. Normally, the amount of these peptides present in skin is nominal. However, in skin of individuals suffering from psoriasis or related inflammatory disease, these peptides concentrate significantly. The investigators compared the amount of LL-37 and human β-defensin 2 (HBD-2) in patients with atopic dermatitis or psoriasis, as compared with otherwise healthy controls. Immunohistochemical assays demonstrated plentiful LL-37 and HBD2 in the epidermis of every patient affected with psoriasis. In contrast, the concentrations of these peptides, and their corresponding mRNA, were significantly decreased in lesions from patients with atopic dermatitis. The authors concluded that reduced levels of these peptides increased host susceptibility to or severity of *Staphylococcus aureus* infection. Similarly, disruption of the gene encoding antimicrobial peptide CRAMP correlates with increased susceptibility to cutaneous infection by *Streptococcus pyogenes* in a murine model (Nizet et al. 2001). These examples are consistent with the concept that antimicrobial peptides are crucial effectors in front-line innate immunity of the skin, and play key protective roles against human colonization or invasion.

2.2
Epithelial Mucosa

Evidence substantiates the relevance of antimicrobial peptides in defense of human oropharyngeal mucosa. Recent investigations have begun to highlight compelling roles of mucosal antimicrobial peptides in defense against human infection.

Nasal Mucosa   Cole et al. showed that *S. aureus* colonization of human nares is more frequent in individuals defective in innate immunity against endogenous nasal carrier isolates of *S. aureus* (1999). Nasal secretions from colonized individuals revealed elevated levels the neutrophil defensins hNP1 and -3, as well as HBD2, indicating a neutrophil-mediated inflammatory response is needed in addition to, or in backgrounds of insufficient antimicrobial peptide efficacy in response to *S. aureus* colonization. Thus, the authors interpreted that nasal carriage of *S. aureus* occurs when nasal fluid is permissive for colonization, or induces a local inflammatory response that fails to clear the organisms.

Oropharyngeal Mucosa or Gingiva   Further insights into antimicrobial peptide defense against invasive infection may be drawn from abnormal innate immunity of oropharyngeal mucosa or gingiva. A very rare congenital condition termed morbus Kostmann exemplifies this fact. Morbus Kostmann patients exhibit severe congenital neutropenia; treatment with recombinant granulocyte-colony stimulating factor restores neutrophil counts, but they continue to have recurrent soft-tissue infections and profound periodontitis. Based on these findings, Putsep and co-workers (2002) hypothesized that antimicrobial peptide effector molecules are deficient in patients with morbus Kostmann. The investigators studied neutrophils, plasma, and saliva from morbus Kostmann patients, as compared with healthy controls, for profiles of antibacterial peptides. Potential defects in neutrophil oxidative killing mechanisms and lactoferrin levels were ruled out, as these parameters appeared normal among affected patients. Neutrophils from patients with morbus Kostmann were deficient in production of the cathelicidin antimicrobial peptide LL-37, and had reduced concentrations of defensins hNP1–3. Additionally, LL-37 was not detectable within patient plasma or saliva. However, bone marrow transplant in patients with morbus Kostmann yielded effectively normal concentrations of LL-37, and most importantly, restored normal periodontia.

Gastrointestinal Mucosa   Insights from experimental models reinforce key roles for antimicrobial peptides in defense of gastrointestinal mucosa. Antimicrobial peptides have been shown to be integral to defense against infection in many animal models studied to date. Fields et al. (1989) originally demonstrated that *Salmonella* mutants resistant to defensins hNP1 in vitro had greater virulence and ability to survive neutrophil antimicrobial mechanisms in mice. Recently, inactivation of the pro-α-defensin activating enzyme matrilysin by gene disruption in transgenic mice abrogated generation of mature α-defensins (Shirafuji et al. 2003). Mice with this defect are relatively ineffective in clearing orally introduced bacteria, which have increased virulence in these animals. Further substantiating this concept, transgenic mice containing human α-defensin HD5 are essentially resistant to enteric infection and ensuing systemic invasion by *Salmonella typhimurium* (Salzman et al. 2003).

Other Integumentary Barriers   Other recent studies indicate that antimicrobial peptides, including defensins, are important to defense against invasive infection of other anatomic and physiologic contexts as well. For example, the recent discoveries of hepcidin and its dual roles in genitourinary defense (Park et al. 2001) and iron metabolism (Ganz 2003) emphasize the importance of effective and coordinated mechanisms of local and systemic host defense. Likewise, human beta defensins HBD1 and HBD2 have been shown to contribute to antimicrobial defense in the intrahepatic biliary tree (Harada et al. 2004). HBD1, HBD3, and LL-37 were found to be up-regulated in the context of infected human synovia (Paulsen et al. 2002a), and joint cartilage was shown to express HBD2 when challenged with microbial pathogens (Varoga et al. 2004). In fact, recent findings illustrate the ubiquitous distribution of antimicrobial peptides in defense against invasive infections of diverse host contexts, including human tears and ocular tissues (Shin et al. 2004; Paulsen et al. 2002b), lung and pulmonary epithelium (Lau et al. 2005), gastric tissue (Isomoto et al. 2004), and other integument-environment interfaces (Fig. 1).

2.3
Vascular Compartment

Neutrophils and platelets share parallels offering new insights into antimicrobial peptide defense vs hematogenous infection. Antimicrobial peptides isolated from mammalian neutrophils are among the most thoroughly studied of all the antimicrobial peptides. Medicine has long been aware of disorders in neutrophil quality that correlate with increased recurrence or severity of infection. Therefore, deficient neutrophil granule composition or function is consistent with the concept that antimicrobial peptides are critical to host defense against infection. A comparison of different types of leukocyte disorders illustrates perspectives of antimicrobial peptides in host defense. On one hand, individuals with Chediak-Higashi syndrome have inherent deficiencies in elastase and cathepsin G (Ganz et al. 1988), consistent with specific roles for antimicrobial peptides in host defense. On the other hand, such inherent conditions manifest multiple neutrophil dysfunctions, obscuring potentially key contributions of antimicrobial peptides to immunity. Alternatively, leukocytes from patients with chronic granulomatous disease or equivalent conditions are deficient in NADPH oxidase systems that are crucial to oxidative mechanisms of microbial killing. However, in certain disorders, definable antimicrobial peptide deficiencies occur in cells that have intact oxidative killing systems. For example, neutrophils isolated from patients with specific granule disorders (e.g., azurophilic or acquired neutrophil disorder) exhibit reduced granule and antimicrobial peptide content, and reduced antimicrobial effi-

118 M. R. Yeaman · A. S. Bayer

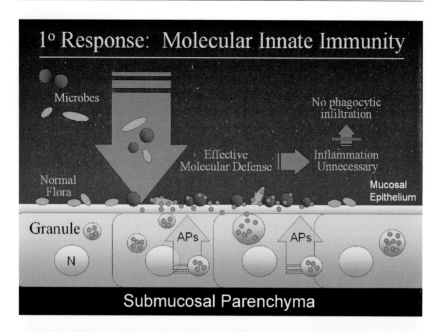

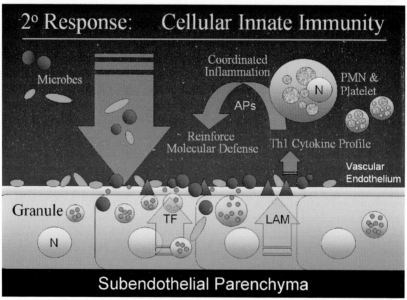

◀──

Fig. 1 Antimicrobial peptides in coordinated molecular and cellular innate immunity. Constant exposure to environmental microorganisms and potential pathogens necessitates constitutive or rapidly inducible host defenses. Ideally, a primary response (*upper panel*) will deploy antimicrobial peptides (*APs*) and other molecular effectors that do not evoke inflammation. However, overwhelming inoculum, breaches in the integument, and pathogens that are refractory to mechanical, chemical, or biological barriers may render molecular defenses ineffective in limiting colonization or invasion. In such cases, pathogens may gain access to and invade deeper tissue parenchyma. In the bloodstream (*lower panel*), pathogens may adhere to the vascular endothelium, prompting endothelial cell injury or expression of tissue factor (*TF*) and leukocyte adhesion molecules (*LAM*) from granules by way of PAMP (pattern-associated molecular patterns) → TLR (toll-like receptor) → NfκB (transcriptional activator) pathways. TF expression promotes thrombin production, triggering platelet activation and ensuing PMP and kinocidin liberation. In turn, platelet and other kinocidins exert direct antimicrobial activity, and facilitate chemonavigation and recruitment of neutrophils (*PMN*), which adhere to LAMs and deposited platelets. These examples illustrate roles of APs in contributing to direct innate immune functions, and recursive coordination of cellular innate immunity

cacy in vitro, as compared with normal cells (Lehrer et al. 1988). Likewise, platelets from humans with congenital α-granule deficiencies are depleted in what are now recognized to be antimicrobial peptide kinocidins. Importantly, recurrent or severe infections are hallmarks of each of the above conditions.

In aggregate, the examples above offer mounting evidence for key functions of antimicrobial peptides as integral to host defense against invasive infection. Clearly, defensins and related peptides are important in these roles. The remainder of this review focuses on new insights relevant to antimicrobial peptides from platelets, and their likely multiple roles in antimicrobial host defense of the vascular compartment and beyond.

3
Platelets as Antimicrobial Host Defense Effector Cells

The affinity of potential microbial pathogens for the bloodstream via invasion or trauma necessitates two critical host responses: (a) rapid and potent antimicrobial systems that function without necessarily evoking inflammation; and (b) prevention of blood loss and initiation of tissue repair. In invertebrates, one cell type (hemocyte) performs both of these functions (Nachum et al. 1980; Yeaman 1997). In mammals, thrombocytes and leukocytes have historically been viewed to mediate hemostasis or inflammation, respectively (Tocantins 1938; Weksler 1992). Although a principal function of platelets is hemostasis, these cells have retained important features of immune effector cells that contribute multiple functions in antimicrobial host defense (Yeaman 1997; Yeaman and Bayer 1999, 2000).

A natural history of endovascular infection, particularly as it pertains to highly virulent organisms such as *S. aureus*, has been considered to include the following series of events (Yeaman and Bayer 2000):

- Microbe access to the vascular compartment

- Adhesion of blood-borne pathogens to normal or abnormal vascular endothelium (e.g., endocardium preconditioned by rheumatic heart disease or effects of prosthetic cardiac valves)

- Tissue expression of soluble mediators or surface ligands that promote platelet activation and deposition

- Ensuing deposition of circulating platelets in response to secondary agonists rendered by initial platelet adhesion and activation (e.g., ADP, platelet-activating factor [PAF]), or GPIIb-IIIa)

- Further platelet deposition, degranulation, and liberation of PMPs and other granule contents.

From this perspective, it follows that platelets among the earliest opportunities for host defense to intervene in establishment and progression of endovascular infection. Hence, the consequences of these initial interactions between microorganisms and platelets may largely determine whether infection is initiated within the cardiovascular compartment. An increasingly compelling body of evidence exists supporting the view that platelets provide a physiologically relevant mechanism of antimicrobial host defense in such settings.

The physiology of mammalian platelets as it relates to antimicrobial host defense has recently been reviewed as compared with other immune effector cells (Yeaman 1997; Yeaman and Bayer 2000). Platelets are small (2–4 μm), short-lived cells originating from megakaryocyte lineage. As platelets are devoid of nuclei, they can only perform limited translation using stable megakaryocyte mRNA templates. Therefore, the bulk of their bioactive constituents are stored in granules. Platelets contain three distinct cytoplasmic granule types. Dense (δ) granules contain mediators of vascular tone, such as serotonin, adenosine diphosphate (ADP), eicosanoids, thromboxane A_2 (TXA$_2$), and ions including calcium and phosphate. By comparison, alpha (α) granules store proteins important to hemostasis, including adhesion (e.g., fibrinogen and von Willebrand factor), coagulation, and endothelial repair (e.g., platelet-derived growth factor, PDGF; transforming growth factors α and β; TGF-α, TGF-β). Lysosomal (λ) granules principally contain lysozyme and precursors of thrombus dissolution (e.g., plasminogen). Interestingly, degranulation is determined by agonist specificity and potency. For example, low levels of thrombin or ADP induce δ and α degranulation, while λ granules only secrete when these agonists are present at considerably higher levels. In addition, platelets are now known to contain a battery of antimicrobial peptides (see below). Thus, platelets are likely the earliest and predominant cells in the bloodstream that respond to microorganisms or tissue infection, and are capable of liberating stored antimicrobial constituents rapidly in these settings.

Beyond their critical role in hemostasis, platelets have structural archetypes of antimicrobial effector cells, supporting the concept that platelets play multiple roles to prevent or limit infection. Furthermore, platelets perform functions that reflect key contributions to immunity. The structures and functions of platelets as they pertain to immunobiology have also been reviewed (Yeaman 1997; Yeaman and Bayer 2000; Klinger and Jelkmann 2002). The ensuing sections will highlight the molecular effectors of platelets in antimicrobial host defense, and how these multifunctional constituents are believed to bridge molecular and cellular mechanisms of innate immunity.

3.1
Classical Studies Implicated Antimicrobial Molecules in Platelets

The exquisite sensitivity of platelets to microbial signals and rapid deployment of bioactive molecules upon interacting with microbial pathogens offer strong teleological arguments for their role in antimicrobial host defense. Pioneering evidence pointed to platelet antimicrobial effects as being mediated by release of antimicrobial constituents. In the nineteenth century, Fodor distinguished a heat-stable bactericidal action of serum, termed β-lysin, from heat-labile or α-lysin complement proteins (Fodor 1887). Fifty years later, Tocantins (1938) published a remarkably insightful review of information relating to the immune functions of platelets. Subsequently, a succession of investigators demonstrated that platelets, not leukocytes, reconstituted the bactericidal activity of serum, and exert bacteriostatic or bactericidal effects against a broad spectrum of microbial pathogens in vitro, including *Bacillus* spp., *Staphylococcus* spp., *Listeria* spp. and *Salmonella* species. Isolation of constituents responsible for platelet antimicrobial activities are first attributed to Myrvik, who found two platelet-derived agents in serum that corresponded to killing of *Bacillus subtilis* (Myrvik 1956). Building upon these findings, Johnson and Donaldson (1968) resolved cationic proteins from rabbit platelets that exerted in vitro bactericidal activity against *B. subtilis* or *S. aureus*. Tew and co-workers (1974) later reported that thrombin stimulation induced the release of a β-lysin from rabbit platelets. Overall, these studies pointed to antimicrobial effector molecules from platelets as being small and cationic, and presumptively identified as platelet factor-4 (PF-4). Thereafter, Darveau et al. (1992) described peptide fragments of human PF-4 that were antimicrobial against Gram-negative bacteria in combination with conventional antibiotics. A detailed chronology of these historical perspectives is available elsewhere (Yeaman 1997).

3.2
Platelet Microbicidal Proteins

Platelets contain an arsenal of antimicrobial peptides. The relationship between platelet structure and function supports their likely integral roles in antimicrobial host defense. Yet, until recently, the molecular effectors contributing to these activities were not known. Over the last decade, contemporary techniques in protein chemistry and molecular biology have been applied to resolve the identification and characteristics of these antimicrobial constituents. Thus, it is now clear that platelets contain a diverse and complementary group of peptides that exert explicit antimicrobial activities.

Characterization of Antimicrobial Peptides from Platelets   Human and other mammalian platelets contain a battery of several antimicrobial peptides, termed platelet microbicidal proteins (PMPs; Yeaman 1997; Yeaman et al. 1997; Tang et al. 2002). These peptides are all cationic, heat-stable, and range in molecular mass from 6.0 to 9.0 kDa, distinguishing them from defensins or lysozyme. Other structural features that distinguish PMPs from defensins include richer composition of lysine residues and the presence of two to four, rather than six, cysteine pairs in the disulfide array of PMPs (Yeaman and Bayer 1999). Recently, a detailed molecular characterization of structure–activity relationships in a prototype PMP kinocidin revealed that differential processing may further extend the structural and antimicrobial repertoire of this group of peptides (Yount et al. 2004). Amino acid sequencing and mass spectrometry showed that distinct N-terminal polymorphism variants of rabbit PMP-1 from unactivated or thrombin-stimulated platelets, respectively, arise from a common PMP-1 propeptide. Cloning from bone marrow and characterization of its cDNA identified that PMP-1 is translated as a 106-amino acid precursor, and processed to 73- (8,053 Da) and 72-residue (7,951 Da) variants. Analysis of sequence motifs revealed PMP-1 is a member of the PF-4 protein family. Moreover, phylogeny, and three-dimensional structures confirmed that PMP-1 has greatest homology with PF-4. Therefore, structural and antimicrobial properties established the identity of PMP-1 as a rabbit analog of human PF-4.

3.3
Distinct Genetic Lineages of PMPs

Although PMPs were initially isolated from rabbit platelets, analogous peptides have been identified in human platelets. Recently, Tang and colleagues (2002) isolated and characterized human platelet antimicrobial peptides, released from human platelets following thrombin stimulation. These human peptides are structural and functional analogs of rabbit PMPs (Yeaman 2004) and include platelet factor 4 (PF-4), platelet basic protein (PBP), and derivatives connective tissue activating peptide-3 (CTAP-3 and a derivative, neutrophil activating peptide-2, NAP-2), RANTES (released upon activation, normal T cell expressed and secreted), thymosin-β-4 (Tβ-4), and fibrinopeptides A and B (FP-A and FP-B). Based on the above, phylogenetically there are five distinct genetic lineages of mammalian PMPs:

- PF-4 and variants

- PBP and proteolytic derivatives CTAP-3 and NAP-2

- RANTES

- Thymosin beta-4 (Tβ-4)

- Fibrinopeptides

While the peptides PF-4, PBP, CTAP-3, and NAP-2 are α- or CXC-chemokines, RANTES is a β-, or CC-chemokine. These findings demonstrate the direct antimicrobial functions of this group of human platelet proteins, even though their structures had been characterized previously. Krijgsveld et al. (2000) found that carboxy-terminal di-amino acid truncated versions of NAP-2 or CTAP-3 exerted antimicrobial activity in vitro. These peptides have been termed thrombocidins 1 and 2, respectively. It is interesting to note that FP-A and FP-B were not detectable in total protein extracts of platelets. This finding suggests these PMPs are generated as a result of thrombin stimulation of platelets. Conceivably, such peptides may be generated via thrombin-cleavable Arg-Gly sites in platelet fibrinogen (Turner et al. 2002). Therefore, thrombin (a serine protease), platelet-derived proteases, proteases generated by tissue injury, phagocytes (e.g., cathepsin G) or inflammation (e.g., plasmin) may process precursor proteins and contribute to generation of multiple antimicrobial peptides from platelets. Alternatively, microbial proteases may also modulate the bioactivity of PMPs and kinocidins. For example, Sieprawska-Lupa et al. (2004) showed that *S. aureus* exoenzymes aureolysin and V8 protease cleave the antimicrobial peptide LL-37. Thus, it is possible that PMPs and kinocidins counter-respond to microbial virulence factors such as proteases, deploying domains with increased or optimized antimicrobial efficacy vs cognate pathogens in corresponding contexts of infection (Yeaman and Yount 2005). Collectively, these relationships may be crucial to the multifunctional roles of platelets in antimicrobial host defense (Yeaman 1997; Yeaman and Bayer 2000; Klinger and Jelkmann 2002).

By convention, the term "platelet microbicidal proteins" (PMPs) is defined in this review to encompass PMPs, thrombin-induced PMPs (tPMPs), HPAPs, thrombocidins, or other antimicrobial polypeptides from platelets. Detailed comparisons of the structures and functions of PMPs with antimicrobial peptides from other sources are reviewed elsewhere (Yeaman 1997, 2004; Yeaman and Bayer 2000; Tang et al. 2002).

3.4
Bridging Molecular and Cellular Defenses: Platelet Kinocidins

Clearly, antimicrobial peptides contribute to the direct inhibition and killing of microbes in host defense against invasive infection. Moreover, as described above, some antimicrobial peptides also perform functions that effectively bridge molecular and cellular immune responses. In addition, we recently proposed a model that posits how different antimicrobial peptides are optimized to defend against infection of specific physiologic and anatomic contexts, or contexts that change over time in the face of infection. This paradigm of immunocoordination has been termed the AEGIS model of antimicrobial peptide immunobiology: Archetype Effectors Governing Immune Syntax (Yeaman and Yount 2005). A key aspect of the AEGIS model is based on the concept of antimicrobial peptide immunorelativity. This concept is a multifunctional concatenation of molecular regulation and physicochemical properties of a host defense peptide, with its particular anatomic or physiologic context, optimized to provide defense against cognate pathogens of that context. From this perspective, it is very likely that distinct PMPs and kinocidins act synergistically with leukocytes and their antimicrobial armamentarium to protect the host against invasive infection. The AEGIS model of immunocoordination is supported by multiple lines of investigation, as illustrated in the following examples:

1. Kinocidins: microbicidal chemokines. Several human and rabbit PMPs have been shown to be antimicrobial peptides, and also known chemokines. Reflecting their dual complementary functions, chemokines

that also exert direct microbicidal activities have been termed kinocidins (Yount and Yeaman 2004; Yeaman and Yount 2005). Importantly, kinocidins represent molecules with roles that bridge and coordinate molecular and cellular mechanisms of antimicrobial host defense. Important reports by Cole et al. (1999) and Yang et al. (2003) have further emphasized this concept. Finally, our recent studies have identified and characterized antimicrobial efficacies of kinocidins from non-platelet sources as well (Yount and Yeaman 2004). Thus, along with their direct antimicrobial properties, kinocidins recruit and potentiate several antimicrobial mechanisms of leukocytes and lymphocytes. Importantly, these kinocidins are released into the bloodstream as a result of platelet degranulation. Consistent with these concepts, meritorious contributions of Dankert and Zaat, Cole and Ganz, Zhang and Oppenheim have enhanced the understanding of direct antimicrobial roles of other kinocidins. Moreover, evidence suggests that distinct kinocidins may function optimally in distinct physiologic contexts. For example, interleukin-8 (IL-8) has direct antimicrobial activity, but has conditional optima and antimicrobial spectra distinct from other kinocidins (Yount and Yeaman, unpublished data). These observations further support the concept of antimicrobial peptide immunorelativity.

2. Kinocidins have complementary and compounded host defense functions. Recently, we described how rabbit and human kinocidins contain autonomous functional motifs that facilitate complementary direct and indirect antimicrobial functions. Direct antimicrobial functions include rapid microbicidal effects of a peptide against its microbial targets (Yeaman et al. 2002). Indirect effects are less obvious, but likely just as important. For example, chemoattraction of leukocytes to sites of infection is an example of an indirect antimicrobial effect of these peptides. Still other functions may further amplify the compounded roles of these molecules in host defense. Our proteomic and functional data illustrate that protease inhibitory, opsonic, and other complementary functions can be integrated along with microbicidal activity within specific disulfide-stabilized antimicrobial peptide configurations (Yount and Yeaman 2004). Elegant studies from the laboratory of Gallo (Zaiou et al. 2003) indicate that some human cathelicidins encode microbicidal and protease inhibitor activities. Previous investigations of classical defensins and other antimicrobial peptides have similarly revealed that these molecules are also capable of leukocyte chemoattraction, consistent with this hypothesis.

3. Multiple and complementary functions encoded in antimicrobial peptides are likely deployed by activation in relevant contexts of infection.

We hypothesize that functional decompression of multiple antimicrobial properties integrated within an antimicrobial peptide may proceed through a coordinated process (Yount and Yeaman 2004; Yeaman and Yount 2005):

– Hallmark signals of infection elicit elaboration of a given antimicrobial peptide in its cognate physiological/anatomic context
– Where effectively intensified, the peptide exerts direct microbicidal activity
– Proteolytic cleavage of peptides at lower concentrations or unassociated with their microbial targets yields a protease inhibitory function of a subset of these molecules, preserving antimicrobial integrity of others
– Diffusion and/or subsequent processing of peptides establishes gradients that govern leukocyte chemonavigation
– decoration by specific dual microbicidal and opsonic peptide motifs promotes opsonophagocytosis of pathogens by arriving leukocytes
– Peptides or domains thereof with enhanced microbicidal activity under acidic conditions are potentiated within the leukocyte phagolysosomes

Compelling data exist substantiating each step in this coordinated pathway. This model further proposes that ensuing steps in this sequence of events would deploy in inverse proportion to success of the native peptides in defending against infection; ensuing cell-mediated immune mechanisms are only necessary if first-line peptides fail to control pathogen(s). Therefore, we believe that certain antimicrobial peptides function by way of coordinated decompression of microbicidal, protease-inhibitory, chemotactic, opsonic, or other mutually complementary functions in an orchestrated sequence in time and place to effect optimal innate immunity, and if needed, coordinate inflammatory mechanisms.

3.5
Elaboration of PMPs and Platelet Kinocidins

The contributions of PMPs to extracellular host defenses against adherent bacteria are related to mechanisms that prompt the mobilization or release of these antibacterial peptides. Stimulation of rabbit or human platelets with thrombin, as naturally occurs at sites of endovascular damage, leads to extracellular accumulation of potent antimicrobial activity due to molecules genetically unrelated to defensins. Recent studies have shown that PMPs and kinocidins are released from platelets exposed to thrombin, staphylococcal α-toxin, or microbial pathogens themselves, including viridans group streptococci, *S. aureus*, and *Candida albicans*. As noted above, thrombin is a potent stimulus for the release of PMPs and kinocidins from rabbit and human platelets. Recent studies have focused on the molecular mechanisms by which platelets detect and respond to tPMP-1-susceptible (ISP479C; tPMP-1^S) or -resistant (ISP479R; tPMP-1^R) strains of *S. aureus* (Trier et al. 2000). At platelet-to-*S. aureus* ratios above or equal to 1,000:1, anti-*S. aureus* effects corresponded to the tPMP-1 susceptibility phenotype of the challenge organism, with greater killing of the tPMP-1^S vs tPMP-1^R strain. Analytic RP-HPLC of staphylocidal supernatants resulting from these platelet–*S. aureus* interactions confirmed tPMP-1 was released from *S. aureus*-stimulated platelets. Below ratios of 1,000:1, equivalent survival of ISP479C or ISP479R strains was seen, and did not differ from respective controls. A panel of specific platelet inhibitors was then used to probe pathways associated with platelet anti-staphylococcal responses. Apyrase (inhibitor of extracellular ADP), ticlopidine (nonspecific inhibitor of platelet ADP receptors), suramin (inhibitor of platelet P2X and P2Y ADP receptors), and pyridoxyl 5′-phosphate derivative (PPND; inhibitor of platelet P2X ADP receptors) each interfered with platelet anti-staphylococcal responses. However, specific inhibition of platelet β-adrenergic (yohimbine), phospholipase C (e.g., propanolol), cyclooxygenase-1 (COX-1) (indomethacin), or thromboxane A2 (SQ29548) pathways each failed to impede platelet anti-*S. aureus* responses. In aggregate, this pattern of results demonstrated that platelet release of PMPs in response to *S. aureus* occurs via an active, rapid, and direct mechanism, amplified through autocrine pathways where platelet ADP release triggering successive waves of platelet degranulation via the platelet P2X-ADP receptor array.

3.6
Antimicrobial Efficacies and Spectra of PMPs and Kinocidins

Human and rabbit PMPs and kinocidins are active against many potential pathogens that have a propensity to access the bloodstream. For example, PMP-1, PF-4, and other kinocidins are efficacious in nanomolar to micromolar concentrations against *S. aureus*, streptococci, and *C. albicans*. It is notable that PMPs from both species exhibit conditional optima. For example, rabbit PMPs 1–5 are more active at acid pH, whereas tPMP-1 and -2 are more active at neutral pH. Similar themes are seen with analogous human PMPs. Of these peptides, the antimicrobial action of tPMP-1 has been studied most intensively. tPMP-1 is microbicidal against most of the clinically relevant Gram-positive pathogens, including *S. aureus*, *Staphylococcus epidermidis*, the viridans streptococci. The Gram-negative bacteria such as *Escherichia coli* and *Pseudomonas aeruginosa*, as well as the fungus *C. albicans* and other candidal species, and *Cryptococcus neoformans* are also susceptible to many PMPs and kinocidins (Yeaman and Bayer 2000). Bacteremia isolates of *S. aureus*, *S. epidermidis*, or viridans streptococci from patients with infective endocarditis are also less susceptible to tPMP-1 in vitro than isolates from infections not associated with infective endocarditis (Wu et al. 1994). An ensuing investigation of a cohort of 60 bacteremic *S. aureus* isolates from a single medical center (Duke University) also discovered that strains from patients with infective endocarditis or vascular catheter infections were more resistant in vitro to low levels of tPMP-1 than were isolates arising from soft tissue abscesses (Fowler et al. 2000). These observations suggest that reduced susceptibility to tPMP-1 provides the invading microbe with a survival advantage at sites of endovascular damage. Extending these themes, Fowler et al. (2004) recently found that methicillin-resistant *S. aureus* (MRSA) with dysfunctions in their accessory gene regulator (*agr*) and reduced tPMP-1 susceptibility in vitro have a greater propensity to cause persistent bacteremia than MRSA lacking these phenotypes. Collectively, these examples illustrate the point that platelets and their antimicrobial peptides are important components of host defense against Gram-positive pathogens in the relevant context of cardiovascular infection.

3.7
Mechanisms of PMP and Kinocidin Antimicrobial Action

Injurious effects of PMPs and kinocidins on whole microbial cells, protoplasts, and lipid bilayers in vitro have been investigated using multiple approaches, including transmission and scanning electron microscopy (TEM, SEM) and biophysical techniques. Generally, these studies reinforce a fundamental theme: PMPs and kinocidins initially target and perturb their microbial target cell membranes, but additional steps are involved in eventual lethal effects. In *S. aureus*, cytoplasmic membrane permeabilization occurs immediately, but membrane depolarization does not necessarily follow. Thereafter, the cytoplasmic membrane appears to condense, corresponding to cell wall hypertrophy after exposure to PMPs for 60–90 min (Yeaman et al. 1998). Typically, perturbations in cell ultrastructure precede significant bactericidal and bacteriolytic effects of these peptides. Fungal pathogens are affected in a similar manner by tPMP-1 in vitro (Yeaman et al. 1993). Protoplasts derived from *S. aureus* whole cells exhibit tPMP-1 susceptibility or resistance characteristics corresponding to those of the phenotype from which they were prepared, suggesting that antimicrobial effects are independent of the cell wall.

PMPs and kinocidins exert rapid microbicidal effects against target pathogens. Within minutes of exposure to peptides such as tPMP-1, perturbations of the *S. aureus* cell membrane are seen by ultrastructural analysis. These results suggest that, similar to other endogenous antimicrobial peptides, tPMP-1 initially targets the microbial cell membrane as part of its lethal pathway (Yeaman and Bayer 1999). Energy auxotroph strains of *S. aureus* with defects in their ability to generate a normal transmembrane electric potential ($\Delta\psi$) are less susceptible to killing by tPMP-1 (and other antimicrobial peptides) than their parental counterparts with normal $\Delta\psi$. Menadione supplementation of the $\Delta\psi$-deficient mutants to restore the $\Delta\psi$ to near parental levels also restores tPMP-1 susceptibility (Yeaman et al. 1998). Likewise, tPMP-1-mediated membrane permeability to propidium iodide is also reduced in the $\Delta\psi$ mutants, and reconstituted by menadione. Related studies (Koo et al. 1999) reinforce that the overall proton motive force ($\Delta\psi$ and ΔpH) is important in tPMP-1-induced microbicidal action. Addition of tPMP-1 to model planar lipid bilayers causes membrane permeabilization and bilayer dysfunction in a voltage-dependent manner, with maximal effects induced at a trans-negative voltage orientation relative to the site of addition of the peptide. Well-defined, voltage-gated pores formed by defensins in similar model membranes are not formed by tPMP-1. The staphylocidal activity of tPMP-1 declines under conditions in vitro in which microbial membrane energetics are reduced (e.g., stationary phase, low temperature).

Cytoplasmic membrane fluidity of tPMP-1-resistant strains of *S. aureus* obtained by serial passage, transposon mutagenesis, or plasmid is distinct from that of genetically related tPMP-1-susceptible counterparts. Recent findings (Mukhopadhyay et al. 2005; Xiong et al. 2005; Weidenmaier et al. 2005) offer new evidence for an interrelationship among the staphylococcal cell wall and cell membrane in terms of antimicrobial peptide susceptibility or resistance. For example, the *mprF* gene in *S. aureus* leads to increased phospholipid lysinylation (cationic charge) and reduced susceptibility to antimicrobial peptides in vitro ostensibly by electrostatic repulsion (Kristian et al. 2003). In comparison, the *S. aureus tagO* gene involved in cell wall teichoic acid profiles does not alter susceptibility to tPMP-1, but does lead to reduced endothelial cell binding and attenuated virulence in the rabbit model of endocarditis (Weidenmaier et al. 2005). Likewise, membrane permeabilization is involved, but not exclusive as a mechanism of tPMP-1 staphylocidal activity (Xiong et al. 2005). Interesting results also reveal that the net cationic charge of the extracellular facet of the *S. aureus* cell membrane also affects susceptibility to PMPs and kinocidins. For example, Mukhopadhyay et al. (2005) found asymmetry of the outer membrane leaflet, with orientation of lysyl-phosphatidylglycerol (LPG) constituents to its exterior, to reduce *S. aureus* susceptibility to tPMP-1 or other antimicrobial peptides in vitro. As LPG is predominant basic lipid species in this respect, asymmetric localization to the outer membrane leaflet likely enhances positive surface charge and may repel cationic antimicrobial peptides.

Although PMPs and kinocidins cause rapid dysfunction of target microbial cell membranes, the 1- to 2-h delay between initial exposure and microbicidal effect implicate other, likely intracellular, targets of action for these peptides. Xiong et al. (1999) demonstrated that pre-exposure of tPMP-1[S] S. aureus with tetracycline, a 30S ribosomal subunit inhibitor, significantly decreased the ensuing staphylocidal effect of tPMP-1 over a concentration range of 0.16–1.25 µg/ml. In these studies, pre-exposure to novobiocin (inhibitor of bacterial DNA gyrase subunit B), or azithromycin, quinupristin, or dalfopristin (inhibitors of 50S ribosomal subunits) mitigated the staphylocidal effect of tPMP-1 over a concentration range of 0.31–1.25 µg/ml. These data suggested that tPMP-1 exerts anti-S. aureus activities, in part, through mechanisms involving inhibition of macromolecular synthesis. More recent studies by Xiong et al. (2002) provide further support for the hypothesis that PMPs and kinocidins access and disrupt intracellular pathways. In tPMP-1[S] S. aureus strains, purified tPMP-1 caused a significant reduction in DNA and RNA synthesis that temporally corresponded to the extent of staphylocidal activity. In contrast, tPMP-1 exerted substantially reduced inhibition of macromolecular synthesis in the isogenic tPMP-1[R] counterpart, mirroring reduced staphylocidal effects. However, tPMP-1 caused equivalent degrees of protein synthesis inhibition in these strains. For example, pre-exposure of tPMP-1-susceptible strains with antibiotics that block either 50S ribosome-dependent protein synthesis or DNA gyrase subunit B actions completely inhibit tPMP-1-induced microbicidal effects. Moreover, S. aureus strains deficient in their autolytic pathway are also less susceptible to tPMP-1, suggesting an important role for this system in the overall lethal mechanism of this peptide (Xiong et al. 2004). Activation of abnormal autolysin functions has also been implicated as a mechanism of cationic antimicrobial peptides, including PMPs and kinocidins (Sakoulas et al. 2005; Ginsburg 1988), but remains to be definitively characterized in this regard. Collectively, these findings support the concept that PMP- or kinocidin-induced lethality involves intracellular targets and perhaps even autolytic pathways subsequent to initial membrane perturbations.

Beyond direct microbicidal effects, these peptides also cause prolonged post-exposure growth-inhibitory effects against staphylococci, similar to those of bacterial cell wall-active antibiotics such as oxacillin and vancomycin (Yeaman et al. 1992). For example, in contrast to the microbicidal effects of tPMP-1, these growth-inhibitory properties of the peptide are observed in both tPMP-1-susceptible and -resistant strains. Simultaneous exposure of staphylococci to tPMP-1 and antibiotics such as oxacillin exerts synergistic bactericidal effects. Moreover, pre-exposure of S. aureus or C. albicans to tPMP-1 reduces the capacity of these pathogens to adhere to platelets in vitro, and this effect can be amplified by exposure to antimicrobial agents (Yeaman et al. 1994a; Yeaman et al. 1994b).

3.8
PMP and Kinocidin Mechanisms of Action Are Distinctive

The antimicrobial mechanisms of PMPs and kinocidins appear to differ from one another, and from traditional classes of antimicrobial peptides. The microbicidal effects of PMPs and kinocidins involve pH-dependent membrane permeabilization, and may occur with (e.g., PF-4) or without (e.g., tPMP-1) membrane depolarization. Moreover, these mechanisms are distinct from defensin hNP-1, or the cationic antibacterial agents protamine or gentamicin. The bactericidal mechanisms of PMP and kinocidin actions vs isogenic tPMP-1-susceptible (tPMP-1^S) and tPMP-1-resistant (tPMP-1^R) S. aureus are distinct from human defensin NP-1 (hNP-1) or these cationic antimicrobial agents in relevant parameters, including roles of transmembrane potential ($\Delta\psi$), permeabilization, and inhibition of macromolecular synthesis on bactericidal activity. For example, rabbit PMP-2 rapidly permeabilizes and depolarizes tPMP-1^S S. aureus strains, with the extent of permeabilization inversely related to pH (Yeaman et al. 1997). However, tPMP-1 did not significantly depolarize the tPMP-1^S strain, but permeabilized this strain directly correlating to exposure pH. Depolarization, permeabilization, and killing of the tPMP-1^R strain by PMP-2 and tPMP-1 were significantly reduced as compared with the tPMP-1^S counterpart. Moreover, culture in menadione reconstituted the tPMP-1^R $\Delta\psi$ to a level equivalent to the tPMP-1^S strain, increased depolarization due to PMP-2 (but not tPMP-1), and restored permeabilization and killing of the tPMP-1^R strain. Hence, a dependence on $\Delta\psi$ for microbicidal efficacy distinguishes certain PMPs/kinocidins from neutrophil type-I defensins, which employ $\Delta\psi$-independent mechanisms of killing.

4
Relevance of PMPs and Kinocidins in Defense Against Invasive Infections

The preceding sections illustrate the actuality that platelets are specialized host defense cells armed with a battery of relatively unique antimicrobial peptides, including multifunctional kinocidins. Extending upon these themes, the following section unifies recent evidence of coordinated roles for PMPs and kinocidins in defense against invasive infections.

4.1
Platelet Deficit and Dysfunction in Relation to Infection

It has long been recognized that normal platelet quantity and quality are important to host defense. Inherited conditions such as Wiscott-Aldrich syndrome, May-Hegglin anomaly, Gray-Platelet syndrome, or related platelet disorders, strongly correlate with increases in morbidity and mortality due to infection (Yeaman 1997). However, these conditions often represent a convergence of multiple disorders in myeloid lineage or cell-mediated immunity, obscuring platelet-specific contributions to host defense or making them difficult to independently define. Nonetheless, specific roles for platelets and their antimicrobial peptides in host defense have now emerged from in vivo studies using complementary methods. Several points can be used to illustrate this theme:

For example, Sullam et al. (1993) examined the role of platelets in host defense against infective endocarditis in vivo using the rabbit model. In this study, a tPMP-1[S] viridans streptococcus strain was inoculated to induce endocarditis in animals with normal platelet counts, or with specific immune thrombocytopenia. There were no differences in these groups regarding leukocyte quantity or quality, or complement profiles. Animals with thrombocytopenia exhibited significantly higher streptococcal densities in vegetations as compared with counterparts having normal platelet counts. Dankert et al. (2001) likewise reported that platelets are active in host defense against infective endocarditis due to viridans group streptococci. These data substantiate the concept that platelets, PMPs, and kinocidins are important host defense effectors that limit establishment and/or evolution of endovascular infections.

In humans, thrombocytopenia is proving to be a significant indicator of worsened morbidity and mortality in contexts of infection. In patients undergoing cytotoxic cancer chemotherapy or for non-neoplastic conditions, thrombocytopenia appears to put patients at increased risk of morbidity and mortality due to bacterial or fungal infection (Feldman et al. 1991; Kirkpatrick et al. 1994; Viscoli et al. 1994; Chang et al. 2000). For example, in patients with normal absolute neutrophil count, thrombocytopenia has been significantly correlated with increased incidence and severity of Gram-negative lobar pneumonia in the elderly (Kirkpatrick et al. 1994). Observations from other human conditions have also implicated platelets as being integral to antimicrobial host defense. For example, Chang et al. (2000) examined the impact of thrombocytopenia on morbidity and mortality due to infection in liver transplant recipients. By multiple measures of outcome, thrombocytopenia was identified to be a significant and independent predictor of infection and related morbidity and mortality. Nadir platelet counts were significantly lower in non-survivors than survivors. Almost 50% of patients with nadir platelet counts of $\leq 30 \times 10^9/l$ presented with a major infection within 30 days after transplantation, as compared with only 17% of patients with nadir platelet counts exceeding this threshold. Similarly, fungal infections were observed in 14% of patients exhibiting nadir platelet counts below this breakpoint vs 0% in those with nadir platelet counts above it. It is striking that every fungal infection occurred in patients with nadir platelet counts below this threshold prior to the presentation of infection, with nadir platelet counts preceding onset of infection by a median of 7 days. Such data strongly implicate thrombocytopenia as predisposing to morbidity and mortality due to infections in these and perhaps other patient populations. Thrombocytopenia has also been implicated in invasive bacterial infection among children with cancer (Santolaya et al. 2002). In such children, five risk factors were evaluated for correlation with invasive bacterial infection: C-reactive protein, hypotension, leukemia relapse, thrombocytopenia (platelet count $\leq 50,000/mm^3$ blood), and chemotherapy. As a result, thrombocytopenia emerged as the sole significant risk factor of invasive bacterial infection in 12% of children in this cohort study (sensitivity, 92%; specificity, 76%; and ± predictive values of 82% and 90%).

The above findings are consistent with important roles for platelet antimicrobial constituents in the defense against infection. Mavrommatis et al. (2000) demonstrated two distinct types of platelet and coagulation responses to Gram-negative infection in humans. In the initial state, uncomplicated sepsis corresponds to substantial increases in blood levels of FBP-A and PF-4, both known PMPs and kinocidins of human platelets. Increases in the release of these antimicrobial peptides were temporally associated with a reduction in platelet count, implying that degranulated platelets are subsequently cleared. However, in severe sepsis, and most notably in the context of septic shock, platelet response is reduced, while coagulation factors are depleted. These observations suggest that, in cases of profound sepsis, platelet responses to microbial challenge may be suppressed or overwhelmed. Inhibitors of platelet functions have also been shown to inhibit bacterial clearance from blood and increase mortality in animal models (Korzweniowski et al. 1979; Sullam et al. 1993; Yeaman 1997). However, anti-platelet therapies that reduce adhesive functions of platelets, without interfering with their antimicrobial responses, may enhance host defense against endovascular infection. For example, Nicolau et al. (1993, 1995) observed that aspirin administered prophylactically or in combination with antibiotics reduced establishment and progression of *S. aureus* infective endocarditis in the rabbit model. However, as aspirin is a global inhibitor of cyclooxygenase function in endothelial cells as well as in platelets, recent studies by Kupferwasser et al. (1999, 2002) have since dissected these findings. These studies demonstrated that acetylsalicylic acid (ASA) exerted a direct antimicrobial efficacy against *S. aureus* in the rabbit model. At 8 mg/kg/day (but not 4 or 12 mg/kg/day), ASA achieved significant decreases in vegetation weight, echocardiographic-confirmed vegetation size, valve and renal *S. aureus* densities, and renal embolic lesions, as compared with untreated controls. Reduced aggregation was seen when platelets were pre-exposed to ASA, or when *S. aureus* was pre-exposed to salicylate. Moreover, *S. aureus* adhesion to sterile vegetations, suspended platelets, fibrin matrices, or fibrin-platelet matrices, was also diminished following bacterial exposure to salicylate. Based on these collective findings, it is highly likely that salicylate suppresses *S. aureus* capacity to bind to platelets, but does not impede the antimicrobial responses of platelets, such as their ability to elaborate PMPs or kinocidins.

4.2
Platelets Recognize and Respond to Settings of Infection

Platelets arguably signify the most rapid and plentiful inflammatory cell force in response to microbial challenge. Hence, platelets may be considered among the earliest of opportunities for cellular responses to intercede in microbial pathogenesis and defend against the establishment and progression of invasive infection. It is clear that platelets target sites of tissue injury and infection, particularly involving the vascular endothelium. Early interpretations of their role in this setting posited that platelets actually promote the establishment and evolution of infection. Several investigators suggested that platelets facilitate microbial adhesion to fibrin matrices or endothelial cells in vitro (for a review, see Yeaman and Bayer 2000). For example, Herzberg et al. (1990) hypothesized that streptococcal binding to and aggregation of platelets is directly correlated with increased virulence of these strains in experimental endocarditis. Uncontrolled platelet aggregation has also been suggested to be detrimental to the host, as large endovascular vegetations may promote clinical events such as emboli and infarcts. Finally, platelet aggregation and internalization may also protect pathogens from antibiotics or other host defenses, such as complement or clearance by neutrophils (Clawson and White 1971).

While it is possible that some pathogens exploit platelets in their pathogenesis, there are no data to substantiate the concept that platelets inherently facilitate pathogen survival, endothelial cell penetration, or invasion into tissue parenchyma). In this regard, host systems that enable detrimental infection would be highly disadvantageous from an evolutionary perspective. In contrast, the considerations above reinforce the concept that platelets recognize and target microbial pathogens. Moreover, a large body of evidence emphasizes that platelets are important in antimicrobial host defense against infection. The following discussion considers roles for specific platelet recognition and response in innate immunity.

Platelets chemonavigate to signals generated in contexts of cardiovascular injury, infection, or complement fixation. Potential settings of infection may include infective endocarditis, suppurative thrombophlebitis, mycotic aneurysm, septic endarteritis, catheter and dialysis access site infections, and infections of vascular prostheses and stents (Yeaman and Bayer 2000). The rapid and numerical abundance of platelets at these sites has been well established. Osler (1886) made the earliest observations of platelets accumulating on filaments introduced into animal veins. Similarly, Cheung and Fischetti (1990) and others showed that platelets are the first cells to adhere to indwelling vascular catheters. Furthermore, platelets rapidly accumulate upon cardiac valve prostheses and endovascular stents, and are the earliest and quantitatively predominant cells in early endocarditis vegetations in rabbits and humans. Thus, platelets rapidly target surfaces, as well as sites of injury to vascular endothelium, vulnerable to infection by organisms that gain access to the bloodstream.

Discoveries over the past several years have demonstrated that platelets interact with microorganisms through specific mechanisms (Yeaman 1997; Bayer et al. 1995). Recent studies by Siboo et al. (2001) and Bensing et. al. (2001) further exemplify such direct receptor–ligand interactions between pathogens and platelets. These facts imply that platelets are sentinel cells continuously monitoring the vascular compartment and beyond for potential pathogens. Microbial infection also prompts indirect interactions with platelets. For example, pathogen-induced injury evokes rapid thromboplastin (tissue factor) elaboration by vascular endothelial cells and monocytes. Tissue factor and ensuing thrombin stimulation increase platelet adhesion to infected vascular endothelial cells, promoting platelet accumulation at these sites (Carney 1992). Endothelial cell ligands also assist in targeting platelets to sites of infection. Platelets possess specific receptors that sense ligands of injured endothelial cells or exposed subendothelial stroma. Such ligands recognized by platelet membrane glycoprotein (GP) receptors include (see Yeaman 1997): collagen (GPIa-IIa, or VLA-2), von Willebrand factor (GPIb-IX-V), laminin (GPIc-IIa, or VLA-6), fibronectin (GPIc-IIa, or VLA-5), vitronectin ($\alpha_V\beta_3$ integrin), and thrombin. At these sites, adherent platelets become activated to liberate granular constituents that have direct antimicrobial functions (e.g., PMPs and kinocidins), and orchestrate further inflammatory defenses against microbial colonization and deeper tissue invasion. A sequence of events may be considered in this regard:

1. Soluble tissue factor liberated from injured endothelium prompts an intrinsic pathway cascade that ultimately generates thrombin.

2. Thrombin is a potent platelet agonist, activating rapid platelet morphogenesis from discoid to amoeboid, as well as microtubule and granule organization in preparation for degranulation.

3. Next, degranulation occurs, liberating bioactive molecules such as ADP, TXA_2 via δ-degranulation, platelet activating factor (PAF) through activation of membrane phospholipase A_2, and PMPs and kinocidins as a result of α-degranulation.

4. Inducible platelet receptors, such as fibrinogen (GPIIb-IIIa) and P-selectin receptors, are expressed upon activated platelet surfaces.

5. Activation products stimulate successive waves of platelet activation and degranulation at sites of endothelial infection.

5
Immunorelativity, PMPs, and Kinocidins in Defense Against Invasive Infection

Two principal aspects of immunobiology distinguish the functional strategies of PMPs and kinocidins from classical antimicrobial peptides: (a) PMPs and kinocidins are released directly into the bloodstream, where they must act without causing concomitant host toxicity; and (b) these peptides may contribute to optimal immune responses by potentiating subsequent defenses if necessary, but must avoid promoting uncontrolled and detrimental inflammation.

To date, the field of antimicrobial peptide discovery and characterization has been focused on peptides evolved to be secreted onto mucosal surfaces, or contained within professional phagocytes. Many such peptides exert potent antimicrobial activity under defined test conditions in vitro, but are inactive in complex media or biomatrices, or cause significant toxicity to mammalian tissues in vitro or in vivo. Such toxicity has been viewed as an impediment to development of these antimicrobial peptides as novel anti-infective agents. The ensuing section considers the functional aspects of PMPs and other kinocidins from the perspective of defending the complex setting of the bloodstream.

5.1
Antimicrobial Efficacy Versus Host Toxicity

Selective toxicity among antimicrobial peptides involves complex and specific interactions between peptide and target pathogen (Yeaman and Yount 2003). However, it is also likely that these peptides may be rendered less harmful to the host simply through strategic localization or expression that minimizes their interaction with potentially vulnerable host tissues. Three paradigms illustrate this conceptual model. Among vertebrates, many antimicrobial peptides are secreted onto relatively inert epithelial surfaces, such as the tracheal, lingual, or intestinal mucosa of mammals, or the skin of amphibians. This localization—along with rapidly inducible expression—places such peptides in key positions to respond to potential pathogens present on mucosal barriers, yet protects more sensitive tissues from host cytotoxicity. A similar, albeit perhaps more complex mechanism, likely contributes to selective toxicity of antimicrobial peptides found in granules of phagocytic leukocytes. The key antimicrobial functions of professional phagocytes include internalization (phagocytosis) of pathogens, subjecting them to the harsh microenvironment of the phagolysosome. Neutrophils, monocytes, and macrophages contain an array of antimicrobial peptides, including defensins. However, defensins may also exhibit poor selective toxicity, exerting membrane permeabilizing and other harmful effects on microorganisms and mammalian cells alike. To protect the host against autocidal effects, phagocytes normally internalize and expose pathogens to lethal concentrations of these peptides in the maturing phagolysosome, rather than degranulating these potentially injurious components into the extracellular milieu. Within these restricted confines of the phagolysome, defensins and other antimicrobial peptides are present in very high relative concentrations, where they may act harshly and synergistically with one another, along with oxidative killing mechanisms. In this way, antimicrobial peptides may be constrained to granules of mammalian phagocytes to minimize their potential for host cytotoxicity. Moreover, the antimicrobial properties of certain PMPs and kinocidins are potentiated in mildly acidic conditions, such as those found in the maturing phagolysosome (see below).

Antimicrobial peptides such as PMPs and kinocidins embody a distinct paradigm likely optimized to function within the vascular compartment, without concomitant host cytotoxicity. First, multiple reports demonstrate that these platelet peptides exert potent microbicidal activities against pathogens that access the bloodstream. Numerous studies in humans have shown that levels of PMPs such as PF-4 increase markedly (four- to sixfold; up to 5 µg/ml) during acute-phase infections with viruses, bacteria, fungi, or protozoa (Eissen and Ebhota 1983; Lorenz and Brauer 1988; Srichaikul et al.

1989; Mezzano et al. 1992; Yamamoto et al. 1997). Thus, PF-4 plasma levels increase during acute phases of cytomegaloviremia, bacterial septicemia, streptococcal nephritis, candidiasis, and malaria. Likewise, PMP levels increase in rabbit plasma as a result of staphylococcal challenge in vivo and in vitro models of infective endocarditis (Yeaman and Bayer 2000; Mercier et al. 2000, 2004). These results are consistent with the observation that staphylococcal cells or α-toxin prompt PMP release from rabbit platelets in vitro (Trier et al. 2000; Bayer et al. 1997). Similarly, Shahan and colleagues (1998) showed that expression of other PF-4 chemokine family proteins (kinocidins) increase due to pathogenic and nonpathogenic fungi. Indeed, transcription of the CXC or α-chemokines is conditionally up-regulated over CC β-chemokines both temporally and quantitatively. Additionally, Wilson et al. (2001) showed that endotoxin prompts dramatic increases in circulating levels of soluble P-selectin, an indicator of platelet degranulation. Collectively, these observations indicate that platelets, kinocidins and PMPs rapidly respond to microbial pathogens.

5.2
Accumulation of PMPs and Kinocidins in Contexts of Infection

The concept that PMPs and kinocidins exert optimal antimicrobial activity in context-specific settings is supported by a body of data in vitro and in vivo. As described above, platelet deposition and degranulation at sites of infection and tissue injury result in locally intensive accumulations of these immunopeptides. Also, PMPs and kinocidins are cationic peptides and likely concentrate on electronegative pathogen surfaces. Thus, measuring their free concentration in blood, plasma, or sera probably underestimates the antimicrobial activities of PMPs at foci of infection (Yeaman 1997; Yeaman and Yount 2003). During bacterial infection, human PF-4 is concentrated within the spleen and liver (Rucinski et al. 1990), most probably as a result of sequestration of pathogens during reticuloendothelial clearance. Reinforcing this view, in both S. aureus and C. albicans, the tPMP-1[S] phenotype negatively influences virulence in experimental models (Yeaman et al. 1996; Dhawan et al. 1998, 1999, 2000; Mercier et al. 2000; Bates et al. 2003). In addition, numerous reports indicate that PMP and kinocidin antimicrobial activities are modulated by pH (e.g., Yeaman et al. 1997; Tang et al. 2002; Yount and Yeaman 2004). The microbicidal efficacy of certain PMPs is also affected by hypernatremic conditions simulating those in the kidney (Koo et al. 1996). PMPs and kinocidins also likely exit the cardiovascular compartment to prevent or limit infection. For example, recent studies by Qiu et al. (2001) show that the chemokine kinocidin CCL22/MDC reaches microgram per gram levels in lung granuloma tissue. Likewise, Frohm et al. (1996) demonstrated that the PMP Tβ-4 and other antimicrobial peptides are detectable within human wound and blister fluid. Thus, the varying abundance, antimicrobial potencies and spectra, along with distinct conditions contributing to optimal antimicrobial activity, suggest context-specific and multifactorial roles for PMPs and kinocidins in antimicrobial host defense.

5.3
Context-Specific Functional Optima of PMPs and Kinocidins

Multiple lines of evidence support the concept that PMPs and kinocidins function in physiologically relevant contexts in response to infection. As detailed above, PMPs and kinocidins exert direct and potent antimicrobial activities in vitro against pathogens that gain access to the bloodstream. The conditional optima of PMPs and kinocidins in vitro also point to complementary host defense functions in relevant contexts in vivo. For example, human PF-4, PBP, and Tβ-4 exert substantially greater in vitro activity against *E. coli* than *S. aureus* (Tang et al. 2002). Except for Tβ-4, each of the human PMPs and kinocidins exert moderately to substantially greater antimicrobial activity at mildly acidic pH than at pH 7.5. It follows that an inflammatory context may optimize antimicrobial functions of such peptides in vivo. For example, the pH of the leukocyte phagolysosomes descends to as low as 4.5 after to phagocytosis of microorganisms (Spitznagel 1984). Moreover, abscess exudate, serum, and purulent interstitial fluid routinely achieve acidic pH. The complementary roles of PMPs and kinocidins in host defense of such contexts has been recently encompassed by the immunorelativity model of antimicrobial peptides (Yeaman and Yount 2005). Thus, PMPs and kinocidins active in mildly acidic conditions may enhance the ability of leukocytes to kill microbial pathogens through nonoxidative mechanisms, particularly as maturing phagolysosomes become acidified (Shafer et al. 1986). By comparison, at pH 7.5, Tβ-4 caused significant and rapid killing of *S. aureus* or *E. coli*, but had no significant microbicidal efficacy at pH 5.5 in vitro. It is significant that Tβ-4 has a net overall anionic charge at neutral pH, while PF-4, CTAP-3, and RANTES are cationic. However, FP-A and FP-B, also anionic, exert greater antimicrobial activities and spectra under slightly acidic conditions. Thus, key structure–activity relationships of individual PMPs and kinocidins may be optimized to specific biochemical or physiological contexts to act against cognate pathogens.

Other potential context-related effects have also been detected for certain PMPs in vitro. For example, when purified most PMPs and kinocidins have enhanced efficacy in solution phase than agar diffusion assays. PMPs may also have greater antimicrobial potency in the fluid context of the vascular compartment (e.g., blood per se, or the blood–endothelial interface) rather than deep in vegetations of infective endocarditis. This concept has been supported in the in vitro model of infective endocarditis (Mercier et al. 2000). Interestingly, PF-4 and certain other kinocidins exhibit bimodal antimicrobial efficacy in vitro (Tang et al. 2002; Yount and Yeaman, unpublished data). For example, at concentrations of less than 5 nmol/ml, human PF-4 exerts significant anti-candidal activity. However, at concentrations greater than 5 nmol/ml, it lacks significant efficacy. Some physicochemical features of PF-4 may provide insights into this result. For example, Mayo and Chen (1989) identified that human PF-4 generates a monomer-dimer-trimer equilibrium in solution. Decreases in peptide concentration, pH, or ionic strength favor the monomeric form, while obverse conditions promote tetramer formation. Hence, the antimicrobial vs chemokine effects of PF-4 may predominate at distinct context-specific concentrations: antimicrobial concentrations could be inversely related to concentration gradients that are optimal potentiation of leukocyte antimicrobial mechanisms (Yeaman 1997; Yeaman and Yount 2005). As importantly, the PMP kinocidins PF-4 and CTAP-3 exert in vitro synergistic killing of *E. coli* (Tang et al. 2002). Synergy among these peptides was seen combined at sublethal concentrations individually. These results support the concept that the elaboration of an array of PMPs and kinocidins in local settings of infection, even at sublethal concentrations of individual peptides, may yield a potent antimicrobial milieu that significantly contributes to host defense.

5.4
Indirect Roles for PMPs and Kinocidins in Antimicrobial Host Defense

PMP and kinocidin modification of microbial surfaces is consistent with a view that these peptides alter microbe–host cell interactions. For example, *S. aureus* clinical isolates exhibit heterogeneity regarding platelet adhesion, aggregation, and susceptibility to tPMP-1 (Yeaman et al. 1994b). However, exposure of tPMP-1^S or tPMP-1^R *S. aureus* strains to sublethal concentrations of tPMP-1 reduces velocity and magnitude of platelet aggregation by *S. aureus* (Bayer et al. 1995). A similar phenomenon has been observed with *C. albicans* adherence to platelets in vitro (Yeaman et al. 1994a). It has not yet been defined whether tPMP-1 exerts this effect by damaging the microbial surface, or by interfering with pathogen-platelet binding or other mechanisms. However, some insights are available that provide a basis for the hypothesis that direct microbicidal effects are independent from the anti-adhesive properties of tPMP-1. For instance, no correlation is seen between platelet adherence or aggregation and in vitro susceptibility to tPMP-1. In addition, many microbes adhere to platelets through rapid, saturable, and reversible interactions, suggesting one or more receptor-ligand mechanisms. Modified Scatchard analyses indicate that the number of binding sites per platelet vary somewhat for distinct *S. aureus* strains (Yeaman et al. 1992). In these studies, binding of individual *S. aureus* cells to platelets was influenced more by the number of binding sites on platelets, than on platelet binding affinities of bacterial cells. These findings further suggest that organism-specific platelet interactions occur. While protease K did not reduce *S. aureus* adherence to rabbit platelets in vitro, periodate or tPMP-1 did significantly reduce staphylococcal adherence. These observations imply that *S. aureus* adhesion to platelets is multimodal, involving tPMP-1-vulnerable and carbohydrate surface ligands (these may be identical). Beyond its direct effect, other studies showed that exposure to tPMP-1 alone or combined with classical anti-infective agents significantly reduces adhesion of microbes to platelets in vitro, regardless of tPMP-1^S or tPMP-1^R phenotype (Yeaman et al. 1994a, 1994b). Evidence also supports a role for platelets, PMPs, and kinocidins in protection against endovascular infections by interfering with pathogen interactions with endothelial cells. In preliminary studies, Filler et al. (1999) showed that platelets protect human umbilical vein endothelial cells (HUVECs) from injury in vitro due to *C. albicans*. In a series of experiments, tPMP-1^S *C. albicans*-induced ^{51}chromium release from HUVECs was reduced by 50% by a platelet-fungus ratio of 20:1. Furthermore, HUVEC protection by platelets corresponded to a 37% reduction in germ tube length in *C. albicans* after a 2-h exposure. In contrast, HUVEC damage by an isogenic tPMP-1^R *C. albicans* strain was not mitigated by platelets at any ratio.

5.5
Immunoenhancing Roles of PMPs and Kinocidins Versus Invasive Infection

Targeting of platelets to local context of infection, intensified PMP and kinocidin elaborating in these contexts, and affinity of these peptides for pathogens likely converge to facilitate effective antimicrobial host defense (Yeaman et al. 1997; Yeaman and Yount 2005). These functions are analogous to other kinocidins, such as interleukin-8 (IL-8), which has direct antimicrobial efficacy (Yount and Yeaman, unpublished data). For example, similar to PF-4, IL-8 can be generated systemically during infection, but exerts local effects that target sites of infection. The following discussion will focus on evidence that emphasizes immunoenhancing functions of PMPs and kinocidins that are believed to limit the establishment and progression of invasive infection.

Beyond the scope of this review, PMP kinocidins are known to perform important functions contributing significantly to antimicrobial host defense, including recruitment of leukocytes to sites of infection, and potentiating their antimicrobial activities (Yeaman 1997; Yeaman and Bayer 2000; Cole et al. 2001; Tang et al. 2002; Yang et al. 2003; Yeaman and Yount 2005). Walz et al. (1989) was among the first to find that chemokine PMPs amplify potential antimicrobial responses in leukocytes. More recently, Cocchi et al. (1995) showed that the β-chemokine RANTES suppresses human immunodeficiency virus proliferation or pathogenesis via direct antiviral effects and/or modulation of T cell function. PMP kinocidins are also members of the intercrine family of chemokines termed alarmins (Oppenheim and Yang 2005). Early studies by Mandell and Hook (1969) showed that activated platelets facilitate phagocytosis of *Salmonella* species by mouse peritoneal macrophages. Subcutaneous injection of PF-4 prompts vigorous neutrophil infiltration at sites of infection in experimental models (Deuel et al. 1981). Christin et al. (1996) demonstrated that platelets and neutrophils act synergistically in vitro to damage and kill *Aspergillus* spp. In turn, molecules produced by activated monocytes or neutrophils may activate platelets. Thus, PMP kinocidins almost certainly play at least two key roles in antimicrobial host defense: (a) direct inhibition or killing of pathogens as platelets target and liberate these peptides at sites of infection; and (b) recruitment and amplification of leukocyte antimicrobial mechanisms.

6
Epidemiological Evidence for PMP and Kinocidin Roles in Host Defense

Even in healthy individuals, bacteria and other potential pathogens access the human bloodstream daily. In comparison to a very high frequency of bloodstream accessibility, the occurrence of invasive infection per se is astonishingly low (Fowler et al. 2004). This very fact illustrates the effectiveness of host defense mechanisms against the establishment and proliferation of invasive infection. Moreover, an abundance of in vitro and in vivo evidence argues that PMPs and kinocidins significantly contribute to these mechanisms. However, the potential for organisms to subvert or withstand exposure to these peptides may provide them with virulence advantages in settings of infection.

Wu et al. (1994) uncovered a correlation between infective endocarditis source, and reduced susceptibility to tPMP-1 in vitro among Gram-positive bacteria. These data suggest that tPMP-1^S strains are less frequently associated with endovascular infection in humans as compared with tPMP-1^R counterparts. Similar observations have correlated *Salmonella* resistance to defensins, antimicrobial peptides present in neutrophils, with increased virulence (Fields et al. 1989, see above). Subsequently, Fowler et al. (2000) examined the relationship between *S. aureus* infective endocarditis and in vitro resistance to tPMP-1. These investigators evaluated in vitro tPMP-1 susceptibility phenotypes of *S. aureus* isolates from prospectively identified patients with infective endocarditis. Multivariate analyses revealed that strains associated with infected vascular devices were significantly more likely to be caused by a tPMP-1^R strain ($P = 0.02$). However, no correlations were detected between in vitro tPMP-1^R phenotype and the severity of endocarditis. Extending upon these findings, Fowler et al. (2004) recently found that persistent bacteremia in methicillin-resistant *S. aureus* (MRSA) infection is correlated with phenotypic *agr* dysfunction and low-level in vitro resistance to thrombin-induced platelet microbicidal protein. The latest studies by Sakoulas et al. (2005) further suggest that vancomycin therapy, *agr* dysfunction, and low-level in vitro resistance to tPMP-1 may be a convergent phenotype associated with infections by glycopeptide intermediate susceptible strains of *S. aureus*. A clear understanding of these potential interrelationships awaits more additional investigation.

The findings above also point to mechanisms by which pathogens may subvert or circumvent the host defense roles of platelets, PMPs, kinocidins, or other antimicrobial peptides. For example, it is possible that certain organisms are capable of exploiting the platelet as an adhesive surface, if they circumvent the antimicrobial functions of platelets. Thus, pathogens may benefit by resisting antimicrobial functions of platelets, or exploit degranulated platelets to induce or evolve invasive infections. Other potential strategies of pathogenesis may be illustrated in viridans group streptococci. By molecular mimicry of collagen structural domains involved in coagulation, *Streptococcus sanguis* triggers platelets to aggregate in vitro (Meyer et al. 1998). This effect was also documented in vivo, where cardiac function and catecholamine abnormalities rapidly ensued inoculation by an aggregation-positive *Strep. sanguis* strain into experimental animals. Aggregation of platelets generated thrombi that were attributed to the observed hemodynamic changes, acute pulmonary hypertension, and cardiac abnormalities. These effects were temporally associated with acute thrombocytopenia and accumulation of [111]indium-labeled platelets in the lungs. On the contrary, a strain incapable of inducing platelet aggregation failed to produce such effects. Thus, it is conceivable that aggregation-positive *Strep. sanguis* strains may avoid or subvert antimicrobial properties of platelets, exploiting these cells as aggregation nuclei in disease pathogenesis.

In light of the above hypotheses, it should be understood that resistance to PMPs and kinocidins differs considerably from classical antibiotic resistance.

For example, to date only low-level in vitro resistance to tPMP-1 in isolation from other peptides and under relatively austere conditions has been observed. Importantly, there does not appear to be a cumulative or high-level resistance phenotype, nor transferability of resistance. For example, unlike logarithmic resistance frequencies often observed when pathogens encounter conventional antibiotics, in vitro tPMP-1^R phenotypes display only modest and arithmetic increases in survival of approximately 10%–40% when tPMP-1 is tested at a concentration of 1–2 µg/ml. Very small increases in tPMP-1 concentration beyond these levels (e.g. 4 µg/ml) achieve complete killing of tPMP^S and tPMP^R phenotypes in vitro (Xiong et al. 1999, 2002). Thus, the in vitro tPMP-1^R phenotype represents a laboratory breakpoint for which relevance to microbial pathogenesis and host defense has not been determined. The above facts emphasize several important points regarding likely functions of PMPs in antimicrobial host defense: (a) a convergence of platelet targeting to sites of infection and local intensification of PMPs or kinocidins upon microbial surfaces likely creates very high concentrations of these peptides; (b) multiple PMPs and kinocidins are released at these sites, and exert synergistic microbicidal effects at concentrations well below those needed for their individual efficacies; and (c) these peptides interact with other host defenses, including leukocytes, and likely soluble factors such as complement. Thus, the context-specific levels and functions of these peptides within infective foci are likely to be measured at a higher level alone in artificial media in vitro. From these perspectives, the observed correlations between tPMP-1^R phenotype and propensity to cause intravascular infection undoubtedly involves multiple factors as yet undefined.

7
Summary and Prospectus

The relentless challenge of microbial exposure has, by necessity, driven evolution of rapid and effective defenses against invasive infection. Antimicrobial peptides found in skin, mucosal epithelia, leukocytes, and platelets represent key molecular effectors of this host defense armamentarium. Recent discoveries have uncovered commonalities among structural and functional archetypes of these peptides that had previously been elusive. In addition, the latest investigations focusing on human or experimental models indicate that antimicrobial peptides from these cells perform critical roles in antimicrobial host defense. Beyond their direct antimicrobial effects, some antimicrobial peptides are multifunctional kinocidins that bridge molecular and cellular immunity. The individual and collective functions of such peptides are only now becoming realized. A clearer understanding of these molecules and their functions has great potential to reveal new insights into host defense, and hasten the development of new anti-infective compounds and strategies against pathogens that resist traditional antibiotics.

References

Bates DM, von Eiff C, McNamara PJ, Peters G, Yeaman MR, Bayer AS, Proctor RA (2003) *Staphylococcus aureus* mutants are as infective as the parent strains but the menadione biosynthetic mutant persists within the kidney. J Infect Dis 187:1654–1661

Bayer AS, Sullam PM, Ramos M, Li C, Cheung AL, Yeaman MR (1995) *Staphylococus aureus* induces platelet aggregation via a fibrinogen-dependent mechanism which is independent of principal platelet GPIIb/IIIa fibrinogen-binding domains. Infect Immun 63:3634–3641

Bayer AS, Ramos MD, Menzies BE, Yeaman MR, Shen AJ, Cheung AL (1997) Hyperproduction of α-toxin by *Staphylococcus aureus* results in paradoxically-reduced virulence in experimental endocarditis: a host defense role for platelet microbicidal proteins. Infect Immun 65:4652–4660

Bensing BA, Rubens CE, Sullam PM (2001) Genetic loci of *Streptococcus mitis* that mediate binding to human platelets. Infect Immun 69:1373–1380

Chang FY, Singh N, Gayowski T, Wagener MM, Mietzner SM, Stout JE, Marino IR (2000) Thrombocytopenia in liver transplant recipients: predictors, impact on fungal infections, and role of endogenous thrombopoietin. Transplantation 69:70–75

Cheung AL, Fischetti VA (1990) The role of fibrinogen in staphylococcal adherence to catheters in vitro. J Infect Dis 161:1177–1186

Christin L, Wyson DR, Meshulam T, Hastey R, Simons ER, Diamond RD (1996) Mechanisms and target sites of damage in killing of *Candida albicans* hyphae by human polymorphonuclear neutrophils. J Infect Dis 176:1567–1578

Clawson CC, White JG (1971) Platelet interaction with bacteria. II. Fate of bacteria. Am J Pathol 65:381–398

Cocchi F, DeVico AL, G-Demo A, Arya SK, Gallo RC, Lusso P (1995) Identification of RANTES, MIP-1α, and MIP-1β as the major HIV-suppressive factors produced by CD8+ T cells. Science 270:1811–1815

Cole AM, Dewan P, Ganz T (1999) Innate antimicrobial activity of nasal secretions. Infect Immun 67:3267–3275

Dankert J, Krijgsveld J, van Der Werff J, Joldersma W, Zaat SA (2001) Platelet microbicidal activity is an important defense factor against viridans streptococcal endocarditis. J Infect Dis 184:597–605

Darveau RP, Blake J, Seachord CL, Cosand WL, Cunninigham MD, Cassiano-Cough L, Maloney G (1992) Peptide related to the carboxy-terminus of human platelet factor IV with antibacterial activity. J Clin Invest 90:447–455

Dhawan VK, Bayer, AS, Yeaman MR (1998) In vitro resistance to thrombin-induced platelet microbicidal protein is associated with enhanced progression and hematogenous dissemination in experimental *Staphylococcus aureus* infective endocarditis. Infect Immun 66:3476–3479

Dhawan VK, Yeaman MR, Bayer AS (1999) Influence of in vitro susceptibility phenotype against thrombin-induced platelet microbicidal protein on treatment and prophylaxis outcomes of experimental *Staphylococcus aureus* endocarditis. J Infect Dis 180:1561–1568

Dhawan VK, Bayer AS, Yeaman MR (2000) Thrombin-induced platelet microbicidal protein susceptibility phenotype influences the outcome of oxacillin prophylaxis and therapy of experimental *Staphylococcus aureus* endocarditis. Antimicrob Agents Chemother 44:3206–3209

Deuel TF, Senior RM, Chang D, Griffith GL, Heinrikson RL, Kaiser ET (1981) Platelet factor-4 is chemotactic for neutrophils and monocytes. Proc Natl Acad Sci U S A 78:4548–4587

Essien EM, Ebhota MI (1983) Platelet secretory activities in acute malaria (*Plasmodium falciparum*) infection. Acta Haematol 70:183–188

Feldman C, Kallenbach JM, Levy H, Thorburn JR, Hurwitz MD, Koornhof HJ (1991) Comparison of bacteraemic community-acquired lobar pneumonia due to *Streptococcus pneumoniae* and *Klebsiella pneumoniae* in an intensive care unit. Respiration 58:265–270

Fields PL, Groisman EA, Heffron F (1989) A *Salmonella* locus that controls resistance to microbicidal proteins from phagocytic cells. Science 243:1059–1062

Filler SG, Joshi M, Phan QT, Diamond RD, Edwards JE Jr, Yeaman MR (1999) Platelets protect vascular endothelial cells from injury due to *Candida albicans*. Abstract 2163, 39th ICAAC, American Society for Microbiology. San Francisco, CA

Fodor J (1887) Die fahigkeit des blutes bakterien zu vernichten. Deutsch Med Wochenschr 13:745–747

Fowler VG, McIntyre LM, Yeaman MR, Peterson GE, Reller LB, Corey GR, Wray D, Bayer AS (2000) In vitro resistance to thrombin-induced platelet microbicidal protein in isolates of *Staphylococcus aureus* from endocarditis patients correlates with an intravascular device source. J Infect Dis 182:1251–1254

Fowler VG, Scheld WM, Bayer AS (2004) Endocarditis and intravascular infections. In: Mandell GL, Bennet JE, Dolin R (eds), Principles and practice of infectious diseases (6th edn.) Chap. 74. Churchill Livingstone, New York

Fowler VG Jr, Sakoulas G, McIntyre LM, Meka V, Arbeit RD, Cabell CH, Stryjewski ME, Eliopoulos GM, Reller LB, Corey GR, Jones T, Lucindo N, Yeaman MR, Bayer AS (2004) Persistent bacteremia due to MRSA infection is associated with *agr* dysfunction and low-level in vitro resistance to thrombin-induced platelet microbicidal protein. J Infect Dis 190:1140–1149

Frohm M, Gunne H, Bergman AC, Agerberth B, Bergman T, Boman A, Lidén S, Jörnvall H, Boman HG (1996) Biochemical and antibacterial analysis of human wound and blister fluid. Eur J Biochem 237:86–92

Ganz T (2003) Hepcidin, a key regulator of iron metabolism and mediator of anemia of inflammation. Blood 102:783–788

Ganz T, Metcalf JA, Gallin JI, Boxer LA, Lehrer RI (1988) Microbicidal/cytotoxic proteins of neutrophils are deficient in two disorders: Chediak-Higashi syndrome and "specific" granule deficiency. J Clin Invest 82:552–556

Ginsburg I (1988) The biochemistry of bacteriolysis: facts, paradoxes, and myths. Microbiol Sci 5:137–142

Harada K, Ohba K, Ozaki S, Isse K, Hirayama T, Wada A, Nakanuma Y (2004) Peptide antibiotic human beta-defensin-1 and -2 contribute to antimicrobial defense of the intrahepatic biliary tree. Hepatology 40:925–932

Herzberg MC, Gong K, MacFarlane GD, Erickson PR, Soberay AH, Krebsbach PH, Gopalraj M, Schilling K, Bowen WH (1990) Phenotypic characterization of *Streptococcus sanguis* virulence factors associated with bacterial endocarditis. Infect Immun 58:515–522

Isomoto H, Mukae H, Ishimoto H, Date Y, Nishi Y, Inoue K, Wada A, Hirayama T, Nakazato M, Kohno S (2004) Elevated concentrations of alpha-defensins in gastric juice of patients with *Helicobacter pylori* infection. Am J Gastroenterol 99:1916–1923

Johnson FB, Donaldson DM (1968) Purification of staphylocidal α-lysin from rabbit serum. J Bacteriol 96:589–595

Kirkpatrick B, Reeves DS, MacGowan AP (1994) A review of the clinical presentation, laboratory features, antimicrobial therapy and outcome of 77 episodes of pneumococcal meningitis occurring in children and adults. J Infect 29:171–182

Klinger MHF, Jelkmann W (2002) Role of blood platelets in infection and inflammation. J Interferon Cytokine Res 22:913–922

Koo SP, Bayer AS, Kagan BL, Yeaman MR (1999) Membrane permeabilization by thrombin-induced PMP-1 is modulated by transmembrane voltage polarity and magnitude. Infect Immun 67:2475–2481

Korzweniowski OM, Scheld WM, Bithell TC, Croft BH, Sande MA (1979) The effect of aspirin on the production of experimental *Staphylococcus aureus* endocarditis 19th ICAAC, American Society for Microbiology, Washington, DC

Krijgsveld J, Zaat SA, Meeldijk J, van Veelen P, Fang G, Poolman B, Brandt E, Ehlert J, Kuijpers A, Engbers G, Feijen J, Dankert J (2000) Thrombocidins, microbicidal proteins from human blood platelets, are C-terminal deletion products of CXC chemokines. J Biol Chem 275:20374–20381

Kristian SA, Durr M, Van Strijp JA, Neumeister B, Peschel A (2003) MprF-mediated lysinylation of phospholipids in *Staphylococcus aureus* leads to protection against oxygen-independent neutrophil killing. Infect Immun 71:546–549

Kupferwasser LI, Yeaman MR, Shapiro SM, Nast CC, Sullam PM, Filler SG, Bayer AS (1999) Acetylsalicylic acid reduces vegetation bacterial density, hematogenous bacterial dissemination, and frequency of embolic events in experimental *Staphylococcus aureus* endocarditis through antiplatelet and antibacterial effects. Circulation 99:2791–2797

Kupferwasser LI, Yeaman MR, Shapiro SM, Nast CC, Bayer AS (2002) In vitro susceptibility to thrombin-induced platelet microbicidal protein is associated with reduced disease progression and complication rates in experimental *Staphylococcus aureus* endocarditis: microbiological, histopathologic, and echocardiographic analyses. Circulation 105:746–752

Lau YE, Rozek A, Scott MG, Goosney DL, Davidson DJ, Hancock RE (2005) Interaction and cellular localization of the human host defense peptide LL-37 with lung epithelial cells. Infect Immun 73:583–591

Lehrer RI, Ganz T, Selsted ME (1988) Oxygen-independent bactericidal systems – mechanisms and disorders. Hematol Oncol Clin North Am 2:159–169

Lorenz R, Brauer M (1988) Platelet factor 4 (PF-4) in septicaemia. Infection 16:273–276

Mandell GL, Hook EW (1969) The interaction of platelets, *Salmonella*, and mouse peritoneal macrophages. Proc Soc Exp Biol Med 132:757–759

Mavrommatis AC, Theodoridis T, Orfanidou A, Roussos C, Christopoulou-Kokkinou V, Zakynthinos S (2000) Coagulation system and platelets are fully activated in uncomplicated sepsis. Crit. Care Med 28:451–457

Mayo KH, Chen MJ (1989) Human platelet factor 4 monomer-dimer-tetramer equilibria investigated by 1H NMR spectroscopy. Biochemistry 28:9469–9478

Mercier RC, Rybak MJ, Bayer AS, Yeaman MR (2000) Influence of platelets and platelet microbicidal protein susceptibility on the fate of *Staphylococcus aureus* in an in vitro model of infective endocarditis. Infect Immun 68:4699–4705

Mercier RC, Dietz RM, Mazzola JL, Bayer AS, Yeaman MR (2004) Beneficial influence of platelets on antibiotic efficacy in an in vitro model of *Staphylococcus aureus* endocarditis. Antimicrob Agents Chemother 48:2551–2557

Meyer MW, Gong K, Herzberg MC (1998) *Streptococcus sanguis*-induced platelet clotting in rabbits and hemodynamic and cardiopulmonary consequences. Infect Immun 66:5906–5914

Mezzano S, Burgos ME, Ardiles L, Olavarria F, Concha M, Caorsi I, Aranda E, Mezzano D (1992) Glomerular localization of platelet factor 4 in streptococcal nephritis. Nephro 61:58–63

Myrvik QN (1956) Serum bactericidins active against Gram-positive bacteria. Ann N Y Acad Sci 66:391–400

Nachum R, Watson SW, Sullivan JD Jr, Seigel SE (1980) Antimicrobial defense mechanisms in the horseshoe crab, Limulus polyphemus: preliminary observations with heat-derived extracts of *Limulus* amoebocyte lysate. J Invert Pathol 32:51–58

Nicolau DP, Freeman CD, Nightingale CH, Quintiliani R, Coe CJ, Maderazo EG, Cooper, BW (1993) Reduction of bacterial titers by low-dose aspirin in experimental aortic valve endocarditis. Infect Immun 61:1593–1595

Nicolau DP, Marangos MN, Nightingale CH, Quintiliani R (1995) Influence of aspirin on development and treatment of experimental *Staphylococcus aureus* endocarditis. Antimicrob Agents Chemother 39:1748–1751

Nizet V, Ohtake T, Lauth X, Trowbridge J, Rudisill J, Dorschner RA, Pestonjamasp V, Piraino J, Huttner K, Gallo RL (2001) Innate antimicrobial peptide protects the skin from invasive bacterial infection. Nature 414:454–457

Ong PY, Ohtake T, Brandt C, Strickland I, Boguniewicz M, Ganz T, Gallo RL, Leung DY (2002) Endogenous antimicrobial peptides and skin infections in atopic dermatitis. N Engl J Med 347:1151–1160

Oppenheim JJ, Yang D (2005) Alarmins: chemotactic activators of immune responses. Curr Opin Immunol 17:359–365

Osler W (1886) On certain problems in the physiology of the blood corpuscles. Med News 48:365–425

Park CH, Valore EV, Waring AJ, Ganz T (2001) Hepcidin, a urinary antimicrobial peptide synthesized in the liver. J Biol Chem 276:7806–7810

Paulsen F, Pufe T, Conradi L, Varoga D, Tsokos M, Papendieck J, Petersen W. (2002a) Antimicrobial peptides are expressed and produced in healthy and inflamed human synovial membranes. J Pathol 198:369–377

Paulsen FP, Pufe T, Schaudig U, Held-Feindt J, Lehmann J, Thale AB, Tillmann BN (2002b) Protection of human efferent tear ducts by antimicrobial peptides. Adv Exp Med Biol 506:547–553

Putsep K, Carlsson G, Boman HG, Andersson M (2002) Deficiency of antibacterial peptides in patients with morbus Kostmann: an observation study. Lancet 360:1144–1149

Qiu B, Frait KA, Reich F, Komuniecki E, Chensue SW (2001) Chemokine expression dynamics in mycobacterial (type-1) and schistosomal (type-2) antigen-elicited pulmonary granuloma formation. Am J Pathol 158:1503–1515

Rucinski B, Niewiarowski S, Strzyzewski M, Holt JC, Mayo KH (1990) Human platelet factor 4 and its C-terminal peptides: heparin binding and clearance from the circulation. Thromb Haemost 63:493–498

Sakoulas G, Eliopoulos GM, Fowler VG, Moellering RC, Novick RP, Lucindo N, Yeaman MR, Bayer AS (2005) *Staphylococcus aureus* accessory gene regulator (*agr*) dysfunction and vancomycin exposure are associated with autolysin defect, glycopeptide intermediate susceptibility (GISA), and reduced antimicrobial peptide susceptibility phenotypes in vitro. Antimicrob Agents Chemother 49:2687–2692

Salzman NH, Ghosh D, Huttner KM, Paterson Y, Bevins CL (2003) Protection against enteric salmonellosis in transgenic mice expressing a human intestinal defensin. Nature 422:522–526

Santolaya ME, Alvarez AM, Aviles CL, Becker A, Cofre J, Enriquez N, O'Ryan M, Paya E, Salgado C, Silva P, Tordecilla J, Varas M, Villarroel M, Viviani T, Zubieta M (2002) Prospective evaluation of a model of prediction of invasive bacterial infection risk among children with cancer, fever, and neutropenia. Clin Infect Dis 35:678–683

Shafer WM, Martin LE, Spitznagel JK (1986) Late intraphagosomal hydrogen ion concentration favors the in vitro antimicrobial capacity of a 37-kilodalton cationic granule protein of human neutrophil granulocytes. Infect Immun 53:651–655

Shahan TA, Sorenson WG, Paulauskis JD, Morey R, Lewis DM (1998) Concentration- and time-dependent upregulation and release of the cytokines MIP-2, KC, TNF, and MIP-1-α in rat alveolar macrophages by fungal spores implicated in airway inflammation. Am J Respir Cell Mol Biol 18:435–440

Shin JS, Kim CW, Kwon YS, Kim JC (2004) Human beta-defensin 2 is induced by interleukin-1beta in the corneal epithelial cells. Exp Mol Med 36:204–210

Shirafuji Y, Tanabe H, Satchell DP, Henschen-Edman A, Wilson CL, Ouellette AJ (2003) Structural determinants of procryptdin recognition and cleavage by matrix metalloproteinase-7. J Biol Chem 278:7910–7919

Siboo IR, Cheung AL, Bayer AS, Sullam PM (2001) Clumping factor A mediates binding of *Staphylococcus aureus* to human platelets. Infect Immun 69:3120–3127

Sieprawska-Lupa M, Mydel P, Krawczyk K, Wojcik K, Puklo M, Lupa B, Suder P, Silberring J, Reed M, Pohl J, Shafer W, McAleese F, Foster T, Travis J, Potempa J (2004) Degradation of human antimicrobial peptide LL-37 by *Staphylococcus aureus*-derived proteinases. Antimicrob Agents Chemother 48:4673–4679

Spitznagel JK (1984) Non-oxidative antimicrobial reactions of leukocytes. Contemp
    Top Immunobiol 14:283–343
Srichaikul T, Nimmannitya S, Sripaisarn T, Kamolsilpa M, Pulgate C (1989) Platelet
    function during the acute phase of dengue hemorrhagic fever. Southeast Asian
    J Trop Med Pub Health 20:19–25
Sullam PM, Frank U, Yeaman MR, Tauber MG, Bayer AS, Chambers HF (1993) Effect
    of thrombocytopenia on the early course of streptococcal endocarditis. J Infect
    Dis 168:910–914
Tang YQ, Yeaman MR, Selsted ME (2002) Antimicrobial peptides from human platelets.
    Infect Immun 70:6524–6533
Tew JG, Roberts RR, Donaldson DM (1974) Release of α-lysin from platelets by throm-
    bin and by a factor produced in heparinized blood. Infect Immun 9:179–186
Tocantins LM (1938) The mammalian blood platelet in health and disease. Medicine
    17:155–257
Trier DA, Bayer AS, Yeaman MR (2000) *Staphylococcus aureus* elicits antimicrobial re-
    sponses from platelets via an ADP-dependent pathway. Abstract 1010, 40th ICAAC,
    American Society for Microbiology. Toronto, Canada
Turner RB, Liu L, Sazonova IY, Reed GL (2002) Structural elements that govern the sub-
    strate specificity of the clot-dissolving enzyme plasmin. J Biol Chem 277:33068–
    33074
Varoga D, Pufe T, Harder J, Meyer-Hoffert U, Mentlein R, Schroder JM, Petersen WJ,
    Tillmann BN, Proksch E, Goldring MB, Paulsen FP (2004) Production of endoge-
    nous antibiotics in articular cartilage. Arth Rheum 50:3526–3534
Viscoli C, Bruzzi P, Castagnola E, Boni L, Calandra T, Gaya H, Meuneir F, Feld R,
    Zinner S, Klastersky J et al (1994) Factors associated with bacteraemia in febrile,
    granulocytopenic patients. The International Antimicrobial Therapy Cooperative
    Group (IATCG) of the European Organization for Research and Treatment of
    Cancer (EORTC). Eur J Cancer 30:430–437
Walz A, Dewald B, von Tscharner V, Baggiolini M (1989) Effects of neutrophil-activating
    peptide NAP-2, platelet basic protein, connective tissue-activating peptide III, and
    platelet factor 4 on human neutrophils. J Exp Med 170:1745–1750
Weksler BB (1992) Platelets. In: Gallin J, Goldstein I, Snyderman R (eds) Inflammation:
    basic principles and clinical correlates, 2nd edn. Raven, New York, pp
Wilson M, Blum R, Dandona P, Mousa S (2001) Effects in humans of intravenously
    administered endotoxin on soluble cell-adhesion molecule and inflammatory
    markers: a model of human diseases. Clin Exp Pharmacol Physiol 28:376–380
Wu T, Yeaman MR, Bayer AS (1994) In vitro resistance to platelet microbicidal protein
    correlates with endocarditis source among staphylococcal isolates. Antimicrob
    Agents Chemother 38:729–732
Weidenmaier C, Peschel A, Xiong YQ, Kristian SA, Dietz K, Yeaman MR, Bayer AS
    (2005) Lack of wall teichoic acids in *Staphylococcus aureus* leads to reduced
    interactions with endothelial cells and to attenuated virulence in a rabbit model
    of endocarditis. J Infect Dis 191:1771–1777
Xiong YQ, Yeaman MR, Bayer AS (1999) In vitro antibacterial activities of platelet
    microbicidal protein and neutrophil defensin against *Staphylococcus aureus* are
    influenced by antibiotics differing in mechanism of action. Antimicrob Agents
    Chemother 43:1111–1117

Xiong YQ, Bayer AS, Yeaman MR (2002) Inhibition of *Staphylococcus aureus* intracellular macromolecular synthesis by thrombin-induced platelet microbicidal proteins. J Infect Dis 185:348–356

Xiong YQ, Bayer AS, Yeaman MR, van Wamel W, Cheung AL (2004) Impact of *sarA* and *agr* in *Staphylococcus aureus* upon fibronectin-binding protein A gene expression and fibronectin adherence capacity in vitro and in experimental infective endocarditis. Infect Immun 72:1832–1836

Xiong YQ, Mukhopadhyay K, Yeaman MR, Adler-Moore J, Bayer AS (2005) Functional interrelationships between cell membrane and wall in antimicrobial peptide-mediated killing of *Staphylococcus aureus*. Antimicrob Agents Chemother 49:3114–3121

Yamamoto Y, Klein TW, Friedman H (1997) Involvement of mannose receptor in cytokine interleukin-1beta (IL-1beta), IL-6, and granulocyte-macrophage colony-stimulating factor responses, but not in chemokine macrophage inflammatory protein 1beta (MIP-1beta), MIP-2, and KC responses, caused by attachment of *Candida albicans* to macrophages. Infect Immun 65:1077–1082

Yang D, Chertov O, Bykovskaia SN, Chen Q, Buffo MJ, Shogan J, Anderson M, Schroder JM, Wang JM, Howard OM, Oppenheim JJ (1999) Beta-defensins: linking innate and adaptive immunity through dendritic and T cell CCR6. Science 286:525–528

Yang D, Chen Q, Hoover DM, Staley P, Tucker KD, Lubkowski J, Oppenheim JJ (2003) Many chemokines including CCL20/MIP-3alpha display antimicrobial activity. J Leukoc Biol 74:448–455

Yeaman MR (1997) The role of platelets in antimicrobial host defense. Clin Infect Dis 25:951–970

Yeaman MR (2004) Antimicrobial peptides from platelets in defense against cardiovascular infections. In: Devine DA, Hancock REW (eds.) Mammalian host defense peptides. Adv Mol Cell Micro. Cambridge University Press, Cambridge, pp 279–322

Yeaman MR, Bayer AS (1999) Antimicrobial peptides from platelets. Drug Resist. Updates 2:116–26

Yeaman MR, Bayer AS (2000) *Staphylococcus aureus*, platelets, and the heart. Curr Infect Dis Rep 2:281–298

Yeaman MR, Yount NY (2003) Mechanisms of antimicrobial peptide action and resistance. Pharm Rev 54:27–55

Yeaman MR, Yount NY (2005) Code among chaos: immunorelativity and the AEGIS model of antimicrobial peptides. ASM News 71:21–27

Yeaman MR, Norman DC, Bayer AS (1992) Platelet microbicidal protein enhances antibiotic-induced killing of and post-antibiotic effect in *Staphylococcus aureus*. Antimicrob Agents Chemother 36:1665–1670

Yeaman MR, Ibrahim A, Filler SG, Bayer AS, Edwards JE Jr, Ghannoum MA (1993) Thrombin-induced rabbit platelet microbicidal protein is fungicidal *in vitro*. Antimicrob Agents Chemother 37:546–553

Yeaman MR, Sullam PM, Dazin PF, Ghannoum MA, Edwards JE Jr, Bayer AS (1994a) Fluconazole and platelet microbicidal protein inhibit *Candida* adherence to platelets in vitro. Antimicrob Agents Chemother 38:1460–1465

Yeaman MR, Sullam PM, Dazin PF, Bayer AS (1994b) Platelet microbicidal protein alone and in combination with antibiotics reduces *Staphylococcus aureus* adherence to platelets in vitro. Infect Immun 62:3416–3423

Yeaman MR, Soldan SS, Ghannoum MA, Edwards JE Jr, Filler SG, Bayer AS (1996) Resistance to platelet microbicidal protein results in increased severity of experimental *Candida albicans* endocarditis. Infect Immun 64:1379–1384

Yeaman MR, Tang Y-Q, Shen AJ, Bayer AS, Selsted ME (1997) Purification and in vitro activities of rabbit platelet microbicidal proteins. Infect Immun 65:1023–1031

Yeaman MR, Bayer AS, Koo SP, Foss W, Sullam PM (1998) Platelet microbicidal proteins and neutrophil defensin disrupt the *Staphylococcus aureus* cytoplasmic membrane by distinct mechanisms of action. J Clin Invest 101:178–187

Yeaman MR, Gank KD, Bayer AS, Brass EP (2002) Synthetic peptides that exert antimicrobial activities in whole blood and blood-derived matrices. Antimicro Agents Chemother 46:3883–3891

Yount NY, Gank KD, Xiong YQ, Bayer AS, Pender T, Welch WH, Yeaman MR (2004) Platelet microbicidal protein-1: structural themes of a multifunctional antimicrobial peptide. Antimicrob Agents Chemother 48:4395–4404

Zaiou M, Nizet V, Gallo RL (2003) Antimicrobial and protease inhibitory functions of the human cathelicidin (hCAP18/LL-37) prosequence. J Invest Dermatol 120:810–816

Zanetti M (2004) Cathelicidins, multifunctional peptides of the innate immunity. J Leukoc Biol 75:39–48

CTMI (2006) 306:153–182

Antimicrobial Peptides in the Airway

D. M. Laube[1] · S. Yim[1] · L. K. Ryan[1] · K. O. Kisich[2] · G. Diamond[1] (✉)

[1]Department of Oral Biology, UMDNJ-New Jersey Dental School,
Newark, NJ 07101, USA
gdiamond@umdnj.edu

[2]Department of Immunology, National Jewish Research Center, Denver, CO, USA

Abstract The airway provides numerous defense mechanisms to prevent microbial colonization by the large numbers of bacteria and viruses present in ambient air. An important component of this defense is the antimicrobial peptides and proteins present in the airway surface fluid (ASF), the mucin-rich fluid covering the respiratory epithelium. These include larger proteins such as lysozyme and lactoferrin, as well as the cationic defensin and cathelicidin peptides. While some of these peptides, such as human β-defensin (hBD)-1, are present constitutively, others, including hBD2 and -3 are inducible in response to bacterial recognition by Toll-like receptor-mediated pathways. These peptides can act as microbicides in the ASF, but also exhibit other activities, including potent chemotactic activity for cells of the innate and adaptive immune systems, suggesting they play a complex role in the host defense of the airway. Inhibition of antimicrobial peptide activity or gene expression can result in increased susceptibility to infections. This has been observed with cystic fibrosis (CF), where the CF phenotype leads to reduced antimicrobial capacity of peptides in the airway. Pathogenic virulence factors can inhibit defensin gene expression, as can environmental factors such as air pollution. Such an interference can result in infections by airway-specific pathogens including *Bordetella bronchiseptica, Mycobacterium tuberculosis,* and influenza virus. Research into the modulation of peptide gene expression in animal models, as well as the optimization of peptide-based therapeutics shows promise for the treatment and prevention of airway infectious diseases.

Abbreviations

ASF	Airway surface fluid
BALF	Bronchoalveolar lavage fluid
CF	Cystic fibrosis
CFTR	Cystic fibrosis transmembrane conductance regulator
CFU	Colony forming units
CMV	Cytomegalovirus
COPD	Chronic obstructive pulmonary disease
hBD	Human β-defensin
HBEC	Human bronchial epithelial cell
hCAP-18	Human cationic antimicrobial protein-18
HD	Human defensin
HIV	Human immunodeficiency virus
HNP	Human neutrophil peptide
HSV	Herpes simplex virus
IFN-γ	Interferon-γ
LAP	Lingual antimicrobial peptide
LPS	Lipopolysaccharide
LTA	Lipotechoic acid
mBD	Mouse β-defensin
MCP-1	Monocyte chemoattractant protein-1
MDDC	Monocyte-derived dendritic cell
MEF	Myeloid ELF-1 like factor
MIC	Minimum inhibitory concentration
NP	Neutrophil peptide
PBEC	Primary bronchial epithelial cells
PBMC	Peripheral blood mononuclear cell
PDC	Plasmacytoid dendritic cell
PMA	Phorbol myristate acetate
PMN	Polymorphonuclear leukocyte
RNI	Reactive nitrogen intermediates
ROFA	Residual oil fly ash
RSV	Respiratory syncytial virus
RV	Rhinovirus
sBD	Sheep β-defensin

SLPI	Serine leukoproteinase inhibitor
SP	Surfactant protein
TAP	Tracheal antimicrobial peptide
TEC	Tracheal epithelial cells
TLR	Toll-like receptor
TNF-α	Tumor necrosis factor-α
VAP	Ventilator associated pneumonia
VSV	Vesicular stomatitis virus

1
Introduction

The airway is constantly barraged by inspired airborne particles and potential pathogens. A complex defense system has evolved to recognize and respond to such threats. The innate immune system is responsible for clearing deposited particles from the airway surface, as well as eliminating pathogens from the alveoli.

Airway surface fluid (ASF) is a slightly viscous, elastic material containing several antimicrobial substances (Singh et al. 2000). The ASF flows continually from the peripheral lung to the trachea and into the esophagus via the mucociliary escalator. The motive force for this flow is the synchronous beating of cilia on the surface of airway epithelial cells. The ASF flows at a rate such that a particle trapped in the distal airway would be transported out of the lung and into the esophagus in less than 24 h (Balashazy et al. 2003; Cheng et al. 2003).

Many microorganisms, however, bypass the defenses of the upper airway. Maintenance of lung sterility against bacterial pathogens falls to resident alveolar macrophages, recruited polymorphonuclear leukocytes (PMNs), and their endogenous antimicrobial molecules. Phagocytosis of microorganisms, followed by nonoxidative antimicrobial processes, is important in eliminating pathogens from the alveolar regions of the lung, and antimicrobial peptides and proteins play an important role in this defense.

2
Antimicrobial Components of Respiratory Secretions

Respiratory secretions are active against a broad range of microorganisms. Most of the components responsible for this activity have low molecular weights and a net positive charge, and are either produced locally in the lung or brought to the site of insult by PMNs. Two classes of molecules that

are important in lung host defense are antimicrobial peptides, like cathelicidins and defensins, and the proteins lactoferrin, lysozyme, and SLPI (serine
leukoproteinase inhibitor).

2.1
Cathelicidins

Cathelicidins have a conserved N-terminal precursor structure of about 100
residues, named the cathelin domain (Zanetti et al. 1995). The C-terminal
domain is highly heterogeneous, and most cathelicidins undergo extracellular
proteolytic cleavage that releases the C-terminus from the precursor, thereby
activating it. Each mammalian species examined has cathelicidins, but each
contains a different set of related genes. The sole human cathelicidin found
to date is LL-37, the C-terminal domain of human cationic antimicrobial
protein 18 (hCAP-18) (Larrick et al. 1995; Gudmundsson et al. 1996). LL-37
expression was first found in the specific granules of neutrophils at about
one-third the abundance of lysozyme and lactoferrin (Sorensen et al. 1997),
and was subsequently demonstrated in monocytes, T cells (Agerberth et al.
2000), the surface epithelia of conducting airways (Bals et al. 1998c), and in
bronchoalveolar lavage fluid (BALF) (Agerberth et al. 1999).

LL-37 has broad-spectrum activity against Gram-positive and Gram-
negative microorganisms. Recently, the cathelin domain of hCAP18 was
found to have distinct antimicrobial activity after the proteolytic processing
that frees LL-37 (Zaiou et al. 2003). Interestingly, the 11-kDa cathelin segment
is active against bacterial strains, including *Staphylococcus aureus* MRSA,
that are resistant to LL-37, demonstrating complimentary antimicrobial
activity between the two domains and allowing the host to respond to an
even broader range of pathogens.

2.2
Defensins

Defensins are a family of antimicrobial peptides found at highest concentrations in cells and tissues responsible for innate host defense. They are 3–5 kDa
in size, and have a characteristic three disulfide-linked secondary structure.
The two main subgroups, the α- and β-defensins, differ in their cysteine pairing and placement in the amino acid sequence. The concentration of defensin
in areas of inflammation varies widely depending on the degree of neutrophil
infiltration and the levels of inflammatory mediators present, which can up-
regulate β-defensin expression (see Sect. 7.4) (Soong et al. 1997; Schnapp and
Harris 1998; Singh et al. 1998; Cole et al. 1999).

The α-defensins are 29–35 residues in length and have broad-spectrum activity against Gram-positive and Gram-negative bacteria (Ganz et al. 1985). Human neutrophil peptides (HNP) 1, 2, and 3 make up approximately 30% of neutrophil azurophilic granule's total protein content (Liu et al. 1997). The presence of neutrophils during airway inflammation can result in the introduction of high concentrations of α-defensins into the lung environment. Transcription of another α-defensin, human defensin 5 (HD5), is seen in nasal and bronchial epithelial cells infrequently and at very low levels (Frye et al. 2000), raising questions about its contribution to airway defense.

Human β-defensins (hBD)1–4, the four human β-defensins characterized to date, are all expressed in the airway. Distribution is highest in the large airways and decreases distally. Ciliated epithelial cells are responsible for the production of all four β-defensins in the nose, trachea, and bronchi. These peptides are slightly larger than the α-defensins, at 36–42 amino acid residues in length. Both hBD1 and -2 have been found in airway secretions at microgram per milliliter concentrations (Bals et al. 1998b; Singh et al. 1998). Both classes of peptide exhibit broad-spectrum activity against Gram-positive and Gram-negative bacteria, mycobacteria, and fungus.

2.3
Lysozyme, Lactoferrin, and SLPI

The most abundant antimicrobial airway proteins are lysozyme, lactoferrin, and SLPI. Lysozyme is found at concentrations around 10 µg/ml in airway lavage fluid (Fleming 1922; Raphael et al. 1989; Thompson et al. 1990) and 1 mg/ml in sputum (Brogan et al. 1975). Lysozyme is a 14-kDa protein that is a component of neutrophil-specific granules and is also produced by monocytes, macrophages, and epithelial cells (Thompson et al. 1990). It targets the $\beta 1 \rightarrow 4$ glycosidic bond between N-acetylglucosamine and N-acetylmuramic residues that make up peptidoglycan, making it highly active against Gram-positive bacteria. To kill Gram-negative bacteria, lysozyme requires a cofactor, such as lactoferrin, that can disrupt the outer membrane, giving the enzyme access to the peptidoglycan cell wall (Ellison and Giehl 1991).

Lactoferrin is an 80-kDa iron-binding protein located in the specific granules of neutrophils and produced by epithelial cells, and is found at 1–10 µg/ml concentrations in airway lavage fluid (Raphael et al. 1989; Thompson et al. 1990) and approximately 1 mg/ml in sputum (Brogan et al. 1975; Jacquot et al. 1983; Harbitz et al. 1984). Lactoferrin sequesters the iron needed for microbial respiration, thereby inhibiting bacterial growth (Arnold et al. 1977), and is also directly microbicidal (Arnold et al. 1982).

SLPI, a 12-kDa protein produced by Clara and goblet cells of the bronchiolar and bronchial lining epithelium, protects tissues from degradation by the serine proteinases released by neutrophils during inflammation (Kramps et al. 1991). It plays a role in epithelial cell proliferation (Zhang et al. 2002) and wound repair (Ashcroft et al. 2000; Zhu et al. 2002). SLPI also has some antimicrobial activity against both Gram-positive and Gram-negative bacteria (Hiemstra et al. 1996). It is found at concentrations of 0.1–2 µg/ml in airway lavage fluid (Vogelmeier et al. 1991; Kouchi et al. 1993) and 2.5 µg/ml in nasal secretions (Lee et al. 1993).

In addition, interactions between the many antimicrobial molecules in the airway have been found to further increase their potency. A high level of synergy has been seen between lysozyme, lactoferrin, and SLPI (Singh et al. 2000). In addition, hBD2 (Bals et al. 1998b) and LL-37 (Bals et al. 1998c) are synergistic with lactoferrin, reducing minimum inhibitory concentrations (MICs) -two- to fourfold against *S. aureus* and *Escherichia coli*.

3
Other Potential Contributions to Airway Host Defense

Defensins and cathelicidins may serve other roles in airway host defense. In general, these functions occur at much lower concentrations of peptide than those required for microbicidal action. Secreted α-defensins may play a role in epithelial repair by enhancing lung epithelial cell proliferation (Aarbiou et al. 2002). LL-37 is chemotactic for monocytes, T cells, neutrophils (Gudmundsson and Agerberth 1999; Ayabe et al. 2000), and mast cells (Niyonsaba et al. 2002). HNP1, -2, and -3 are chemotactic for human T cells (Territo et al. 1989; Chertov et al. 1996), monocytes (D. Yang et al. 2000; Y.S. Yang et al. 2000), and PMNs. Cells producing hBD-1 or -2 could also influence chemotaxis, as both of these β-defensins have been shown to bind to HEK293 cells expressing CCR6, and hBD2 attracts immature monocyte-derived dendritic cells (MDDC) and memory T cells via CCR6 in vitro (Yang et al. 1999). HBD3 was shown to attract macrophages (Garcia et al. 2001a).

Low concentrations of LL-37 were also shown to modulate epithelial cell response to lipopolysaccharide (LPS). The LPS stimulated human lung epithelial cell line A549 had reduced release of IL-8 and monocyte chemoattractant protein (MCP)-1 (Scott et al. 2002) when co-incubated with as low as 1 µg/ml LL-37. A differential response is seen in vitro when A549 cells are co-incubated with LPS and LL-37; the peptide suppresses IL-8 production at lower concentrations, while LPS-independent IL-8 stimulation occurs as the levels of LL-37 increase.

RAW cells were also used to study the ability of LL-37 to affect macrophage response to LPS and other bacterial products. Inhibition of tumor necrosis factor (TNF)-α release was seen in RAW cells co-incubated with as low as 1 μg/ml LL-37 and 100 ng/ml *E. coli* or *Burkholderia cepacia* LPS, or 1 μg/ml *S. aureus* lipotechoic acid (LTA). In vivo, LL-37 partially protected mice against lethal endotoxin shock after a 3-μg injection of *E. coli* LPS. A separate experiment using 20 μg/ml HNP1 or hBD1 in place of LL-37 also showed some reduction in TNF-α release, indicating that cationic peptides may contribute to the immune response by limiting damage induced by bacterial and viral infections in addition to recruiting cells of the immune system to the site of infection (Scott et al. 2000).

SLPI is regulated by neutrophil defensins (van Wetering et al. 2000). Primary bronchial epithelial cells (PBECs) incubated for 24–48 h with 100 μg/ml HNP1 released significantly higher levels of SLPI protein than control cells. The presence of neutrophil defensins could signal the need for more SLPI to neutralize the concurrently released proteases by degranulating neutrophils. However, adding neutrophil elastase to cultures of PBECs actually decreases defensin-induced SLPI release. The physiological relevance of this relationship is, therefore, currently unknown.

Taken together, these results indicate that antimicrobial peptides may play a dual function in host defense and response to infection. In addition to direct antimicrobial activity, these released peptides may exert other effects on airway epithelial cells like those observed in vitro, including the induction of cytokine release (van Wetering et al. 2002), and the stimulation of cell proliferation (Aarbiou et al. 2002). At the site of insult, they are present at microbicidal concentrations. The quantity of peptide could diffuse outward from this area of higher concentration, until it falls below the level of bacterial killing and becomes a chemotactic agent, directing immune cells to where they are needed and containing the extent of inflammation

4
Expression and Regulation of β-Defensins

Isolated from the bovine tracheal mucosa, tracheal antimicrobial peptide (TAP) was the first described β-defensin (Diamond et al. 1991). TAP mRNA is expressed primarily in the adult conducting airway, but not in tissue other than airway mucosa (Diamond et al. 1993). Another bovine β-defensin, lingual antimicrobial peptide (LAP), was later discovered to be expressed in respiratory epithelial cells (Schonwetter et al. 1995). Both TAP and LAP mRNA expression is induced in bovine tracheal epithelial cells (TEC) upon stimulation with heat-killed Gram-negative *Pseudomonas aeruginosa*, LPS, and TNF-α (Russell et al. 1996). Exposure of IL-1β, LTA, and muramyl dipeptide to bovine TEC also induced TAP mRNA expression while IL-6 did not (Diamond et al. 2000a). DNA analysis of TAP gene indicated that the 5′ flanking region, including NF-κB and NF-IL-6 transcription factor binding sites, was responsible for mediating transcription. Additionally, stimulation of bovine TEC with LPS led to NF-κB binding activity in these cells.

In humans, β-defensins in the airway exhibit more complex patterns of regulation. HBD1 gene expression is seen throughout the respiratory epithelia (Goldman et al. 1997) and hBD1 peptide can be found in BALF (Singh et al. 1998; Agerberth et al. 1999). It is constitutively expressed, as it is not inducible by inflammatory mediators or bacterial products, including LPS, IL-6, phorbol myristate acetate (PMA), interferon-γ (IFN-γ), TNF-α (Zhao et al. 1996), or IL-1α (Singh et al. 1998), and does not have the transcription factor binding sites needed for response to these stimuli. Because of this basal level of expression, hBD1 is believed to act as a baseline host defense molecule in the absence of injury or inflammation.

In contrast, hBD2 can be induced by these stimuli, due in part to NF-κB consensus binding sites in the 5′ flanking region of its promoter. Recognition of bacteria, bacterial products, and proinflammatory cytokines results in NF-κB translocation and hBD2 transcription in cultured airway epithelial cells (Becker et al. 2000; Diamond et al. 2000a). Endogenous proteins such as neutrophil elastase and myeloid ELF-1 like factor (MEF) have also been shown to stimulate hBD2 expression in A549 cells and human bronchial epithelial cells (HBECs), respectively (Griffin et al. 2003; Lu et al. 2004) The amount of stimulus required to elicit equivalent increases in hBD2 expression in vivo increases distally. Nasal epithelial cells produce the highest levels of hBD2 mRNA in response to 10^4 colonies/ml, while tracheal and bronchial epithelial cells require 10^5 colonies/ml to reach the same level of stimulation (Harder et al. 2000).

The hBD2 promoter region also contains other transcription binding sites, including AP-1 and NF-IL6 (Harder et al. 2000), suggesting additional levels of regulation. Recently, Kao et al. (2004) showed that IL-17 is the most potent cytokine studied to date in stimulating hBD2 expression in human airway epithelial cells. This suggests that the IL-17R-dependent JAK signaling pathway may also be involved in the transcriptional regulation of hBD2 in this cell type, in addition to NF-κB (Kao et al. 2004).

HBD3 was initially purified from human skin. Later investigation revealed that A549 cells co-incubated with *P. aeruginosa* also produced hBD3 peptide (Harder et al. 2000). Similar to hBD2, hBD3 gene expression is induced by proinflammatory cytokines and bacterial pathogens. In fetal lung explants, hBD3 mRNA was expressed in response to IL-1β treatment (Jia et al. 2001), and TNF-α induced the mRNA expression of hBD3 in A549 cells (Harder et al. 2000). This defensin has activity under both low salt concentrations and physiological saline (Harder et al. 2001; Lehrer and Ganz 2002).

HBD4 mRNA expression was induced by *P. aeruginosa*, *Streptococcus pneumoniae*, and PMA in human respiratory epithelial cells, but not proinflammatory cytokines (Garcia et al. 2001b). A novel hBD6 was shown to be expressed in human airway using an ORFeome-based Hidden Markov Model search (Kao et al. 2003).

4.1
Toll-Like Receptors

Induction of β-defensins in the airway epithelium occurs as part of a coordinated response to pathogen-associated molecular patterns via pattern recognition receptors. These include, but are not limited to, the Toll-like receptors (TLRs), of which 11 have been identified in human cells. Among these, TLRs 1–10 are expressed in cultured human airway epithelial cells, with TLR-2, -3, -5 and -6 expressed in the greatest abundance (Sha et al. 2004). Induction of host defense gene expression in the airway was observed in response to TLR ligands peptidoglycan (TLR-2), double-stranded RNA (TLR-3), LPS (TLR-4), and flagellin (TLR-5). Activation of these receptors can induce the NF-κB-mediated transcription of the β-defensins via a common MyD88-mediated pathway. Thus, in vitro, airway β-defensin gene expression is induced by LPS (Becker et al. 2000), LTA (Wang et al. 2003), flagellin (G. Diamond, unpublished observations) and CpG DNA (TLR-9, G. Diamond, unpublished observations).

While the response to LPS in bovine airway epithelial cells is robust (Russell et al. 1996), it is much less so in human cells, probably due to the lack of MD2, a TLR-4 adapter protein, in the trachea (Jia et al. 2004), although it is expressed in the A549 alveolar epithelial cell line (M. Klein-Patel and G. Diamond, unpublished observations). However, this response can be significantly enhanced by preincubating with respiratory syncytial virus (RSV), which up-regulates TLR-4 expression in these cells (Monick et al. 2003). Together, the published and preliminary data strongly support the hypothesis that the airway epithelium maintains an active host defense strategy that utilizes TLRs to respond to pathogenic infection by the up-regulation of β-defensins.

5
Antimicrobial Peptides in Human Airway Diseases

5.1
Cystic Fibrosis

CF is the most well examined disease with respect to antimicrobial peptides, as a lack of antimicrobial activity has been proposed to play a role in CF pathogenesis (Smith et al. 1996; Goldman et al. 1997). The respiratory tract of CF patients is colonized very early in life by bacteria such as *Haemophilus influenzae*, *S. aureus*, and *Klebsiella pneumoniae*. Later, almost all CF patients are colonized with *P. aeruginosa*, and some with *B. cepacia*. Thus, decreased levels of antimicrobial peptides or activity may play a role in both the initial colonization as well as the subsequent infections seen in this disease. For example, only mucoid strains of *P. aeruginosa* (such as those associated with CF airway infections) but not non-mucoid strains induce hBD2 expression in airway epithelial cells (Harder et al. 2000), and *B. cepacia* is resistant to antimicrobial peptides from airway cells (Baird et al. 1999).

The defect found in CF patients is in the expression or activity of the cystic fibrosis transmembrane conductance regulator (CFTR), although the reason why a defect in this channel-like molecule leads to colonization of the airway is under debate. Smith et al. (1996) hypothesized that antimicrobial peptides play a role in CF pathogenesis, showing that the ASF from CF patients had reduced microbicidal activity compared with ASF from normal individuals. They also observed that the salt concentration in CF ASF is elevated in comparison to normal ASF, and by adjusting the concentration of salt to that of healthy individuals its antimicrobial activity was restored.

Since both hBD1 and -2 are sensitive to elevated salt concentrations (Goldman et al. 1997; Bals et al. 1998b), it has been suggested that the reduced activity of these peptides may be at least partly responsible for the increased colonization of bacterial species. However, as techniques to measure the salt concentration of ASF differ, others have found no difference in the salt concentration between CF and normal airways (Knowles et al. 1997). Also, since neither hBD3 nor LL-37 are sensitive to elevated salt concentrations (Turner et al. 1998; Harder et al. 2001), other hypotheses to explain the role of peptides in CF pathogenesis have emerged.

In contrast to the high-salt hypothesis, a second group hypothesized that differences in CF and normal ASF were not due to this altered salt concentration, but rather resulted from CF epithelium hyperabsorbing water, due to the CFTR defect, thus changing the fluid makeup of the ASF (Matsui et al. 1998). Even with equivalent salt concentrations, this difference in the ASF from CF patients could have an effect on antimicrobial activity.

Supporting this salt-independent hypothesis, Bals et al. (2001) showed that while ASF from both CF and normal epithelium exhibited increased activity in low-salt conditions, ASF from CF cells exhibited reduced microbicidal activity compared with normals when the fluid was desalted. An examination of the peptide makeup of the ASF showed no difference in the levels of hBD1 and -2, LL-37, lysozyme, or lactoferrin. This and other data from the study suggest that another protein in the ASF may be responsible for increased colonization in CF airways.

To address this issue, a human bronchial xenograft model has been used to study the role of β-defensins in the CF lung. The xenografts are generated using primary human respiratory epithelial cells on denuded tracheas obtained from Fisher 344 rats. These tracheas are ligated to tubing, and then implemented subcutaneously in the flank of nu/nu BALB/c mice (Engelhardt et al. 1992; Goldman et al. 1997; Bals et al. 1999b). Both hBD1 and -2 and LL-37 peptides are detectable in ASF from both normal and CF human bronchial xenografts (Bals et al. 1998b, 1998c).

Using this model, Goldman and colleagues demonstrated that CF xenograft ASF contained abnormally high NaCl concentrations and failed to kill bacteria (Goldman et al. 1997). Correction of the CF defect with an adenoviral construct expressing normal CFTR restored antibacterial activity, suggesting a link between the CF defect and antimicrobial peptide activity. Furthermore, Bals and colleagues overexpressed the LL-37/hCAP-18 gene in human CF xenografts, also restoring antimicrobial activity (Bals et al. 1999b). This experiment implies that the inhibition of antibacterial activity in CF can be overcome by increased levels of peptide. Thus, while the mechanism of inhibition of antimicrobial peptide activity in CF airway is still under examination, signs point to a role for these peptides in the prevention of airway infection.

The role of chronic inflammation has been under investigation with regards to CF pathogenesis as well. The observation that hBD2 peptide levels are increased in BALF from CF patients (Singh et al. 1998), but not normal subjects, is consistent with the fact that β-defensin genes such as hBD2 and -3 are induced in response to inflammatory mediators (Kaiser and Diamond 2000). A hypothesis that they may be inactive, however, is derived from the observation that degradative enzymes such as cathepsins, produced by inflammatory cells, and present in BALF as a result of the chronic inflammation seen in CF, can cleave and inactivate β-defensins hBD2 and -3 (Taggart et al. 2003). This could result in a reduced capacity to contain a growing infection. Supporting this is the observation that in later, severe disease, lower levels of hBD2 are found in BALF (Chen et al. 2004). Therefore, it appears that both in early and late stages, bacterial infections associated with CF may be facilitated by inhibition of antimicrobial peptide activity.

5.2
Viral Infections

The first indication that viruses could be susceptible to antimicrobial peptides came from a 1985 study by the Lehrer laboratory with rabbit antimicrobial peptides MCP-1 and -2, isolated from rabbit alveolar macrophages (Lehrer et al. 1985). These peptides, which are also found in rabbit granulocytes, inactivated herpes simplex virus (HSV)-1 and -2, vesicular stomatitis virus (VSV), and influenza virus A/WSN, but not cytomegalovirus (CMV), echovirus type 11, or reovirus type 3. Further studies showed that these viruses could also be inactivated by the human neutrophil defensins HNP1–3 (Daher et al. 1986).

In both studies, the enveloped viruses were inactivated by these α-defensins, whereas the nonenveloped echovirus and reovirus were not. Indolicidin (Robinson et al. 1998), a tridecapeptide amide isolated from the cytoplasmic granules of bovine neutrophils, and defensins, both natural (Nakashima et al. 1993) and synthetic (Cole et al. 2002), also exhibit in vitro activity against human immunodeficiency virus (HIV), an enveloped virus. Therefore, possessing an envelope may be an important feature of susceptibility to the direct antiviral activity of antimicrobial peptides. Interestingly, LL-37 effectively kills vaccinia virus, a nonenveloped virus, in vitro and in vivo, but HNP1, hBD1, and hBD2 do not inhibit its viral replication, despite strong activity against *E. coli* (Howell et al. 2004).

Although there are many examples of direct antiviral activity, few published studies examine the mechanism of inactivation of viruses and viral infections by human defensins. One study examined the mechanism of the inactivation of HSV-2 by rabbit neutrophil peptide (NP)-1 (MCP-1) described above. The results confirmed the direct inactivation of HSV-2 by NP-1, and also showed that pre-treating target Vero (monkey kidney epithelial) and CaSki (human cervical epithelial) cells with NP-1 also prevents infection. NP-1 inhibited cell-to-cell spread of the virus, prevented viral entry, and stopped virally mediated fusion events, namely the transport of VP16 to the nucleus and subsequent fusion (Sinha et al. 2003). HNP1–3 also inhibited this transport mechanism, but had little effect on HSV-2 binding (Yasin et al. 2004).

Other factors besides the possession of a lipid envelope appear to be important in determining the antiviral activity of defensins, for both α- and β-defensins were described to inhibit infectivity of adenovirus (Gropp et al. 1999; Bastian and Schafer 2001) and adeno-associated virus (Virella-Lowell et al. 2000) in vitro. Modulation of receptors for viral entry may be another mechanism for inhibition of viral replication, as hBD2 and -3 have been shown to alter the surface expression of CXCR4, an HIV coreceptor, in peripheral blood mononuclear cells (PBMCs) (Quiñones-Mateu et al. 2003). HNP1–3 and six θ-defensins were shown to bind the gp120 protein of HIV and HNP1–3 bound human soluble CD4 with relatively high affinity (Wang et al. 2004).

Enhancement of antiviral immunity may occur when these cationic peptides act on dendritic cells to influence uptake and antigen presentation to T lymphocytes, an essential step in the initiation of an acquired immune response. LL-37 was shown to induce the uptake of antigen, enhance the maturation and antigen-presenting ability of MDDCs, and influence the development of a T-helper type 1 (T_H1) response by driving the development and proliferation of T_H1 lymphocytes, a response essential for viral immunity (Davidson et al. 2004). Lung dendritic cells, which are primarily type 1 myeloid-derived, have been shown to be the primary antigen-presenting cells initiating immune responses in the lung (Holt 2000). Defensins and cathelicidins may also influence innate immunity against viral infections via the attraction of immune cells to the site of infection. Taken together, these data provide strong evidence that antimicrobial peptides may play an important role in the immune response to respiratory viruses.

Few studies have investigated the role of antimicrobial peptides in innate immunity against viral infections in the respiratory tract. Viruses that affect the respiratory epithelium include influenza, rhinovirus, adenovirus, RSV, and parainfluenza virus. As mentioned above, direct inactivation of respiratory viruses by defensins has been demonstrated. Adenovirus infectivity was reduced when α- (HNP1) or β-defensins (hBD2) were either directly mixed with virus (Bastian and Schafer 2001) or transfected into HT-29 cells (HD5, an α-defensin or hBD1, a β-defensin) (Gropp et al. 1999). MCP-1 (NP-1) and MCP-2 (NP-2) from rabbit alveolar macrophages could inactivate influenza virus A/WSN (Lehrer et al. 1985).

Viruses were shown to affect β-defensin gene expression in bronchial epithelial cells in vitro. Human rhinovirus (RV-16) induced hBD2 and -3, but not hBD1 mRNA expression in PBECs, measured 24 h following infection (Duits et al. 2003). This study also found that these epithelial cells express TLR-3 and that ultraviolet light inactivation of RV-16 did not induce the increase in these β-defensins, indicating that viral replication and the presence of dsRNA is required and implying that the TLR-3 pathway is involved. In vivo infection of normal human subjects with human RV induced hBD2 mRNA expression, and hBD2 protein levels were increased in nasal lavages (Proud et al. 2004). Together, these results suggest that hBD-2 might play a role in host defense to human rhinovirus infection.

In another in vivo model of viral respiratory infection, a parainfluenza type 3 virus infection in sheep, mRNA levels of sBD-1 and two surfactant proteins known to bind virus, SP-A and SP-D, were increased in lung homogenates 3, 6, and 17 days following infection. This increased expression coincided with decreased viral replication, suggesting a role for these factors in response to viral respiratory infections (Grubor et al. 2004). The increase in sBD-1 is interesting especially when most studies describe the ineffectiveness of other stimuli to increase BD-1 in tracheal epithelial cells (Zhao et al. 1996; Singh et al. 1998).

Therefore, one possibility for the observed increase in sBD-1 is the recruitment of defensin-producing cells to the lung. One candidate cell may be the plasmacytoid or type 2 dendritic cell (PDC), also known as the natural interferon-producing cell. The role of PDCs in influenza virus infection is to provide IFN-α activity to aid the antiviral response (Fitzgerald-Bocarsly 2002). PDCs are present in human lung during tuberculosis infection, an infection that induces a T_H1-type of immune response (L.K. Ryan et al., unpublished data). Therefore, since influenza infection initiates a T_H1 immune response, PDCs may also migrate to the lung during viral infection.

We also have shown that PDCs stimulated with influenza virus increase both hBD1 mRNA and peptide in response to the virus as early as 2 h and as late as 18 h after infection. This increase is inhibited by ultraviolet inactivation of the virus and the magnitude of the increase was minimal in other cells of the immune system (Ryan et al. 2006, submitted). When influenza and PDCs are co-incubated and IFN-α is induced, little virus protein is found inside the PDCs, indicating that something is protecting the cell from viral replication (Fonteneau et al. 2003). Data in our laboratory has also shown that replication of HSV containing a green fluorescent protein construct does not occur in PDCs, but occurs in MDDCs, a cell that does not up-regulate hBD1 in response to virus (L.K. Ryan et al., unpublished data). One role of hBD1 may be to protect this cell from viral replication and destruction so that IFN-α can be induced via a replication-independent mechanism. Further research will elucidate the role of hBD1 in the innate immune response to viral infection in the respiratory tract.

5.3
Tuberculosis

Mycobacterium tuberculosis, the causative agent for tuberculosis, typically creates a primary lesion in the peripheral lung. The primary lesion, or Gohn focus, is thought to initiate most often in an alveolus (Canetti 1955). In order for particles of expectorated material containing viable *M. tuberculosis* to penetrate down human airways to the alveoli, they must be of such a small size that they could contain no more than about five bacilli (Fennelly et al. 2004). However, most of the particles emitted from an infectious tuberculosis patient appear to be much larger than this, containing tens to hundreds of tubercle bacilli (Fennelly et al. 2004). This would suggest that substantial numbers of *M. tuberculosis* land in the airways of potential victims, and fail to cause disease. There are several potential mechanisms through which airways might resist infection by M. tuberculosis.

M. tuberculosis landing in the airways would initially be confronted with ASF. M. tuberculosis is in part inhibited from replication at the airway surface by expulsion from the lungs. Additionally, the combined antimicrobial components of the ASF may inhibit mycobacterial growth. The activity of ASF against M. tuberculosis has not yet been reported, but some of its constituents such as β-defensins and LL-37 exhibit in vitro antimycobacterial activity (Kisich et al. 2001; Kisich et al., manuscript in preparation).

Airway epithelial cells are a key link in the innate immune response to M. tuberculosis that are frequently neglected. They are one of the first cell types whose surface receptors come into contact with both M. tuberculosis and inflammatory substances derived from it. As such, both their direct antimicrobial effects and the proinflammatory signals they produce in response to M. tuberculosis must be considered. TLR-2 and -4 have been reported to signal in response to M. tuberculosis (Brightbill et al. 1999; Thoma-Uszynski et al. 2001). However, little is known about the consequences of M. tuberculosis binding to these receptors on airway epithelial cells. Other agonists for these two TLRs have been reported to stimulate synthesis and secretion of several antimycobacterial molecules including hBD2 (Becker et al. 2000) and reactive nitrogen intermediates (RNI) via iNOS/eNOS (Jones et al. 2001a, 2001b; Means et al. 2001; Firmani and Riley 2002). Both RNI (Firmani and Riley 2002) and hBD2 (Kisich et al. 2001) have been shown to possess activity against M. tuberculosis, while LL-37 potentiates the anti-tuberculosis activity of neutrophil defensins (Kisich et al., manuscript in preparation). Therefore, while the ability of human airway epithelial cells to kill M. tuberculosis has not been reported, they appear to display anti-tuberculosis mechanisms which could be hypothesized.

In addition to the potential for direct anti-tuberculosis activity, the inflammatory consequences of TLR-2 and -4 stimulation in airway epithelial cells must be considered. Airway epithelial cells synthesize and secrete several cytokines and chemokines in response to TLR-2 and -4 agonists, including IL-18 (Cameron et al. 1999), IL-6, IL-8 (Palmberg et al. 1998; Wickremasinghe et al. 1999), GM-CSF, GROα/γ, and MCP-1 (Becker et al. 1994). These proinflammatory cytokines combine with those secreted from resident macrophages in the airway lumen and walls, including TNF-α, βIL-1α/β, to enhance the adhesive and hemodynamic properties of nearby vascular endothelium, resulting in a rapid influx of several waves of inflammatory cells, beginning with neutrophils (Wang et al. 2002).

While macrophages have long been ascribed a primary role in protection of the human lung from M. tuberculosis, cultured human neutrophils are far more potent anti-tuberculosis effector cells, particularly in the presence of TNF-α (Kisich et al. 2002). The anti-tuberculosis activity of human neutrophils is dependent on the neutrophil α-defensins (HD1, -2, and -3) working synergistically with LL-37 (Kisich et al., submitted). The rapid recruitment of neutrophils in response to *M. tuberculosis* has been observed following intraperitoneal introduction of tubercle bacilli in mice (Bloch 1948), and the rapid influx of neutrophils into the airways in response to bacterial challenge has also been well documented (Wang et al. 2002).

Therefore, *M. tuberculosis* that are not eliminated by mucociliary clearance or the antimicrobial activity of the ASF may provoke airway epithelial cells to synthesize and secrete additional antimicrobial compounds, including hBD2 and LL-37, which are known to possess anti-tuberculosis activity. The consequence of activation of airway macrophages and epithelial cells is a rapid influx of neutrophils, which are potent anti-tuberculosis effector cells. This response, which operates within the time required for a single cell cycle of *M. tuberculosis*, may account for a degree of protection of humans from infection by *M. tuberculosis*. As with other lung infections, susceptibility to pulmonary mycobacterial infections may be increased by factors which impair normal function of the airway epithelium, including cigarette smoke, chronic exposure to dusty environments, and air pollution (Marras and Daley 2002). Impairment of the inflammatory response may also increase the risk of infection. This may occur due to underlying conditions such as the neutropenia associated with malnutrition (Drenick and Alvarez 1971; Gotch et al. 1975; Cegielski and McMurray 2004), X-linked chronic granulomatous disease (Hodsagi et al. 1986; Gonzalez et al. 1989; Casanova et al. 1995), IL-12 axis deficiencies (Picard et al. 2002), or HIV infection (Aaron et al. 2004).

5.4
Other Airway Diseases

An increase in antimicrobial peptide mRNA or protein levels has been observed in other airway disorders such as sarcoidosis (Agerberth et al. 1999), pneumonia (Hiratsuka et al. 1998; Schaller-Bals et al. 2002), empyema (Ashitani et al. 1998b), and diffuse panbronchiolitis (Ashitani et al. 1998a). As these diseases are characterized by an increase in neutrophil influx and other markers of inflammation, the actual role of these peptides in disease is unclear. However, if defensins and other peptides are important in preventing infection, it is likely that inhibition of expression could lead to a reduction in this defense. This could occur due to a genetic defect in defensin expression, or by an inhibition of activity or expression by endogenous mediators, pathogenic virulence factors, or environmental toxins.

For example, a single nucleotide polymorphism in the hBD1 coding region that results in a Val-Ile substitution is associated with chronic obstructive pulmonary disease (COPD) (Matsushita et al. 2002). While this disease is not apparently associated with microbial infection, it is hypothesized that the amino acid substitution may affect some other defensin function such as leukocyte chemotaxis. Even in the absence of genetic mutation, modification of α-defensin peptide in the airway by ADP-ribosylation under certain conditions is observed to inhibit their activity (Paone et al. 2002), leading to reduced antibacterial capacity or chemotactic activity.

At least one pathogen has evolved a strategy to evade induction of defensin expression. Challenge of bovine TECs with the animal respiratory pathogen *Bordetella bronchiseptica* indicates that a type III secretion factor inhibits the activation of NF-κB induced by the bacterium (Legarda et al. 2004). This results in a suppression of β-defensin gene induction, and potentially a concomitant reduction in the host defense capability of the airway. The same results were observed with human airway epithelial cells (S. Yim and G. Diamond, unpublished data). As mentioned earlier, influenza virus may also depress β-defensin expression in lung epithelium in order to evade the host defense response and infect and replicate in the epithelial cell (S. Yim, G. Diamond, and L.K. Ryan, unpublished data).

High levels of air pollutant particles in ambient air also have detrimental effects on the respiratory tract, leading to serious health problems, including increased susceptibility to airway infections. Initial results from our laboratory demonstrated a dose-dependent inhibition of LPS-induced β-defensin gene expression in airway epithelial cells with increased levels of residual oil fly ash (ROFA), a particulate air pollutant (Diamond et al. 2000b). Further research has suggested that this effect may be due to the inhibition of NF-κB activation by soluble metal components of the particle, such as vanadium (M. Klein-Patel et al., submitted). Such an inhibition can suppress both the antibacterial capability of the airway as well as other functions of defensins, leading to an overall impaired defense.

6
Animal Models

6.1
Mice

Murine models are useful for investigating the role of antimicrobial peptides in airway host defense. Six β-defensins have been found in mice, five of which are fully characterized and expressed in the airway (Lehrer and Ganz

2002). Mouse β-defensin (mBD)-1, the mouse homolog of hBD1, possesses salt-sensitive antimicrobial activity against both *P. aeruginosa* and *E. coli* (Morrison et al. 1998; McCray et al. 1999) and is constitutively expressed in the airway (Huttner et al. 1997; Bals et al. 1998a; Morrison et al. 1998). mBD4 is also constitutively expressed, and is found in the esophagus, tongue, and trachea (Jia et al. 2000).

MBD2 and -6 expression is up-regulated in response to LPS at the mucosal surface of the airway (Morrison et al. 1999; Yamaguchi et al. 2001). Gene expression of mBD3, the mouse hBD2 homolog, is induced by exposure to *P. aeruginosa* in the airway (Bals et al. 1999a). A recent study demonstrated that IL-17 stimulates mBD3, but not mBD4, expression, in mouse tracheal epithelial cells (Kao et al. 2004). This mirrors the response seen with hBD2 in humans, suggesting that, in addition to their structural similarity, there might also be functional homology between these two peptides.

In order to define the in vivo role of β-defensins in the airway, two groups independently developed mBD1-deficient mice (Morrison et al. 2002; Moser et al. 2002). Surprisingly, while the knockout mice spontaneously exhibited an increased number of Staphylococcal bacteria in the urine, no statistically significant differences were observed in recovered *S. aureus* when 10^4–10^5 CFU of bacteria were introduced into the trachea (Morrison et al. 2002). However, when mBD1$^{-/-}$ mice were inoculated with larger doses of *H. influenza* (10^6–10^7 CFU), they exhibited a decrease in bacterial clearance from the lungs when compared with wild type or heterozygotes (Moser et al. 2002). The differences, while statistically significant, were small, and both strains of mice were able to eventually clear bacteria from the lungs. Furthermore, no differences were observed with *S. pneumonia* infection. Taken together, the results suggest that β-defensins selectively participate in airway defense, probably as part of a defense system with redundant components. The development of more β-defensin knockout mice, or preferably, a strain in which all β-defensin activity is inhibited (similar to the matrilysin knockout for intestinal defensin activity; Wilson et al. 1999) will help further our understanding of this defense.

6.2
Large Animals

While small animal models such as the mouse are useful due to their size, expense, and ease in manipulation, larger animals are often a better model for the study of human airway defense and disease due to greater similarity of airway morphology and the large quantities of tissue obtainable. The first example of in vivo bacterial induction of defensins was shown by Stolzenberg et al., whereby intratracheal instillation of *Manheimia (Pasteurella) haemolyt-*

ica into a single lobe of a cow lung caused an increase in LAP expression in the airway epithelium localized to the site of the infection (Stolzenberg et al. 1997).

Furthermore, inoculation of neonatal calves with *M. haemolytica* resulted in the NF-κB-mediated gene expression of several markers of inflammation, including TAP (Caverly et al. 2003). Interestingly, the degree of this induction varied greatly between individual animals, suggesting that large animals exhibit a complex innate immune response similar to that seen in humans. Surprisingly, introduction of high levels of the same bacterium into the lungs of sheep resulted in a reduction in the levels of sBD1, which lacks an NF-κB element (Ackermann et al. 2004).

In contrast, inoculation of ovine parainfluenza virus-3 into sheep lungs resulted in an increase in sBD1 mRNA levels (Meyerholz et al. 2004). These data suggest the differential roles of β-defensins in the lung, potentially distinguishing between those involved in viral defense and those for bacterial defense. It is hoped that with the recent identification of β-defensin sequences in primates homologous to hBD1−4 (Del Pero et al. 2002; Boniotto et al. 2003a, 2003b; Ventura et al. 2004), these animals will also provide a useful model for future studies on human diseases and gene therapy trials.

7
The Therapeutic Potential of Peptides in Airway Infectious Disease

Due to their broad spectrum of activity and the inability of bacteria to develop resistance to them, antimicrobial peptides are seen as having great potential for use as therapeutic antibiotics against lung infections. Several companies have been formed based on the initial discoveries of antimicrobial peptides, in order to develop these initial sequences into lead compounds for the production of antimicrobial therapies in the airway. Peptides based on indolicidins, histatins, protegrins, and other α-helical peptides have been undergoing preclinical and clinical trials for the treatment of CF (see Table 1; Falla and Zhang 2004). Similarly, Intrabiotics, Inc. has been testing one type of compound for the treatment of ventilator-associated pneumonia (VAP).

Due to the natural expression of peptides such as β-defensins and LL-37 in the respiratory tract, research into the control of their gene expression in this tissue is ongoing. A more comprehensive understanding of this regulation may lead to therapies such as the introduction of small, stable molecules that can increase the concentration or activity of the naturally occurring peptides in the airway to treat or prevent infection. One company, Inimex, Inc., has been developing peptide-based compounds toward this end.

Table 1 Potential therapeutic antimicrobial peptides in development

Company	Peptide type	Target disease	Website
Micrologix, Inc.	Indolicidin	CF	Mbiotech.com
Helix Biomedix, Inc.	α-helical	CF	Helixbiomeix.com
Intrabiotics, Inc.	Protegrin	CF, VAP	Intrabiotics.com
Demegen, Inc.	Histatin	CF	Demegen.com
Inimex, Inc.	Inducer of innate immunity	Nosocomial pneumonia	Inimexpharma.com

In order to produce viable drugs, however, major hurdles must be overcome. These include development of methods for the rapid, inexpensive production of pure, endotoxin-free, protease-resistant peptides. One area that may be promising is the development of peptide mimetics—small molecules whose structure mimics that of the peptides, but shares none of the aforementioned difficulties of peptides (for review see Patch and Barron 2002).

References

Aarbiou J, Ertmann M, van Wetering S, van Noort P, Rook D, Rabe KF, Litvinov SV, van Krieken JH, de Boer WI, Hiemstra PS (2002) Human neutrophil defensins induce lung epithelial cell proliferation in vitro. J Leukoc Biol 72:167–174

Aaron L, Saadoun D, Calatroni I, Launay O, Memain N, Vincent V, Marchal G, Dupont B, Bouchaud O, Valeyre D, Lortholary O (2004) Tuberculosis in HIV-infected patients: a comprehensive review. Clin Microbiol Infect 10:388–398

Ackermann MR, Gallup JM, Zabner J, Evans RB, Brockus CW, Meyerholz DK, Grubor B, Brogden KA (2004) Differential expression of sheep beta-defensin-1 and -2 and interleukin 8 during acute Mannheimia haemolytica pneumonia. Microb Pathog 37:21–27

Agerberth B, Grunewald J, Castanos-Velez E, Olsson B, Jornvall H, Wigzell H, Eklund A, Gudmundsson GH (1999) Antibacterial components in bronchoalveolar lavage fluid from healthy individuals and sarcoidosis patients. Am J Respir Crit Care Med 160:283–290

Agerberth B, Charo J, Werr J, Olsson B, Idali F, Lindbom L, Kiessling R, Jornvall H, Wigzell H, Gudmundsson GH (2000) The human antimicrobial and chemotactic peptides LL-37 and alpha-defensins are expressed by specific lymphocyte and monocyte populations. Blood 96:3086–3093

Arnold RR, Cole MF, McGhee JR (1977) A bactericidal effect for human lactoferrin. Science 197:263–265

Arnold RR, Russell JE, Champion WJ, Brewer M, Gauthier JJ (1982) Bactericidal activity of human lactoferrin: differentiation from the stasis of iron deprivation. Infect Immun 35:792–799

Ashcroft GS, Lei K, Jin W, Longenecker G, Kulkarni AB, Greenwell-Wild T, Hale-Donze H, McGrady G, Song XY, Wahl SM (2000) Secretory leukocyte protease inhibitor mediates non-redundant functions necessary for normal wound healing. Nat Med 6:1147–1153

Ashitani J, Mukae H, Nakazato M, Ihi T, Mashimoto H, Kadota J, Kohno S, Matsukura S (1998a) Elevated concentrations of defensins in bronchoalveolar lavage fluid in diffuse panbronchiolitis. Eur Respir J 11:104–111

Ashitani J, Mukae H, Nakazato M, Taniguchi H, Ogawa K, Kohno S, Matsukura S (1998b) Elevated pleural fluid levels of defensins in patients with empyema. Chest 113:788–794

Ayabe T, Satchell DP, Wilson CL, Parks WC, Selsted ME, Ouellette AJ (2000) Secretion of microbicidal alpha-defensins by intestinal Paneth cells in response to bacteria. Nat Immunol 1:113–118

Baird RM, Brown H, Smith AW, Watson ML (1999) Burkholderia cepacia is resistant to the antimicrobial activity of airway epithelial cells. Immunopharmacology 44:267–272

Balashazy I, Farkas A, Szoke I, Hofmann W, Sturm R (2003) Simulation of deposition and clearance of inhaled particles in central human airways. Radiat Prot Dosimetry 105:129–132

Bals R, Goldman MJ, Wilson JM (1998a) Mouse beta-defensin 1 is a salt-sensitive antimicrobial peptide present in epithelia of the lung and urogenital tract. Infect Immun 66:1225–1232

Bals R, Wang X, Wu Z, Freeman T, Bafna V, Zasloff M, Wilson JM (1998b) Human beta-defensin 2 is a salt-sensitive peptide antibiotic expressed in human lung. J Clin Invest 102:874–880

Bals R, Wang X, Zasloff M, Wilson JM (1998c) The peptide antibiotic LL-37/hCAP-18 is expressed in epithelia of the human lung where it has broad antimicrobial activity at the airway surface. Proc Natl Acad Sci U S A 95:9541–9546

Bals R, Wang X, Meegalla RL, Wattler S, Weiner DJ, Nehls MC, Wilson JM (1999a) Mouse beta-defensin 3 is an inducible antimicrobial peptide expressed in the epithelia of multiple organs. Infect Immun 67:3542–3547

Bals R, Weiner DJ, Meegalla RL, Wilson JM (1999b) Transfer of a cathelicidin peptide antibiotic gene restores bacterial killing in a cystic fibrosis xenograft model. J Clin Invest 103:1113–1117

Bals R, Weiner DJ, Meegalla RL, Accurso F, Wilson JM (2001) Salt-independent abnormality of antimicrobial activity in cystic fibrosis airway surface fluid. Am J Respir Cell Mol Biol 25:21–25

Bastian A, Schafer H (2001) Human α-defensin 1 (HNP-1) inhibits adenoviral infection in vitro. Regul Pept 101:157–161

Becker MN, Diamond G, Verghese MW, Randell SH (2000) CD14-dependent lipopolysaccharide-induced beta-defensin-2 expression in human tracheo-bronchial epithelium. J Biol Chem 275:29731–29736

Becker S, Quay J, Koren HS, Haskill JS (1994) Constitutive and stimulated MCP-1, GRO alpha, beta, and gamma expression in human airway epithelium and bronchoalveolar macrophages. Am J Physiol 266:L278–L286

Bloch HM (1948) The relationship between phagocytic cells and human tubercle bacilli. Am Rev Tuberc 57:662–670

Boniotto M, Antcheva N, Zelezetsky I, Tossi A, Palumbo V, Verga Falzacappa MV, Sgubin S, Braida L, Amoroso A, Crovella S (2003a) A study of host defence peptide beta-defensin 3 in primates. Biochem J 374:707–714

Boniotto M, Tossi A, DelPero M, Sgubin S, Antcheva N, Santon D, Masters J, Crovella S (2003b) Evolution of the beta defensin 2 gene in primates. Genes Immun 4:251–257

Brightbill HD, Libraty DH, Krutzik SR, Yang RB, Belisle JT, Bleharski JR, Maitland M, Norgard MV, Plevy SE, Smale ST, Brennan PJ, Bloom BR, Godowski PJ, Modlin RL (1999) Host defense mechanisms triggered by microbial lipoproteins through toll-like receptors. Science 285:732–736

Brogan TD, Ryley HC, Neale L, Yassa J (1975) Soluble proteins of bronchopulmonary secretions from patients with cystic fibrosis, asthma, and bronchitis. Thorax 30:72–79

Cameron LA, Taha RA, Tsicopoulos A, Kurimoto M, Olivenstein R, Wallaert B, Minshall EM, Hamid QA (1999) Airway epithelium expresses interleukin-18. Eur Respir J 14:553–559

Canetti G (1955) The tubercle bacillus in the pulmonary lesion of man; histobacteriology and its bearing on the therapy of pulmonary tuberculosis. Springer Verlag, New York

Casanova JL, Jouanguy E, Lamhamedi S, Blanche S, Fischer A (1995) Immunological conditions of children with BCG disseminated infection. Lancet 346:581

Caverly JM, Diamond G, Gallup JM, Brogden KA, Dixon RA, Ackermann MR (2003) Coordinated expression of tracheal antimicrobial peptide and inflammatory-response elements in the lungs of neonatal calves with acute bacterial pneumonia. Infect Immun 71:2950–2955

Cegielski JP, McMurray DN (2004) The relationship between malnutrition and tuberculosis: evidence from studies in humans and experimental animals. Int J Tuberc Lung Dis 8:286–298

Chen CI, Schaller-Bals S, Paul KP, Wahn U, Bals R (2004) Beta-defensins and LL-37 in bronchoalveolar lavage fluid of patients with cystic fibrosis. J Cyst Fibros 3:45–50

Cheng YS, Yazzie D, Gao J, Muggli D, Etter J, Rosenthal GJ (2003) Particle characteristics and lung deposition patterns in a human airway replica of a dry powder formulation of polylactic acid produced using supercritical fluid technology. J Aerosol Med 16:65–73

Chertov O, Michiel DF, Xu L, Wang JM, Tani K, Murphy WJ, Longo DL, Taub DD, Oppenheim JJ (1996) Identification of defensin-1, defensin-2, and CAP37/azurocidin as T-cell chemoattractant proteins released from interleukin-8-stimulated neutrophils. J Biol Chem 271:2935–2940

Cole AM, Dewan P, Ganz T (1999) Innate antimicrobial activity of nasal secretions. Infect Immun 67:3267–3275

Cole AM, Hong T, Boo LM, Nguyen T, Zhao C, Bristol G, Zack JA, Waring AJ, Yang OO, Lehrer RI (2002) Retrocyclin: a primate peptide that protects cells from infection by T- and M-tropic strains of HIV-1. PNAS 99:1813–1818

Daher KA, Selsted ME, Lehrer RI (1986) Direct inactivation of viruses by human granulocyte defensins. J Virol 60:1068–1074

Davidson DJ, Currie AJ, Reid GS, Bowdish DM, MacDonald KL, Ma RC, Hancock RE, Speert DP (2004) The cationic antimicrobial peptide LL-37 modulates dendritic cell differentiation and dendritic cell-induced T cell polarization. J Immunol 172:1146–1156

Del Pero M, Boniotto M, Zuccon D, Cervella P, Spano A, Amoroso A, Crovella S (2002) Beta-defensin 1 gene variability among non-human primates. Immunogenetics 53:907–913

Diamond G, Zasloff M, Eck H, Brasseur M, Maloy WL, Bevins CL (1991) Tracheal antimicrobial peptide, a cysteine-rich peptide from mammalian tracheal mucosa: peptide isolation and cloning of a cDNA. Proc Natl Acad Sci U S A 88:3952–3956

Diamond G, Jones DE, Bevins CL (1993) Airway epithelial cells are the site of expression of a mammalian antimicrobial peptide gene. Proc Natl Acad Sci U S A 90:4596–4600

Diamond G, Kaiser V, Rhodes J, Russell JP, Bevins CL (2000a) Transcriptional regulation of beta-defensin gene expression in tracheal epithelial cells. Infect Immun 68:113–119

Diamond G, Legarda D, Ryan LK (2000b) The innate immune response of the respiratory epithelium. Immunol Rev 173:27–38

Drenick EJ, Alvarez LC (1971) Neutropenia in prolonged fasting. Am J Clin Nutr 24:859–863

Duits LA, Nibbering PH, van Strijen E, Vos JB, Mannesse-Lazeroms SP, van Sterkenburg MA, Hiemstra PS (2003) Rhinovirus increases human beta-defensin-2 and -3 mRNA expression in cultured bronchial epithelial cells. FEMS Immunol Med Microbiol 38:59–64

Ellison RT 3rd, Giehl TJ (1991) Killing of Gram-negative bacteria by lactoferrin and lysozyme. J Clin Invest 88:1080–1091

Engelhardt JF, Yankaskas JR, Wilson JM (1992) In vivo retroviral gene transfer into human bronchial epithelia of xenografts. J Clin Invest 90:2598–2607

Falla TJ, Zhang L (2004) Therapeutic potential and applications of innate immunity peptides. In: Devine DA, Hancock REW (eds) Mammalian host defense peptides. Cambridge University Press, Cambridge, pp 69–110

Fennelly KP, Martyny JW, Fulton KE, Orme IM, Cave DM, Heifets LB (2004) Cough-generated aerosols of Mycobacterium tuberculosis: a new method to study infectiousness. Am J Respir Crit Care Med 169:604–609

Firmani MA, Riley LW (2002) Reactive nitrogen intermediates have a bacteriostatic effect on Mycobacterium tuberculosis in vitro. J Clin Microbiol 40:3162–3166

Fitzgerald-Bocarsly PA (2002) Natural interferon-α producing cells, the plasmacytoid dendritic cells. Biotechniques 33:S16–S29

Fleming A (1922) On a remarkable bacteriolytic element found in tissues and secretions. Proc R Soc Lond B Biol Sci 93: 306–317

Fonteneau J-F, Gilliet M, Larsson M, Dasilva I, Münz C, Liu Y-J, Bhardwaj N (2003) Activation of influenza virus-specific CD4$^+$ and CD8$^+$ T cells: a new role for plasmacytoid dendritic cells in adaptive immunity. Blood 101: 3520–3526

Frye M, Bargon J, Dauletbaev N, Weber A, Wagner TO, Gropp R (2000) Expression of human alpha-defensin 5 (HD5) mRNA in nasal and bronchial epithelial cells. J Clin Pathol 53:770–773

Ganz T, Selsted ME, Szklarek D, Harwig SS, Daher K, Bainton DF, Lehrer RI (1985) Defensins. Natural peptide antibiotics of human neutrophils. J Clin Invest 76:1427–1435

Garcia JR, Jaumann F, Schulz S, Krause A, Rodriguez-Jimenez J, Forssmann U, Adermann K, Kluver E, Vogelmeier C, Becker D, Hedrich R, Forssmann WG, Bals R (2001a) Identification of a novel, multifunctional beta-defensin (human beta-defensin 3) with specific antimicrobial activity. Its interaction with plasma membranes of Xenopus oocytes and the induction of macrophage chemoattraction. Cell Tissue Res 306:257–264

Garcia JR, Krause A, Schulz S, Rodriguez-Jimenez FJ, Kluver E, Adermann K, Forssmann U, Frimpong-Boateng A, Bals R, Forssmann WG (2001b) Human beta-defensin 4: a novel inducible peptide with a specific salt-sensitive spectrum of antimicrobial activity. FASEB J 15:1819–1821

Goldman MJ, Anderson GM, Stolzenberg ED, Kari UP, Zasloff M, Wilson JM (1997) Human beta-defensin-1 is a salt-sensitive antibiotic in lung that is inactivated in cystic fibrosis. Cell 88:553–560

Gonzalez B, Moreno S, Burdach R, Valenzuela MT, Henriquez A, Ramos MI, Sorensen RU (1989) Clinical presentation of Bacillus Calmette-Guerin infections in patients with immunodeficiency syndromes. Pediatr Infect Dis J 8:201–206

Gotch FM, Spry CJ, Mowat AG, Beeson PB, Maclennan IC (1975) Reversible granulocyte killing defect in anorexia nervosa. Clin Exp Immunol 21:244–249

Griffin S, Taggart CC, Greene CM, O'Neill S, McElvaney NG (2003) Neutrophil elastase up-regulates human beta-defensin-2 expression in human bronchial epithelial cells. FEBS Lett 546:233–236

Gropp R, Frye M, Wagner TOF, Bargon J (1999) Epithelial defensins impair adenoviral infection: implication for adenovirus-mediated gene therapy. Human Gene Ther 10:957–964

Grubor B, Gallup JM, Meyerholz DK, Crouch EC, Evans RB, Brogden KA, Lehmkuhl HD, Ackermann MR (2004) Enhanced surfactant protein and defensin mRNA levels and reduced viral replication during parainfluenza virus type 3 pneumonia in neonatal lambs. Clin Diag Lab Immunol 11:599–607

Gudmundsson GH, Agerberth B (1999) Neutrophil antibacterial peptides, multifunctional effector molecules in the mammalian immune system. J Immunol Methods 232:45–54

Gudmundsson GH, Agerberth B, Odeberg J, Bergman T, Olsson B, Salcedo R (1996) The human gene FALL39 and processing of the cathelin precursor to the antibacterial peptide LL-37 in granulocytes. Eur J Biochem 238:325–332

Harbitz O, Jenssen AO, Smidsrod O (1984) Lysozyme and lactoferrin in sputum from patients with chronic obstructive lung disease. Eur J Respir Dis 65:512–520

Harder J, Meyer-Hoffert U, Teran LM, Schwichtenberg L, Bartels J, Maune S, Schroder JM (2000) Mucoid Pseudomonas aeruginosa, TNF-alpha, and IL-1beta, but not IL-6, induce human beta-defensin-2 in respiratory epithelia. Am J Respir Cell Mol Biol 22:714–721

Harder J, Bartels J, Christophers E, Schroder JM (2001) Isolation and characterization of human beta -defensin-3, a novel human inducible peptide antibiotic. J Biol Chem 276:5707–5713

Hiemstra PS, Maassen RJ, Stolk J, Heinzel-Wieland R, Steffens GJ, Dijkman JH (1996) Antibacterial activity of antileukoprotease. Infect Immun 64:4520–4524

Hinds W (1982) Aerosol technology: properties, behavior, and measurement of airborne particles. John Wiley & Sons, New York

Hiratsuka T, Nakazato M, Date Y, Ashitani J, Minematsu T, Chino N, Matsukura S (1998) Identification of human beta-defensin-2 in respiratory tract and plasma and its increase in bacterial pneumonia. Biochem Biophys Res Commun 249:943–947

Hodsagi M, Uhereczky G, Kiraly L, Pinter E (1986) BCG dissemination in chronic granulomatous disease (CGD). Dev Biol Stand 58:339–346

Holt PG (2000) Antigen presentation in the lung. Am J Respir Crit Care Med 162:S151–S156

Howell MD, Jones JF, Kisich KO, Streib JE, Gallo RL, Leung DYM (2004) Selective killing of vaccinia virus by LL-37: implications for eczema vaccinatum. J Immunol 172:1763–1767

Huttner KM, Kozak CA, Bevins CL (1997) The mouse genome encodes a single homolog of the antimicrobial peptide human beta-defensin 1. FEBS Lett 413:45–49

Jacquot J, Tournier JM, Carmona TG, Puchelle E, Chazalette JP, Sadoul P (1983) [Proteins of bronchial secretions in mucoviscidosis. Role of infection]. Bull Eur Physiopathol Respir 19:453–458

Jia HP, Wowk SA, Schutte BC, Lee SK, Vivado A, Tack BF, Bevins CL, McCray PB Jr (2000) A novel murine beta -defensin expressed in tongue, esophagus, and trachea. J Biol Chem 275:33314–33320

Jia HP, Schutte BC, Schudy A, Linzmeier R, Guthmiller JM, Johnson GK, Tack BF, Mitros JP, Rosenthal A, Ganz T, McCray PB Jr (2001) Discovery of new human beta-defensins using a genomics-based approach. Gene 263:211–218

Jia HP, Kline JN, Penisten A, Apicella MA, Gioannini TL, Weiss J, McCray PB Jr (2004) Endotoxin responsiveness of human airway epithelia is limited by low expression of MD-2. Am J Physiol Lung Cell Mol Physiol 287:L428–L437

Jones BW, Heldwein KA, Means TK, Saukkonen JJ, Fenton MJ (2001a) Differential roles of Toll-like receptors in the elicitation of proinflammatory responses by macrophages. Ann Rheum Dis 60 Suppl 3:iii6–iii12

Jones BW, Means TK, Heldwein KA, Keen MA, Hill PJ, Belisle JT, Fenton MJ (2001b) Different Toll-like receptor agonists induce distinct macrophage responses. J Leukoc Biol 69:1036–1044

Kaiser V, Diamond G (2000) Expression of mammalian defensin genes. J Leukoc Biol 68:779–784

Kao CY, Chen Y, Zhao YH, Wu R (2003) ORFeome-based search of airway epithelial cell-specific novel human [beta]-defensin genes. Am J Respir Cell Mol Biol 29:71–80

Kao CY, Chen Y, Thai P, Wachi S, Huang F, Kim C, Harper RW, Wu R (2004) IL-17 markedly up-regulates beta-defensin-2 expression in human airway epithelium via JAK and NF-kappaB signaling pathways. J Immunol 173:3482–3491

Kisich KO, Heifets L, Higgins M, Diamond G (2001) Antimycobacterial agent based on mRNA encoding human beta-defensin 2 enables primary macrophages to restrict growth of Mycobacterium tuberculosis. Infect Immun 69:2692–2699

Kisich KO, Higgins M, Diamond G, Heifets L (2002) Tumor necrosis factor alpha stimulates killing of Mycobacterium tuberculosis by human neutrophils. Infect Immun 70:4591–4599

Knowles MR, Robinson JM, Wood RE, Pue CA, Mentz WM, Wager GC, Gatzy JT, Boucher RC (1997) Ion composition of airway surface liquid of patients with cystic fibrosis as compared with normal and disease-control subjects. J Clin Invest 100:2588–2595

Kouchi I, Yasuoka S, Ueda Y, Ogura T (1993) Analysis of secretory leukocyte protease inhibitor (SLPI) in bronchial secretions from patients with hypersecretory respiratory diseases. Tokushima J Exp Med 40:95–107

Kramps JA, Rudolphus A, Stolk J, Willems LN, Dijkman JH (1991) Role of antileukoprotease in the human lung. Ann N Y Acad Sci 624:97–108

Larrick JW, Hirata M, Zhong J, Wright SC (1995) Anti-microbial activity of human CAP18 peptides. Immunotechnology 1:65–72

Lee CH, Igarashi Y, Hohman RJ, Kaulbach H, White MV, Kaliner MA (1993) Distribution of secretory leukoprotease inhibitor in the human nasal airway. Am Rev Respir Dis 147:710–716

Legarda D, Klein-Patel ME, Yim S, Yuk MH, Diamond G (2005) Suppression of NF-kappaB-mediated beta-defensin gene expression in the mammalian airway by the Bordetella type III secretion system. Cell Microbiol 7:489–497

Lehrer RI, Daher K, Ganz T, Selsted M (1985) Direct inactivation of viruses by MCP-1 AND MCP-2, natural peptide antibiotics from rabbit leukocytes. J Virol 54:467–472

Lehrer RI, Ganz T (2002) Defensins of vertebrate animals. Curr Opin Immunol 14:96–102

Liu L, Zhao C, Heng HH, Ganz T (1997) The human beta-defensin-1 and alpha-defensins are encoded by adjacent genes: two peptide families with differing disulfide topology share a common ancestry. Genomics 43:316–320

Lu Z, Kim KA, Suico MA, Shuto T, Li JD, Kai H (2004) MEF up-regulates human beta-defensin 2 expression in epithelial cells. FEBS Lett 561:117–121

Marras TK, Daley CL (2002) Epidemiology of human pulmonary infection with non-tuberculous mycobacteria. Clin Chest Med 23:553–567

Matsui H, Grubb BR, Tarran R, Randell SH, Gatzy JT, Davis CW, Boucher RC (1998) Evidence for periciliary liquid layer depletion, not abnormal ion composition, in the pathogenesis of cystic fibrosis airways disease. Cell 95:1005–1015

Matsushita I, Hasegawa K, Nakata K, Yasuda K, Tokunaga K, Keicho N (2002) Genetic variants of human beta-defensin-1 and chronic obstructive pulmonary disease. Biochem Biophys Res Commun 291:17–22

McCray PB Jr, Zabner J, Jia HP, Welsh MJ, Thorne PS (1999) Efficient killing of inhaled bacteria in DeltaF508 mice: role of airway surface liquid composition. Am J Physiol 277:L183–L190

Means TK, Jones BW, Schromm AB, Shurtleff BA, Smith JA, Keane J, Golenbock DT, Vogel SN, Fenton MJ (2001) Differential effects of a Toll-like receptor antagonist on Mycobacterium tuberculosis-induced macrophage responses. J Immunol 166:4074–4082

Meyerholz DK, Grubor B, Gallup JM, Lehmkuhl HD, Anderson RD, Lazic T, Ackermann MR (2004) Adenovirus-mediated gene therapy enhances parainfluenza virus 3 infection in neonatal lambs. J Clin Microbiol 42:4780–4787

Monick MM, Yarovinsky TO, Powers LS, Butler NS, Carter AB, Gudmundsson G, Hunninghake GW (2003) Respiratory syncytial virus up-regulates TLR4 and sensitizes airway epithelial cells to endotoxin. J Biol Chem 278:53035–53044

Morrison GM, Davidson DJ, Kilanowski FM, Borthwick DW, Crook K, Maxwell AI, Govan JR, Dorin JR (1998) Mouse beta defensin-1 is a functional homolog of human beta defensin-1. Mamm Genome 9:453–457

Morrison GM, Davidson DJ, Dorin JR (1999) A novel mouse beta defensin, Defb2, which is upregulated in the airways by lipopolysaccharide. FEBS Lett 442:112–116

Morrison G, Kilanowski F, Davidson D, Dorin J (2002) Characterization of the mouse beta defensin 1, Defb1, mutant mouse model. Infect Immun 70:3053–3060

Moser C, Weiner DJ, Lysenko E, Bals R, Weiser JN, Wilson JM (2002) Beta-defensin 1 contributes to pulmonary innate immunity in mice. Infect Immun 70:3068–3072

Nakashima H, Yamamoto N, Masuda M, Fujii N (1993) Defensins inhibit HIV replication in vitro. AIDS 7:1129

Niyonsaba F, Iwabuchi K, Matsuda H, Ogawa H, Nagaoka I (2002) Epithelial cell-derived human beta-defensin-2 acts as a chemotaxin for mast cells through a pertussis toxin-sensitive and phospholipase C-dependent pathway. Int Immunol 14:421–426

Palmberg L, Larsson BM, Malmberg P, Larsson K (1998) Induction of IL-8 production in human alveolar macrophages and human bronchial epithelial cells in vitro by swine dust. Thorax 53:260–264

Paone G, Wada A, Stevens LA, Matin A, Hirayama T, Levine RL, Moss J (2002) ADP ribosylation of human neutrophil peptide-1 regulates its biological properties. Proc Natl Acad Sci U S A 99:8231–8235

Patch JA, Barron AE (2002) Mimicry of bioactive peptides via non-natural, sequence-specific peptidomimetic oligomers. Curr Opin Chem Biol 6:872–877

Picard C, Fieschi C, Altare F, Al-Jumaah S, Al-Hajjar S, Feinberg J, Dupuis S, Soudais C, Al-Mohsen IZ, Genin E, Lammas D, Kumararatne DS, Leclerc T, Rafii A, Frayha H, Murugasu B, Wah LB, Sinniah R, Loubser M, Okamoto E, Al-Ghonaium A, Tufenkeji H, Abel L, Casanova JL (2002) Inherited interleukin-12 deficiency: IL12B genotype and clinical phenotype of 13 patients from six kindreds. Am J Hum Genet 70:336–348

Proud D, Sanders SP, Wiehler S (2004) Human rhinovirus infection induces airway epithelial cell production of human beta-defensin 2 both in vitro and in vivo. J Immunol 172:4637–4645

Quiñones-Mateu ME, Lederman MM, Feng Z, Chakraborty B, Weber J, Rangel HR, Marotta ML, Mirza M, Jiang B, Kiser P, Medvik K, Sieg SF, Weinberg A (2003) Human epithelial β-defensins 2 and 3 inhibit HIV-1 replication. AIDS 17:F39–F48

Raphael GD, Jeney EV, Baraniuk JN, Kim I, Meredith SD, Kaliner MA (1989) Pathophysiology of rhinitis. Lactoferrin and lysozyme in nasal secretions. J Clin Invest 84:1528–1535

Robinson WE Jr, McDougall B, Tran D, Selsted ME (1998) Anti-HIV-1 activity of indolicidin, an antimicrobial peptide from neutrophils. J Leukoc Biol 63:94–100

Russell JP, Diamond G, Tarver AP, Scanlin TF, Bevins CL (1996) Coordinate induction of two antibiotic genes in tracheal epithelial cells exposed to the inflammatory mediators lipopolysaccharide and tumor necrosis factor alpha. Infect Immun 64:1565–1568

Schaller-Bals S, Schulze A, Bals R (2002) Increased levels of antimicrobial peptides in tracheal aspirates of newborn infants during infection. Am J Respir Crit Care Med 165:992–995

Schnapp D, Harris A (1998) Antibacterial peptides in bronchoalveolar lavage fluid. Am J Respir Cell Mol Biol 19:352–356

Schonwetter BS, Stolzenberg ED, Zasloff MA (1995) Epithelial antibiotics induced at sites of inflammation. Science 267:1645–1648

Scott MG, Vreugdenhil AC, Buurman WA, Hancock RE, Gold MR (2000) Cutting edge: cationic antimicrobial peptides block the binding of lipopolysaccharide (LPS) to LPS binding protein. J Immunol 164:549–553

Scott MG, Davidson DJ, Gold MR, Bowdish D, Hancock RE (2002) The human antimicrobial peptide LL-37 is a multifunctional modulator of innate immune responses. J Immunol 169:3883–3891

Sha Q, Truong-Tran AQ, Plitt JR, Beck LA, Schleimer RP (2004) Activation of airway epithelial cells by toll-like receptor agonists. Am J Respir Cell Mol Biol 31:358–364

Singh PK, Jia HP, Wiles K, Hesselberth J, Liu L, Conway BA, Greenberg EP, Valore EV, Welsh MJ, Ganz T, Tack BF, McCray PB Jr (1998) Production of beta-defensins by human airway epithelia. Proc Natl Acad Sci U S A 95:14961–14966

Singh PK, Tack BF, McCray PB Jr, Welsh MJ (2000) Synergistic and additive killing by antimicrobial factors found in human airway surface liquid. Am J Physiol Lung Cell Mol Physiol 279:L799–L805

Sinha S, Cheshenko N, Lehrer RI, Herold BC (2003) NP-1, a rabbit alpha-defensin, prevents the entry and intercellular spread of herpes simplex virus type 2. Antimicrob Agents Chemother 47:494–500

Smith JJ, Travis SM, Greenberg EP, Welsh MJ (1996) Cystic fibrosis airway epithelia fail to kill bacteria because of abnormal airway surface fluid. Cell 85:229–236

Soong LB, Ganz T, Ellison A, Caughey GH (1997) Purification and characterization of defensins from cystic fibrosis sputum. Inflamm Res 46:98–102

Sorensen O, Arnljots K, Cowland JB, Bainton DF, Borregaard N (1997) The human antibacterial cathelicidin, hCAP-18, is synthesized in myelocytes and metamyelocytes and localized to specific granules in neutrophils. Blood 90:2796–2803

Stolzenberg ED, Anderson GM, Ackermann MR, Whitlock RH, Zasloff M (1997) Epithelial antibiotic induced in states of disease. Proc Natl Acad. Sci U S A 94:8686–8690

Taggart CC, Greene CM, Smith SG, Levine RL, McCray PB Jr, O'Neill S, McElvaney NG (2003) Inactivation of human beta-defensins 2 and 3 by elastolytic cathepsins. J Immunol 171:931–937

Territo MC, Ganz T, Selsted ME, Lehrer R (1989) Monocyte-chemotactic activity of defensins from human neutrophils. J Clin Invest 84:2017–2020

Thoma-Uszynski S, Stenger S, Takeuchi O, Ochoa MT, Engele M, Sieling PA, Barnes PF, Rollinghoff M, Bolcskei PL, Wagner M, Akira S, Norgard MV, Belisle JT, Godowski PJ, Bloom BR, Modlin RL (2001) Induction of direct antimicrobial activity through mammalian toll-like receptors. Science 291:1544–1547

Thompson AB, Bohling T, Payvandi F, Rennard SI (1990) Lower respiratory tract lacto-
    ferrin and lysozyme arise primarily in the airways and are elevated in association
    with chronic bronchitis. J Lab Clin Med 115:148–158

Turner J, Cho Y, Dinh NN, Waring AJ, Lehrer RI (1998) Activities of LL-37,
    a cathelin-associated antimicrobial peptide of human neutrophils. Antimicrob
    Agents Chemother 42:2206–2214

Van Wetering S, van der Linden AC, van Sterkenburg MA, de Boer WI, Kuijpers AL,
    Schalkwijk J, Hiemstra PS (2000) Regulation of SLPI and elafin release from
    bronchial epithelial cells by neutrophil defensins. Am J Physiol Lung Cell Mol
    Physiol 278:L51–L58

Van Wetering S, Mannesse-Lazeroms SP, van Sterkenburg MA, Hiemstra PS (2002)
    Neutrophil defensins stimulate the release of cytokines by airway epithelial cells:
    modulation by dexamethasone. Inflamm Res 51:8–15

Ventura M, Boniotto M, Pazienza M, Palumbo V, Cardone MF, Rocchi M, Tossi A,
    Amoroso A, Crovella S (2004) Localization of beta-defensin genes in non human
    primates. Eur J Histochem 48:185–190

Virella-Lowell I, Poirer A, Chesnut KA, Brantly M, Flotte TR (2000) Inhibition of
    recombinant adeno-associated virus (rAAV) transduction by bronchial secretions
    from cystic fibrosis patients. Gene Therapy 7:1783–1789

Vogelmeier C, Hubbard RC, Fells GA, Schnebli HP, Thompson RC, Fritz H, Crys-
    tal RG (1991) Anti-neutrophil elastase defense of the normal human respiratory
    epithelial surface provided by the secretory leukoprotease inhibitor. J Clin Invest
    87:482–488

Wang W, Owen SM, Rudolph DL, Cole AM, Hong T, Waring AJ, Lal RB, Lehrer RI
    (2004) Activity of alpha- and theta-defensins against primary isolates of HIV-1.
    J Immunol 173:515–520

Wang X, Moser C, Louboutin JP, Lysenko ES, Weiner DJ, Weiser JN, Wilson JM (2002)
    Toll-like receptor 4 mediates innate immune responses to Haemophilus influenzae
    infection in mouse lung. J Immunol 168:810–815

Wang X, Zhang Z, Louboutin JP, Moser C, Weiner DJ, Wilson JM (2003) Airway epithelia
    regulate expression of human beta-defensin 2 through Toll-like receptor 2. FASEB
    J 17:1727–1729

Wickremasinghe MI, Thomas LH, Friedland JS (1999) Pulmonary epithelial cells are
    a source of IL-8 in the response to Mycobacterium tuberculosis: essential role of
    IL-1 from infected monocytes in a NF-kappa B-dependent network. J Immunol
    163:3936–3947

Wilson CL, Ouellette AJ, Satchell DP, Ayabe T, Lopez-Boado YS, Stratman JL, Hult-
    gren SJ, Matrisian LM, Parks WC (1999) Regulation of intestinal alpha-defensin
    activation by the metalloproteinase matrilysin in innate host defense. Science
    286:113–117

Yamaguchi Y, Fukuhara S, Nagase T, Tomita T, Hitomi S, Kimura S, Kurihara H, Ouchi Y
    (2001) A novel mouse beta-defensin, mBD-6, predominantly expressed in skeletal
    muscle. J Biol Chem 276:31510–31514

Yang D, Chertov O, Bykovskaia SN, Chen Q, Buffo MJ, Shogan J, Anderson M,
    Schroder JM, Wang JM, Howard OM, Oppenheim JJ (1999) Beta-defensins: link-
    ing innate and adaptive immunity through dendritic and T cell CCR6. Science
    286:525–528

Yang D, Chen Q, Chertov O, Oppenheim JJ (2000) Human neutrophil defensins selectively chemoattract naive T and immature dendritic cells. J Leukoc Biol 68:9–14

Yang YS, Mitta G, Chavanieu A, Calas B, Sanchez JF, Roch P, Aumelas A (2000) Solution structure and activity of the synthetic four-disulfide bond Mediterranean mussel defensin (MGD-1). Biochemistry 39:14436–11447

Yasin B, Wang W, Pang M, Cheshenko N, Hong T, Waring AJ, Herold BC, Wagar EA, Lehrer RI (2004) Theta defensins protect cells from infection by herpes simplex virus by inhibiting viral adhesion and entry. J Virol 78:5147–5156

Zaiou M, Nizet V, Gallo RL (2003) Antimicrobial and protease inhibitory functions of the human cathelicidin (hCAP18/LL-37) prosequence. J Invest Dermatol 120:810–816

Zanetti M, Gennaro R, Romeo D (1995) Cathelicidins: a novel protein family with a common proregion and a variable C-terminal antimicrobial domain. FEBS Lett 374:1–5

Zhang D, Simmen RC, Michel FJ, Zhao G, Vale-Cruz D, Simmen FA (2002) Secretory leukocyte protease inhibitor mediates proliferation of human endometrial epithelial cells by positive and negative regulation of growth-associated genes. J Biol Chem 277:29999–30009

Zhao C, Wang I, Lehrer RI (1996) Widespread expression of beta-defensin hBD-1 in human secretory glands and epithelial cells. FEBS Lett 396:319–322

Zhu J, Nathan C, Jin W, Sim D, Ashcroft GS, Wahl SM, Lacomis L, Erdjument-Bromage H, Tempst P, Wright CD, Ding A (2002) Conversion of proepithelin to epithelins: roles of SLPI and elastase in host defense and wound repair. Cell 111:867–878

CTMI (2006) 306:183–198

Hepcidin—A Peptide Hormone at the Interface of Innate Immunity and Iron Metabolism

T. Ganz (✉)

Departments of Medicine and Pathology, David Geffen School of Medicine,
University of California, Los Angeles, Los Angeles, CA 90095–1690, USA
tganz@mednet.ucla.edu

Abstract Hepcidin is a cationic amphipathic peptide made in the liver, released into plasma and excreted in urine. Hepcidin is the homeostatic regulator of intestinal iron absorption, iron recycling by macrophages, and iron mobilization from hepatic stores, but it is also markedly induced during infections and inflammation. Under the influence of hepcidin, macrophages, hepatocytes, and enterocytes retain iron that would otherwise be released into plasma. Hepcidin acts by inhibiting the efflux of iron through ferroportin, the sole known iron exporter that is expressed in the small intestine, and in hepatocytes and macrophages. As befits an iron-regulatory hormone, hepcidin synthesis is increased by iron loading, and decreased by anemia and hypoxia. Hepcidin is also rapidly induced by cytokines, including IL-6. The resulting decrease in plasma iron levels eventually limits iron availability to erythropoiesis and contributes to the anemia associated with infection and inflammation. The decrease in extracellular iron concentrations due to hepcidin probably limits iron availability to invading microorganisms, thus contributing to host defense.

1
Introduction

1.1
Iron: An Essential Trace Element

Iron (Andrews 1999) is an essential element for microbes, plants, and higher animals. It is a component of heme and iron-sulfur centers in many key redox enzymes, and is an essential component of oxygen-storage and -transporting proteins such as hemoglobin and myoglobin. Iron in humans is strictly conserved by efficient recycling of iron from hemoglobin of senescent red blood cells (about 20 mg/day). The recycled iron is mostly used in the bone marrow to produce new hemoglobin for red cells. Smaller amounts of iron from myoglobin and various redox enzymes are also recycled.

1.2
Links Between Iron Metabolism and Host Defense

Iron metabolism is importantly linked to host defense responses. Several mechanisms have been identified that may deprive microbes of iron, and thereby decrease their rate of growth (Jurado 1997; Schaible and Kaufmann 2004). Lactoferrin is an abundant constituent of neutrophil granules and epithelial secretions, and a homolog of the plasma iron transport protein, transferrin. It differs from transferrin by its ability to bind iron even in the acidic environment characteristic of infected tissue. The growth of some bacteria is inhibited by iron-poor but not iron-saturated lactoferrin, indicating that lactoferrin could act to deprive some bacteria of iron (Arnold et al. 1977, 1980). However, many microbes faced with iron starvation can induce the secretion of small organic molecules (siderophores) that chelate environmental iron and are then actively reimported by microbes. More recently, another abundant constituent of neutrophils and epithelial secretions, lipocalin/NGAL (now also called siderocalin) has been shown to sequester enterochelin, a siderophore from *Escherichia coli* (Flo et al. 2004). Mice made deficient in siderocalin had much lower survival after *E. coli* sepsis than control mice but were equally resistant to infections with bacteria that employed unrelated siderophores. Another mechanism of host resistance linked to iron metabolism involves the metal transporters in the membranes of phagocytic vacuoles of macrophages. Macrophages express a divalent metal transporter Nramp1 that may function to deplete the phagosome of essential trace metals, including iron and manganese (Fortier et al. 2005). Disruptive mutations in Nramp1 cause increased susceptibility to several intracellular pathogens, manifested most prominently during the early phases of infection. This chapter is focused on yet another iron-linked host defense mechanism: a systemic response to infection manifested as an acute decrease in plasma iron concentration (hypoferremia of inflammation).

1.3
Hypoferremia of Inflammation

Hypoferremia develops within hours after the onset of inflammatory responses to infection or other proinflammatory stimuli (Nemeth et al. 2004a). Despite hypoferremia, macrophages in patients or animals with infection or inflammation contain abundant stainable iron. The rapid drop in plasma iron is due to the inhibition of iron release from macrophages that take up and degrade senescent erythrocytes, as was demonstrated nearly 50 years ago by comparing the incorporation of radiolabeled iron into new erythrocytes when iron was infused in the form of labeled senescent erythrocytes or labeled transferrin (Freireich et al. 1957). In control animals, both sources were reutilized rapidly but in animals with abscesses there was substantial delay in reutilization of erythrocyte-derived iron as compared to iron-transferrin. It has recently become clear that the newly discovered iron-regulatory hormone hepcidin plays a central role in this response.

2
Hepcidin

2.1
Amino Acid Sequences and Distribution

Human hepcidin (Fig. 1) is a 25-amino acid peptide first identified in human urine (Park et al. 2001) and plasma (Krause et al. 2000). In addition to the 25-amino acid form, the urine also contains minor 20- and 22-amino acid forms

```
hHEP     DTHFPICIFCCGCCHRSKCGMCCKT
pHEP*    DTHFPICIFCCGCCRKAICGMCCKT
rHEP*    DTNFPICLFCCKCCKNSSCGLCCIT
mHEP*    DTNFPICIFCCKCCNNSQCGICCKT
dHEP*    DTHFPICIFCCGCCKTPKCGLCCKT
```

Fig. 1 Sequences of mammalian hepcidins (Park et al. 2001). Starred sequences were deduced from cDNA sequences in est databases and the N-terminus of the peptide form(s) is uncertain. The mammalian species are *h*, human, *p*, pig, *r*,rat, *m*,mouse (mHepcidin-1, the functional homolog of human hepcidin), and *d*,dog. The conserved cysteines are *boxed*

truncated at the N-terminus, probably the products of degradation of the 25-amino acid form (Park et al. 2001). The smaller peptides are now known to be nearly inactive. Hepcidin is predominantly synthesized in the liver (Krause et al. 2000; Park et al. 2001; Pigeon et al. 2001). The human hepcidin gene contains three exons that encode a 72-amino-acid preprohepcidin with a characteristic furin cleavage site immediately N-terminal to the 25-amino acid peptide. Closely related hepcidin genes and peptides (Fig. 1) were found across different species, some of which contain two hepcidin genes (mouse).

2.2
Structure of Hepcidin

Mass spectrometric studies of the human 25-amino acid hepcidin peptide showed that it contains four disulfide bonds (Park et al. 2001), and circular dichroism spectrometry showed urinary hepcidin is rich in β-sheet. The subsequent NMR spectroscopy confirmed hepcidin forms a simple hairpin stabilized by four disulfide bonds (Fig. 2), but one of these is an unusual vicinal disulfide bond in the turn (Hunter et al. 2002), whose functional contribution remains to be determined.

2.3
Evolutionary Origin

Hepcidin genes and peptides are abundant in fish and their expression is further enhanced during infections (Shike et al. 2002; Douglas et al. 2003; Shike et al. 2004). Multiple hepcidin genes in each fish species and the presence of hepcidin mRNAs in other organs in addition to the liver suggests that hepcidins in fish may have roles not only in iron metabolism but also in other host defense capacities.

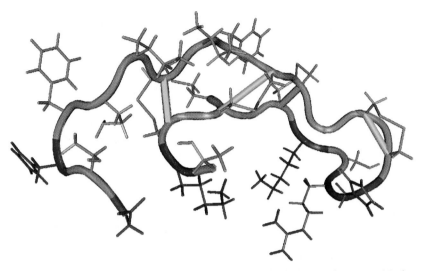

Fig. 2 NMR structure of hepcidin. The backbone and side chains are shown. Positively-charged residues are in *blue*, negatively charged in *red*, disulfides in *yellow*. The amphipathic segregation of residues is clearly seen in this view, as is the vicinal disulfide bond in the turn

2.4
Direct Antimicrobial Activity

In vitro, human hepcidin exerts antibacterial and antifungal activities (Krause et al. 2000; Park et al. 2001) at 10- to 30-µM concentrations. As is the case with many other cationic peptides, the antimicrobial activity is favored by low-ionic-strength media. Urinary hepcidin concentrations are typically in the 3- to 30-nM range (10–100 ng/ml) and can be at least tenfold higher during infections (Nemeth et al. 2003). It is thus unlikely that hepcidin can exert antimicrobial activity in urine. The concentrations of hepcidin in its other potential sites of activity, liver and plasma, are not yet known with certainty.

2.5
Iron-Regulatory Activity

The involvement of hepcidin in iron metabolism was first suggested by the observation that hepcidin synthesis is induced in mice by dietary iron over-load (Pigeon et al. 2001). The specific role of hepcidin was then examined by assessing the effects of its deficiency or excess in transgenic mouse models. Fortuitously, a mouse deficient in hepcidin already existed (Nicolas et al. 2001) as an accidental byproduct of targeting a nearby gene, USF2. The hepcidin-deficient mouse was found to have hemochromatosis with iron deposition in the liver and pancreas and sparing of the macrophage-rich spleen. This phenotype indicated that hepcidin normally controlled intestinal iron uptake and the retention of iron in macrophages. In the absence of hepcidin, iron absorption in the small intestinal was maximal and unregulated, leading to systemic iron overload. The phenotype was not due to USF2 disruption since an independent USF2 knockout line expressed normal amounts of hepcidin mRNA and had normal iron metabolism (Nicolas et al. 2002a). Importantly, human patients with homozygous disruption of the hepcidin gene were soon identified and were found to suffer from the most severe form of hemochromatosis, juvenile hemochromatosis(Roetto et al. 2003). At the other extreme, mice that overexpressed hepcidin-1 under the control of a liver-specific promoter were born with severe iron deficiency, suggesting that hepcidin inhibited placental transport of iron (Nicolas et al. 2002a). The mice died of iron deficiency unless supplemented with parenteral iron, suggesting that hepcidin also blocked intestinal iron uptake. Taken together, the mouse models indicated hepcidin was a negative regulator of iron absorption in the intestine, iron transport across the placenta, and iron release from (mainly splenic) macrophages recycling senescent erythrocytes. Hepcidin-producing tumors in mice cause anemia and hypoferremia accompanied by higher than normal amounts of iron in hepatocytes (Rivera et al. 2004). This suggests that iron export from hepatocytes is also controlled by hepcidin. The regulatory effects of hepcidin are summarized in Fig. 3. Mice have a second hepcidin gene that encodes a peptide less similar to human hepcidin. Overexpression of hepcidin-2 in mice, however, had no effect on iron metabolism and the function of the second hepcidin gene is still unclear (Lou et al. 2004).

3
Regulation of Hepcidin Synthesis

3.1
Regulation by Iron and Oxygen

Most of the iron absorbed from diet or recycled from hemoglobin is used for hemoglobin synthesis in developing erythrocytes in the bone marrow. It is therefore not surprising that hepcidin production is homeostatically regulated by anemia and hypoxemia (Nicolas et al. 2002b). When oxygen delivery is inadequate, the homeostatic response is to produce more erythrocytes. Thus, in anemia, hepcidin levels decrease, its inhibitory effects diminish, and more

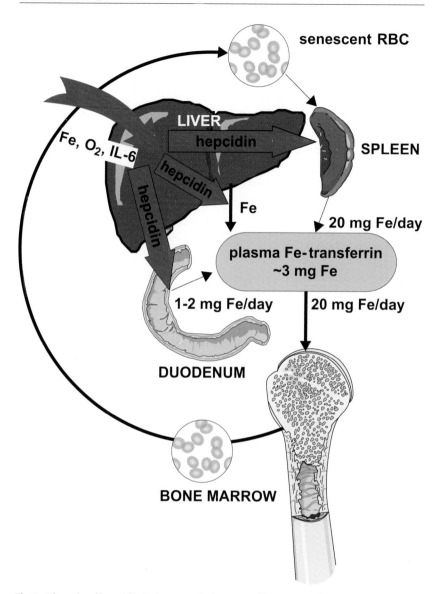

Fig. 3 The role of hepcidin in iron regulation. Hepcidin negatively regulates intestinal iron absorption, iron recycling by macrophages, iron release from hepatic stores and, during pregnancy, iron transfer in placenta. In turn, hepcidin secretion is regulated by iron stores, oxygenation and inflammatory signals, chiefly IL-6

iron is made available from diet and from the storage pool in macrophages and hepatocytes. The human possible hepcidin promoter contains several binding sites for hypoxia-inducible factor (HIF) and it is that the mechanism of hypoxic regulation of hepcidin will turn out to be transcriptional, via the common oxygen-sensing regulatory pathway (Safran and Kaelin 2003).

Hepcidin is also homeostatically regulated by iron loading. Dietary iron or transfusions increase hepcidin synthesis (Nemeth et al. 2004a); however, the molecular mechanisms of this regulation are still unclear. The hepcidin gene and mRNA lack any canonical binding sites for iron-regulatory proteins, and the study of patients with hemochromatosis whose hepcidin regulation is defective suggest the involvement of a previously uncharacterized pathway.

3.2
Regulation by Inflammation

Hepcidin, as an iron-regulatory hormone, constitutes an important link between host defense, inflammation, and iron metabolism. In mice, fish, and humans, hepcidin synthesis is markedly induced by infection and inflammation (Pigeon et al. 2001; Nemeth et al. 2003; Nicolas et al. 2002b; Shike et al. 2002). The cytokine IL-6 is an important inducer of hepcidin synthesis during inflammation (Nemeth et al. 2004a) since (a) IL-6, but not IL-1α or TNF-α, induces hepcidin synthesis in human hepatocytes, (b) anti-IL-6 antibodies block the induction of hepcidin mRNA in primary human hepatocytes treated with bacterial endotoxins LPS or peptidoglycan, (c) anti-IL-6 antibodies block the induction of hepcidin mRNA in human hepatocyte cell lines treated with supernatants of LPS- or peptidoglycan-stimulated macrophages, (d) IL-6 knockout mice (unlike control mice) fail to induce hepcidin in response to turpentine injection, and (e) in human volunteers, urinary hepcidin excretion is increased an average of 7.5-fold 2 h after IL-6 infusion. However, we and others (Lee et al. 2005; Lou et al. 2005) have observed that IL-6-deficient mice retain limited responsiveness to inflammatory stimuli other than turpentine, suggesting that alternative cytokines may also stimulate hepcidin production.

3.3
IL-6, Hepcidin, and Hypoferremia of Inflammation

During inflammation induced by subcutaneous injections of turpentine, normal mice show a marked decrease in serum iron (hypoferremia) (Nicolas et al. 2002b; Nemeth et al. 2004a). This response is completely ablated in hepcidin-deficient mice and in IL-6-deficient mice. In humans, the hepcidin increase elicited by IL-6 infusion is accompanied by a more than 30% decrease in serum iron and in transferrin saturation (Nemeth et al. 2004a). It therefore appears that the IL-6-hepcidin axis is important for this response and that hepcidin is the main mediator of hypoferremia of inflammation.

It is worthwhile to consider briefly how and why the hypoferremia develops so rapidly (within hours of the inflammatory stimulus). The plasma transferrin compartment contains roughly 3 mg of iron and functions as a transit compartment through which flows about 20 mg of iron each day, largely generated by recycling of senescent erythrocytes and mostly destined for the production of new erythrocytes. This means that plasma iron turns over every 3–4 h. If hepcidin could completely block iron recycling, this would result in a 25% drop in plasma iron in 1 h. It is hard to attribute any role to this effect other than that of an acute response aimed at eliminating microbial infections.

4
The Mechanism of Hepcidin Action

4.1
Cellular Iron Transport

Depending on the cell type, iron can be taken up by several distinct pathways. In the intestinal lumen, bioavailable iron in the diet is mostly present either in its ferric (Fe^{3+}) form or as heme. The uptake of ferric iron is mediated by a combination of a ferric reductase (duodenal cytochrome B), which reduces iron to its ferrous (Fe^{2+}) form and a ferrous iron transporter DMT1 that moves iron across the cell membrane (Hentze et al. 2004). The absorption of heme is less well characterized. In macrophages, recycling of iron from senescent erythrocytes starts with erythrocyte phagocytosis and lysis, followed by the extraction of iron from heme by heme oxygenase. Other cells import iron using transferrin receptors that capture and endocytose diferric transferrin. Under the low pH in endocytic vacuoles, iron dissociates from the transferrin-transferrin receptor complex. The transport of iron across vacuolar membranes of macrophages and other cells probably involves DMT1. In the cytoplasm, iron is stored bound to ferritin. The export of iron from different cell types (enterocytes, macrophages, hepatocytes, placental trophoblast) involves ferroportin, the sole known exporter of iron in vertebrates (Donovan et al. 2005), and it also requires a ferroxidase (hephaestin in enterocytes and ceruloplasmin in macrophages) to deliver ferric iron to transferrin.

4.2
Hepcidin Binds to Ferroportin and Induces Its Internalization from Cell Membranes

Recent studies (Nemeth et al. 2004b) indicate that hepcidin directly binds to ferroportin, and that the binding of hepcidin causes ferroportin to be internalized and degraded in lysosomes. The loss of ferroportin from cell membranes ablates cellular iron export. This posttranslational mechanism is sufficient to explain the regulation of iron absorption, since absorptive enterocytes only perform their function for 2 days before being shed from the tips of the villi into the intestinal lumen. Therefore, the transport of iron by ferroportin across the basolateral membrane determines whether the iron is delivered to plasma transferrin or removed from the body with the shed enterocytes.

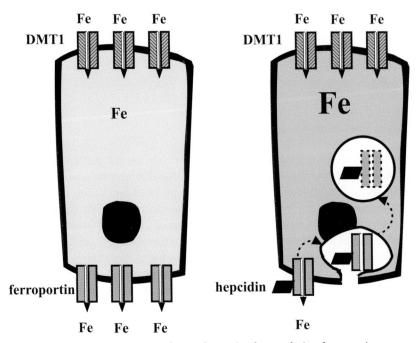

Fig. 4 Hepcidin controls intestinal iron absorption by regulating ferroportin expression on the basolateral membrane of enterocytes. The *left panel* illustrates iron deficiency, with hepcidin secretion suppressed and ferroportin strongly expressed on the basolateral membrane. Iron absorption is maximal. The *right panel* indicates iron excess or the effects of inflammation. The liver secretes hepcidin, which interacts with ferroportin molecules on the basolateral membrane, causing ferroportin to be endocytosed and degraded. Iron export from enterocytes is decreased and the cells fill with iron. In less than 2 days, the iron-filled enterocytes will be shed into the lumen of the intestine

When iron stores are adequate or high, the liver produces hepcidin, which circulates to the proximal small intestine. There, hepcidin causes ferroportin to be internalized, blocking the sole pathway for the transfer of iron from the enterocytes to plasma (Fig. 4). When iron stores are low, hepcidin production is suppressed, ferroportin molecules are displayed on basolateral membranes of enterocytes, transporting iron from the enterocyte cytoplasm to plasma transferrin. Similarly, the hepcidin–ferroportin interaction also explains how macrophage recycling of iron is regulated, and accounts for the characteristic finding of iron-containing macrophages in inflammatory states where hepcidin production is high. In the presence of hepcidin, ferroportin is internalized, iron export is blocked and iron is trapped within macrophages. To recover the ability of the cell to export iron, ferroportin must be resynthesized, a step whose rate could be independently regulated. There is evidence that ferroportin mRNA levels are also regulated by iron (McKie et al. 2000; Frazer et al. 2002). In addition to direct effects on ferroportin and, thus, iron export, hepcidin would be expected to have secondary effects on cellular iron intake. A block of iron export by hepcidin would result in a rise of intracellular iron and the suppression of synthesis of the DMT1 splice variant containing the iron-regulatory element (IRE) (Frazer et al. 2002, 2003), thus reducing iron uptake.

5
Anemia of Inflammation

Anemia of inflammation (AI) is a common consequence of chronic infections (Bush et al. 1956; Cartwright 1966) including HIV, tuberculosis, bacterial endocarditis, and osteomyelitis, but AI can also develop within days during sepsis (Jurado 1997; Corwin and Krantz 2000). AI is also seen in noninfectious generalized inflammatory disorders, including rheumatologic diseases, inflammatory bowel disease, multiple myeloma, and other malignancies. These anemias are characterized by decreased iron and iron-binding capacity (transferrin), increased ferritin, and the presence of iron in bone marrow macrophages, indicating impaired mobilization of iron from stores. Recent studies indicate that inflammation causes hypoferremia through the cytokine-mediated increase in hepcidin production (Nemeth et al. 2004a). Hypoferremia results from hepcidin-mediated inhibition of iron transport into plasma from macrophages involved in iron recycling, from iron stores in hepatocytes and from enterocytes that absorb dietary iron. Because most of the iron in the transferrin compartment is destined for the bone marrow, hypoferremia resulting from excess hepcidin should diminish the amount of

iron available for hemoglobin synthesis and erythrocyte production. Therefore, AI develops as a side effect of the hypoferremic response to infection and
inflammation. Indeed, clinical and experimental situations in which hepcidin
is overproduced are commonly associated with anemia. In addition to the
transgenic mice that overproduce hepcidin-1 and suffer from lethal anemia
(Nicolas et al. 2002a), severe anemia is seen in rare patients with liver tumors
that autonomously produce hepcidin (Weinstein et al. 2002). Also, increased
urinary hepcidin excretion has been observed in patients with hypoferremia
and anemia due to infections or inflammatory disorders (Nemeth et al. 2003).
It is interesting to note that IL-6 excess, which will stimulate hepcidin production, is also commonly associated with anemia. Thus transgenic mice that
overexpress IL-6 become anemic (Katsume et al. 2002), as do patients on experimental IL-6 treatment (van Gameren et al. 1994; Nieken et al. 1995), as well
as patients with Castleman's disease or multiple myeloma, both conditions
associated with IL-6 excess. Based on these observations, we have proposed
that the pathogenic (as well as host defense) cascade that produces anemia
of inflammation leads from IL-6 to hepcidin to hypoferremia, and then to
anemia of inflammation (Nemeth et al. 2004a). The specific contribution that
this response makes to innate immunity is not yet known, but iron excess iron
(even in the absence of clinical iron overload) has been linked to increased
susceptibility to tuberculosis in animals and humans (Gangaidzo et al. 2001;
Schaible et al. 2002; Cronje and Bornman 2005).

6
Hereditary Hemochromatosis

Humans and other mammals lack the capacity to excrete excess iron, so iron
balance is achieved almost exclusively by regulating iron uptake. Hereditary
hemochromatosis (Fleming and Sly 2002) is a group of disorders characterized
by dysregulated, excessive iron uptake from the diet, eventually leading to the
saturation of transferrin, ferritin, and other iron-binding proteins and the
deposition of iron in vital organs. Free iron is toxic, probably due to its ability
to catalyze the production of reactive oxygen products. Hemochromatosis
may progress to liver failure, cardiomyopathy, destruction of endocrine glands
and damage to joints. Moreover, patients with hemochromatosis appear to be
selectively susceptible to sepsis and liver abscess formation with *Vibrio* spp.
and *Yersinia* spp. (Ashrafian 2003), possibly due to the stimulating effects of
excess iron on the pathogenicity of these bacteria.

The most common form of hemochromatosis (type 1) is due to mutations in
the *HFE* gene, resulting in an autosomal recessive disorder of low penetrance

that clinically affects predominantly older men. Mutations in transferrin receptor 2, *TfR2*, are much rarer but cause a similar phenotype (type 3). The autosomal recessive diseases due to mutations in the hepcidin (antimicrobial peptide) gene *HAMP* or the hemojuvelin gene *HJV* cause a much more severe phenotype (juvenile hemochromatosis, type 2) affecting young men and women equally. Autosomal dominant hemochromatosis due to mutations in the ferroportin gene (type 4) differs from other hemochromatoses by usually causing early iron overload in the Kupffer cells (liver macrophages) rather than hepatocytes. Several models have been proposed to account for the role of these molecules in the normal and aberrant regulation of intestinal iron absorption (Pietrangelo 2004). It now appears that all major forms of hemochromatosis are either due to hepcidin deficiency or to ferroportin mutations that render the molecule resistant to the effects of hepcidin.

7
Summary

Hepcidin is the long-anticipated hormone responsible for the regulation of iron recycling and iron balance. It may have evolved from an typical amphipathic ancestral antimicrobial peptide. Hepcidin is produced in the liver, where its synthesis is stimulated by iron excess and inflammation, and inhibited by anemia and hypoxia. Hepcidin acts by binding to the cellular iron exporter ferroportin and inducing its internalization and degradation, thus trapping iron in enterocytes, macrophages, and hepatocytes. The net effect of hepcidin is the diminished absorption of dietary iron, sequestration of iron in macrophages and sequestration of iron in hepatic stores. Hepcidin excess may be the key pathogenic feature of anemia of inflammation, and hepcidin deficiency may be responsible for most cases of familial hemochromatosis. Hepcidin may participate in host defense by decreasing extracellular iron concentrations and slowing the growth of those microbes whose growth is highly dependent on this iron source.

References

Andrews NC (1999) Disorders of iron metabolism. N Engl J Med 341:1986–1995
Arnold RR, Cole MF, McGhee JR (1977) A bactericidal effect for human lactoferrin. Science 197:263–265
Arnold RR, Brewer M, Gauthier JJ (1980) Bactericidal activity of human lactoferrin: sensitivity of a variety of microorganisms. Infect Immun 28:893–898

Ashrafian H (2003) Hepcidin: the missing link between hemochromatosis and infections. Infect Immun 71:6693–6700

Bush JA, Ashenbrucker H, Cartwright GE, Wintrobe MM (1956) The anemia of infection. XX. The kinetics of iron metabolism in the anemia associated with chronic infection. J Clin Invest 35:89–97

Cartwright GE (1966) The anemia of chronic disorders. Semin Hematol 3:351–375

Corwin HL, Krantz SB (2000) Anemia of the critically ill: "acute" anemia of chronic disease. Crit Care Med 28:3098–3099

Cronje L, Bornman L (2005) Iron overload and tuberculosis: a case for iron chelation therapy. Int J Tuberc Lung Dis 9:2–9

Donovan A, Lima CA, Pinkus JL, Pinkus GS, Zon LI, Robine S, Andrews NC (2005) The iron exporter ferroportin/Slc40a1 is essential for iron homeostasis. Cell Metabolism 1:191–200

Douglas SE, Gallant JW, Liebscher RS, Dacanay A, Tsoi SC (2003) Identification and expression analysis of hepcidin-like antimicrobial peptides in bony fish. Dev Comp Immunol 27:589–601

Fleming RE, Sly WS (2002) Mechanisms of iron accumulation in hereditary hemochromatosis. Annu Rev Physiol 64:663–680

Flo TH, Smith KD, Sato S, Rodriguez DJ, Holmes MA, Strong RK, Akira S, Aderem A (2004) Lipocalin 2 mediates an innate immune response to bacterial infection by sequestrating iron. Nature 432:917–921

Fortier A, Min-Oo G, Forbes J, Lam-Yuk-Tseung S, Gros P (2005) Single gene effects in mouse models of host: pathogen interactions. J Leukoc Biol 77:868–877

Frazer DM, Wilkins SJ, Becker EM, Murphy TL, Vulpe CD, McKie AT, Anderson GJ (2003) A rapid decrease in the expression of DMT1 and Dcytb but not Ireg1 or hephaestin explains the mucosal block phenomenon of iron absorption. Gut 52:340–346

Frazer DM, Wilkins SJ, Becker EM, Vulpe CD, McKie AT, Trinder D, Anderson GJ (2002) Hepcidin expression inversely correlates with the expression of duodenal iron transporters and iron absorption in rats. Gastroenterology 123:835–844

Freireich EM, Miller A, Emerson CP, Ross JF (1957) The effect of inflammation on the utilization of erythrocyte and transferrin-bound radio-iron for red cell production. Blood 12:972

Gangaidzo IT, Moyo VM, Mvundura E, Aggrey G, Murphree NL, Khumalo H, Saungweme T, Kasvosve I, Gomo ZA, Rouault T, Boelaert JR, Gordeuk VR (2001) Association of pulmonary tuberculosis with increased dietary iron. J Infect Dis 184:936–939

Hentze MW, Muckenthaler MU, Andrews NC (2004) Balancing acts: molecular control of mammalian iron metabolism. Cell 117:285–297

Hunter HN, Fulton DB, Ganz T, Vogel HJ (2002) The solution structure of human hepcidin, a peptide hormone with antimicrobial activity that is involved in iron uptake and hereditary hemochromatosis. J Biol Chem 277:37597–37603

Jurado RL (1997) Iron, infections, and anemia of inflammation. Clin Infect Dis 25:888–895

Katsume A, Saito H, Yamada Y, Yorozu K, Ueda O, Akamatsu K, Nishimoto N, Kishi-moto T, Yoshizaki K, Ohsugi Y (2002) Anti-interleukin 6 (IL-6) receptor antibody suppresses Castleman's disease like symptoms emerged in IL-6 transgenic mice. Cytokine 20:304–311

Krause A, Neitz S, Magert HJ, Schulz A, Forssmann WG, Schulz-Knappe P, Ader-mann K (2000) LEAP-1, a novel highly disulfide-bonded human peptide, exhibits antimicrobial activity. FEBS Lett 480:147–150

Lee P, Peng H, Gelbart T, Wang L, Beutler E (2005) Regulation of hepcidin transcription by interleukin-1 and interleukin-6. Proc Natl Acad Sci U S A 102:1906–1910

Lou DQ, Nicolas G, Lesbordes JC, Viatte L, Grimber G, Szajnert MF, Kahn A, Vaulont S (2004) Functional differences between hepcidin 1 and 2 in transgenic mice. Blood 103:2816–2821

Lou DQ, Lesbordes JC, Nicolas G, Viatte L, Bennoun M, van Rooijen N, Kahn A, Renia L, Vaulont S (2005) Iron- and inflammation-induced hepcidin gene expression in mice is not mediated by Kupffer cells in vivo. Hepatology 41:1056–1064

McKie AT, Marciani P, Rolfs A, Brennan K, Wehr K, Barrow D, Miret S, Bomford A, Peters TJ, Farzaneh F, Hediger MA, Hentze MW, Simpson RJ (2000) A novel duodenal iron-regulated transporter, IREG1, implicated in the basolateral transfer of iron to the circulation. Mol Cell 5:299–309

Nemeth E, Valore EV, Territo M, Schiller G, Lichtenstein A, Ganz T (2003) Hepcidin, a putative mediator of anemia of inflammation, is a type II acute-phase protein. Blood 101:2461–2463

Nemeth E, Rivera S, Gabayan V, Keller C, Taudorf S, Pedersen BK, Ganz T (2004a) IL-6 mediates hypoferremia of inflammation by inducing the synthesis of the iron regulatory hormone hepcidin. J Clin Invest 113:1271–1276

Nemeth E, Tuttle MS, Powelson J, Vaughn MB, Donovan A, Ward DM, Ganz T, Kaplan J (2004b) Hepcidin regulates cellular iron efflux by binding to ferroportin and inducing its internalization. Science 306:2090–2093

Nicolas G, Bennoun M, Devaux I, Beaumont C, Grandchamp B, Kahn A, Vaulont S (2001) Lack of hepcidin gene expression and severe tissue iron overload in up-stream stimulatory factor 2 (USF2) knockout mice. Proc Natl Acad Sci U S A 98:8780–8785

Nicolas G, Bennoun M, Porteu A, Mativet S, Beaumont C, Grandchamp B, Sirito M, Sawadogo M, Kahn A, Vaulont S (2002a) Severe iron deficiency anemia in trans-genic mice expressing liver hepcidin. Proc Natl Acad Sci U S A 99:4596–4601

Nicolas G, Chauvet C, Viatte L, Danan JL, Bigard X, Devaux I, Beaumont C, Kahn A, Vaulont S (2002b) The gene encoding the iron regulatory peptide hepcidin is regulated by anemia, hypoxia, and inflammation. J Clin Invest 110:1037–1044

Nieken J, Mulder NH, Buter J, Vellenga E, Limburg PC, Piers DA, de Vries EG (1995) Re-combinant human interleukin-6 induces a rapid and reversible anemia in cancer patients. Blood 86:900–905

Park CH, Valore EV, Waring AJ, Ganz T (2001) Hepcidin, a urinary antimicrobial peptide synthesized in the liver. J Biol Chem 276:7806–7810

Pietrangelo A (2004) Hereditary hemochromatosis—a new look at an old disease. N Engl J Med 350:2383–2397

Pigeon C, Ilyin G, Courselaud B, Leroyer P, Turlin B, Brissot P, Loreal O (2001) A new mouse liver-specific gene, encoding a protein homologous to human antimicrobial peptide hepcidin, is overexpressed during iron overload. J Biol Chem 276:7811–7819

Rivera S, Liu L, Nemeth E, Gabayan V, Sorensen OE, Ganz T (2005) Hepcidin excess induces the sequestration of iron and exacerbates tumor-associated anemia. Blood 105:1797–1802

Roetto A, Papanikolaou G, Politou M, Alberti F, Girelli D, Christakis J, Loukopoulos D, Camaschella C (2003) Mutant antimicrobial peptide hepcidin is associated with severe juvenile hemochromatosis. Nat Genet 33:21–22

Safran M, Kaelin WG Jr (2003) HIF hydroxylation and the mammalian oxygen-sensing pathway. J Clin Invest 111:779–783

Schaible UE, Kaufmann SHE (2004) Iron and microbial infection. Nat Rev Micro 2:946–953

Schaible UE, Collins HL, Priem F, Kaufmann SH (2002) Correction of the iron overload defect in beta-2-microglobulin knockout mice by lactoferrin abolishes their increased susceptibility to tuberculosis. J Exp Med 196:1507–1513

Shike H, Lauth X, Westerman ME, Ostland VE, Carlberg JM, Van Olst JC, Shimizu C, Bulet P, Burns JC (2002) Bass hepcidin is a novel antimicrobial peptide induced by bacterial challenge. Eur J Biochem 269:2232–2237

Shike H, Shimizu C, Lauth X, Burns JC (2004) Organization and expression analysis of the zebrafish hepcidin gene, an antimicrobial peptide gene conserved among vertebrates. Dev Comp Immunol 28:747–754

Van Gameren MM, Willemse PH, Mulder NH, Limburg PC, Groen HJ, Vellenga E, de Vries EG (1994) Effects of recombinant human interleukin-6 in cancer patients: a phase I-II study. Blood 84:1434–1441

Weinstein DA, Roy CN, Fleming MD, Loda MF, Wolfsdorf JI, Andrews NC (2002) Inappropriate expression of hepcidin is associated with iron refractory anemia: implications for the anemia of chronic disease. Blood 100:3376–3381

CTMI (2006) 306:199–230

Innate Host Defense of Human Vaginal and Cervical Mucosae

A. M. Cole (✉)

Department of Molecular Biology and Microbiology, Biomolecular Science Center,
University of Central Florida, 4000 Central Florida Blvd., Bldg. 20, Rm. 136,
Orlando, FL 32816–2364, USA
acole@mail.ucf.edu

Abstract Host defense responses of the human female genital tract mucosa to pathogenic microbes and viruses are mediated in part by the release of antimicrobial substances into the overlying mucosal fluid. While host defense has long been considered a prominent function of vaginal and cervical mucosae, evidence that cationic antimicrobial peptides and proteins have fundamental roles in the innate host defense of this tissue has only recently become available. This chapter explores elements of the physical and chemical defense barriers of the cervicovaginal mucosa, which protect against infections of the lower genital tract. Cationic antimicrobial and antiviral polypeptide components of cervicovaginal fluid are discussed in detail, with special emphasis placed on the defensin family of peptides as well as polypeptides that are active against viruses such as HIV-1. The reader should be cognizant that each polypeptide by itself does not provide complete protection of the genital tract. On the contrary, the abundance and multiplicity of antimicrobial peptides and proteins suggest protection of the cervicovaginal mucosa may be best realized from the aggregate effector molecules.

Abbreviations

AIDS	Acquired immunodeficiency syndrome
BV	Bacterial vaginosis
CCR5	CC chemokine receptor 5
CXCR4	CXC chemokine receptor 4
DEFT	θ-Defensin gene
gp120	Envelope glycoprotein of 120 kDa
gp41	Envelope glycoprotein of 41 kDa
H_2O_2	Hydrogen peroxide
HAART	Highly-active anti-retroviral therapy
HBD	Human β-defensin
HD	Human (α-)defensin
HIV	Human immunodeficiency virus
HNP	Human neutrophil peptide
HSV	Herpes simplex virus
IgG	Immunoglobulin G
LPS	Lipopolysaccharides
MALT	Mucosal-associated lymphoreticular tissue
NMR	Nuclear magnetic resonance
PG	Protegrin
RTD	Rhesus θ-defensin
sIgA	Secretory immunoglobulin A
SLPI	Secretory leukocyte protease inhibitor
STD	Sexually transmitted disease
TLR	Toll-like receptor

1
Overview

1.1
Introduction to Antimicrobial Defense of the Female Genital Tract

The female genital tract is divided into three major compartments: the lower genital tract (vagina and ectocervix), the endocervix, and the upper genital tract (endometrium and fallopian tubes). The ectocervix is the vaginal portion of the cervix that is structurally and immunologically similar to the vagina, and both are nonsterile. In healthy women, the upper genital tract is considered a sterile environment under normal conditions. The endocervix serves as an interface between the upper and lower tracts, and its sterility is likely related to the phase of the menstrual cycle (Quayle 2002). The lower genital tract is lined with nonkeratinized, stratified, squamous epithelium (150–200 μm in thickness) that sits atop a lamina propria and vascular submucosa. Proliferation and maturation of the epithelial surface is under hormonal control, with maximum thickness occurring during peak circulating levels of estrogen (Patton et al. 2000). Apical epithelial cells are covered by a glycocalyx layer that hydrates the luminal surface and may act to prevent microbial attachment.

Commensal microbes blanket the surface of the vagina and ectocervix, and are thought to play a particularly important role in host defense of the lower genital tract. *Lactobacillus* spp. are the predominant resident bacteria of healthy cervicovaginal mucosa. Lactobacilli metabolize glycogen released by vaginal epithelial cells into lactic acid, which in turn renders vaginal fluid acidic (normally ~pH 3.5–4.7). Lactic acid and the low pH of vaginal fluid have been shown to exert selective antimicrobial activity against nonresident species of bacteria while sparing the commensal microbiota (Valore et al. 2002). The presence of H_2O_2-producing lactobacilli has also been associated with homeostasis of the normal vaginal mucosa (Hillier 1999). The biological concentrations of H_2O_2 that have been measured in vaginal fluid are toxic to many nonresident microbiota, which suggests these lactobacilli may be more beneficial to the host than lactobacilli that do not generate H_2O_2. Another mechanism by which lactobacilli suppress the proliferation of nonadvantageous bacterial species (e.g., *Gardnerella vaginalis*, *Enterococcus* spp., and *Escherichia coli*) is through the production of broad-spectrum antimicrobial peptides called bacteriocins (Aroutcheva et al. 2001; Reid 2001). Conversely, some clinicians have ascribed vaginal diseases associated with increased levels of lactobacilli (e.g., lactobacillus vaginosis and cytolytic vaginosis; Cibley and Cibley 1991; Horowitz et al. 1994). While these finding are controversial, they underscore the importance of maintaining concentrations of vaginal lactobacilli in equilibrium. Likewise, if homeostasis of the cervicovaginal mucosa is altered, for example as a result of bacterial vaginosis, perturbation of the resident vaginal flora leads to an increase in pH, an increase in nonresident or underrepresented microbes, and the ascension of pathogens into the sterile upper genital tract. In total, the extreme environment of the vaginal mucosa may explain its intrinsic resistance to colonization by pathogens.

Although the vaginal mucosa serves as the portal for entry of infectious and sexually transmitted diseases (STDs), knowledge about immune mechanisms of the vagina in health and disease continues to be fragmentary. An organized mucosal-associated lymphoreticular tissue (MALT) is situated within the lamina propria of the cervix, and consists primarily of T lymphocytes and monocytes/macrophages. Langerhans (dendritic) cells are also abundant components of both the vaginal and cervical mucosa, and play important roles in presenting antigen to T lymphocytes in regional lymph nodes and activating the adaptive immune response. Mucosal secretions overlay the entire genital tract and form a mechanical and chemical barrier, constituting a first-line host defense against microbial and viral invaders. Among the components of the secretions, mediators of adaptive immunity include secretory immunoglobulin A (sIgA) and IgG, produced by plasma cells adjacent to submucosal glands (Meredith et al. 1989). Immunoglobulins are thought to act in part by providing beacons for the recruitment and activation of innate effector cells, preventing the attachment and invasion of pathogenic organisms, and acting as opsonins for pathogen destruction. Mediators of innate vaginal host defense include substances that sequester microbial nutrients, selectively disrupt bacterial cell walls and membranes, or act as decoys for the attachment of microbes. As the physical barrier of the cervical mucus plug is absent during ovulation and menstruation, the importance of the innate host defense molecules of cervical mucus and vaginal secretions is emphasized. Host-derived antimicrobial peptides and proteins are principal innate effector molecules of the cervicovaginal mucosa, and are the central subjects hereinafter.

1.2
Infectious Diseases of the Lower Female Genital Tract

This chapter is focused on innate host defense functions of the *lower* genital tract in women. In the sections to follow, general background is provided for three of the most frequent conditions of the lower genital tract: vulvovaginal candidiasis, bacterial vaginosis/vaginitis, and trichomoniasis. HIV-1 will also be discussed given the recent attention that the innate host response to HIV-1 infection has received in the literature, and the causal relationship that vaginosis and vaginitis have with HIV-1 infection. *Neisseria gonorrhoeae*, *Chlamydia trachomatis*, and other causative agents of pelvic inflammatory disease and other *upper* genital tract sequelae are not discussed in the current chapter. Instead, the reader is referred to several recent reviews on these pathogens (Shafer et al. 2001; Edwards and Apicella 2004; Ghosh et al. 2004; Wang et al. 2005; Brunham and Rey-Ladino 2005).

1.2.1
Vulvovaginal Candidiasis (Candida Vaginitis)

Vulvovaginal candidiasis is an opportunistic yeast infection that affects nearly three-quarters of otherwise healthy US women of reproductive age (Sobel 1992, 2004). The most common causative agent of yeast infections is *Candida albicans*; however, a significant proportion of yeast vaginitis results from other species of *Candida* (Edwards 2004). Cutaneous vulvar candidiasis involves superficial yeast invasion of the epithelium (Edwards 2004), which leads to inflammation and a characteristic itching of the affected tissues. Non-*albicans* candidiasis often produces painful burning sensations and can lead to more severe disease. Factors that contribute to the development of vulvovaginal candidiasis include elevated hormone levels and the use of immunosuppressive agents such as corticosteroids. Intravaginal or systemic antibiotics increase the risk for *Candida* infections due to adverse effects on protective commensal microbiota of the vagina.

Recurrent vulvovaginal candidiasis, classified as three or more episodes per year, manifests in 5%–10% of cases. A majority of recurrent *Candida* infections are a result of predisposing factors (e.g., diabetes mellitus, hormone replacement therapy; reviewed in Fidel 2004). Although the secondary form of recurrent vulvovaginal candidiasis is idiopathic, dysregulation of adaptive and/or innate immunity have been proposed. Until recently, T helper cell type 1 (Th1)-mediated cellular immunity was considered the primary host defense mechanism against infections by *Candida* spp. (reviewed in Fidel 2005). By contrast, Th2 responses were associated with susceptibility to *Candida* infection, and a shift from Th1 to Th2 immunity was thought to trigger episodes of candidiasis. Recently, innate immune factors have been purported to be effectors of anti-*Candida* host defense. Innate anti-*Candida* host defense of vaginal epithelium requires contact with a cell-associated carbohydrate moiety and has been shown for the human, mouse, and rhesus macaque (Fidel 2005). A study by Barousse and colleagues revealed that while vaginal epithelial cells from healthy donors prevented *C. albicans* colonization, epithelia from women with recurrent vulvovaginal candidiasis presented markedly reduced activity against *C. albicans* (Barousse et al. 2001). Although cationic antimicrobial peptides such as the β-defensins are putative candidates, further study will be necessary to determine which dysregulated components contribute to epithelial colonization by *Candida*.

An enticing study by Fidel and colleagues monitored host defense responses in humans following controlled intravaginal challenge with live *Candida* (Fidel et al. 2004). Approximately 10% of women with no prior history of vaginitis became symptomatic following *Candida* challenge. By contrast, in women with documented prior *Candida* vaginitis, nearly 50% became symptomatic with the remainder asymptomatically carrying the yeast (Fidel et al. 2004). Protection was not associated with an inflammatory response. Conversely, symptomatic infection resulted in vaginal infiltration of neutrophils and a high vaginal burden, suggesting that neutrophils were responsible for the infection-related symptoms (Fidel et al. 2004). This paradigm shift should spark novel research into innate factors responsible for the susceptibility and resistance to *Candida* vaginitis.

1.2.2
Bacterial Vaginosis/Bacterial Vaginitis

Bacterial vaginosis (BV) is the most prevalent polymicrobial condition of women of reproductive age. BV is caused by overgrowth of nonresident bacteria in the vagina, or often resident bacteria that under normal conditions are underrepresented members of the endogenous vaginal microbiota. BV is associated with a number of serious obstetrical and gynecological complications, including preterm birth and increased susceptibility to STDs (Taha et al. 1998, 1999; Cherpes et al. 2003; Wiesenfeld et al. 2003). Most women with BV have either a markedly reduced concentration of vaginal lactobacilli or are completely devoid of lactobacilli (Agnew and Hillier 1995), both of which result in an elevation in vaginal fluid pH. It is uncertain whether vaginal pH contributes to the initiation of BV or becomes elevated as BV becomes established. Additionally, while nearly a quarter of women with BV are populated with H_2O_2-producing lactobacilli (Agnew and Hillier 1995), the number of these lactobacilli or amount of H_2O_2 liberated are presumably insufficient to prevent colonization by non-lactobacilli microbes.

BV is frequently asymptomatic, and although excessive vaginal discharge can help with diagnosis of BV, it is imperative to substantiate this finding with microscopic evaluation of the fluid. Noninvasive tests of BV involve positive tests for 16S rRNA/DNA or Gram stains of vaginal fluid, an absence of leukocytes on vaginal wet mounts, and the presence of Clue cells—epithelial cells blanketed with coccobacilli and short rod bacteria (e.g., *G. vaginalis*, *Bacteroides* spp., *Mobiluncus* spp.) that offer a "clue" to the diagnosis of BV. Amino acid metabolism by anaerobic bacteria releases volatile amines as the vaginal pH rises (reviewed in McGregor and French 2000). The characteristic fishy odor in BV is a result of amines reacting with alkaline solutions, such as semen or experimentally added potassium hydroxide (Whiff test). Bacterial vaginitis is often coupled with BV, perhaps because of the similarity in name. However, it is important to note that bacterial vaginitis is relatively uncommon, and is typically caused by outgrowth of a specific bacterium resulting in symptomatic inflammation. The most common etiological agents of bacterial vaginitis are Streptococcus, α-hemolytic *Streptococcus*, and *Staphylococcus aureus*. Unfortunately, for both bacterial vaginosis and bacterial vaginitis, there is a paucity of studies related to mechanisms of innate host defense dysregulation.

1.2.3
Trichomoniasis

The parabasalid protozoan *Trichomonas vaginalis* is the causative agent of trichomoniasis. While trichomoniasis is one of the most common STDs in the US that causes vaginitis, it is not a reportable infection and is given scant attention from public health programs. The World Health Organization (WHO) estimates that trichomoniasis accounts for nearly half of the curable infections worldwide (reviewed in Cates 1999; Schwebke and Burgess 2004). As with many other STDs, there is a firm association between the rate of *T. vaginalis* infection and preterm birth, low birth weight, and incidence of HIV-1 infection. The last is postulated to occur as a consequence of macroscopic or microscopic tears in the vaginal mucosa caused by *T. vaginalis* infection (Brown 2004). Symptoms usually present in the vagina; however, the urethra is often affected (Edwards 2004). Trichomoniasis results in an elevation of the vaginal fluid pH greater than 4.7, adversely affecting lactobacilli and other protective commensal microbiota. Frothy yellow or green vaginal discharge is a hallmark of *Trichomonas* infection, with 10% of cases presenting profuse discharge. *T. vaginalis* infection can cause significant inflammation and influx of leukocytes into the female genital tract; neutrophils are abundant on wet mounts, and thus other types of purulent vaginitis are often confused with trichomoniasis. Whereas research for this disease has been focused on acquired immunity, recent studies revealed that *T. vaginalis* activates cells through toll-like receptor 4 (TLR-4) (Zariffard et al. 2004) and inhibits proinflammatory cytokine production in macrophages by suppressing NF-κB activation (Chang et al. 2004), suggesting an innate host defense component to this infection.

1.2.4
HIV-1

In 2004, an estimated 4.9 million people were newly infected with HIV, 39.4 million were living with HIV/AIDS, and 3.1 million lost their lives due to AIDS (UNAIDS/WHO 2004). Heterosexual transmission is the most common route of spread of HIV-1 worldwide, with more than half of the reported cases of HIV/AIDS being women. HIV prevalence is still increasing in several regions of the world, including sub-Saharan Africa and South and Southeast Asia, which together harbor 85% of persons living with HIV/AIDS. Annually, over 300 million people are infected with STDs (UNAIDS/WHO 2004), and the presence of an STD can dramatically increase the risk of HIV-1 infection. In women, both ulcerative and nonulcerative STDs have been shown to increase the risk of HIV-1 transmission approximately three- to fivefold, with the incidence of HIV-1 infection reaching eightfold for syphilis or herpes (Wasserheit 1992).

Mucosal surfaces of the vagina and cervix are the portals for heterosexual transmission of HIV-1 and therefore play a fundamental role in the pathogenesis of primary infection. In recent years, substantial progress has been made toward the understanding of anti-HIV immunity in the vagina. While many studies have focused on the adaptive, clonal responses and antibody production, there has been an increased interest in innate immune mechanisms of antiviral host defense. Vaginal subepithelial stromal tissues are densely populated with dendritic cells (DCs), monocytes/macrophages and T lymphocytes that express CD4 and the HIV-1 co-receptors, CXCR4 and/or CCR5 (Zhang et al. 1998; Patterson et al. 1998). Mechanisms whereby HIV-1 journeys across the mucosal epithelia are not completely understood, but may directly involve the vaginal epithelial cells. Once the virus reaches the lamina propria, it can either directly infect macrophages or T lymphocytes or adhere to (or infect) dendritic cells, whose traffic to the regional lymph nodes converts them into sites of vigorous viral replication (Grouard and Clark 1997; Steinman and Inaba 1999). Mucosal effector cells and molecules that modulate steps in the process of viral tissue invasion, cell adsorption, or cell entry should reduce the incidence of HIV-1 transmission. In the next section, we investigate the putative role of antimicrobial peptides and proteins in anti-HIV-1 host defense of the cervicovaginal mucosa.

2
Antimicrobial Peptides and Proteins of Cervicovaginal Fluid

2.1
Overview of Antimicrobial Peptides and Proteins of Mucosal Secretions

As early as the turn of the twentieth century, its was revealed that human tissues could produce substances inimical to microorganisms. It was later realized that the antimicrobial properties of mucosal tissues were largely due to the contributions of bioactive polypeptides. In 1922, Alexander Fleming attributed the antimicrobial properties of human nasal secretions to an enzyme he named lysozyme (Fleming 1922). Since the discovery of lysozyme, many other antimicrobial components of mucosal secretions have been identified, including defensins, lactoferrin, secretory leukocyte protease inhibitor (SLPI), secretory phospholipase A2, cathelicidins, calprotectin, chemokines, and histatins. Lysozyme and lactoferrin, two polypeptides expressed at appreciable concentrations in many mucosae, are stored in and secreted from serous cells in submucosal glands, and in inflamed secretions are also released from neutrophils. SLPI, a serine protease inhibitor, is also present at potentially antimicrobial concentrations (Hiemstra et al. 1996). The human β-defensin-1 (HBD1) is an antimicrobial peptide constitutively expressed at low levels at many mucosal sites, while a structurally similar peptide, human β-defensin-2 (HBD2), is predominantly induced at sites of inflammation (McCray and Bentley 1997; Singh et al. 1998). The α-defensin, HNP1 (Ganz et al. 1985) is likely a degranulation product of neutrophils that extravasate into mucosal tissues during infection. Other antimicrobial (poly)peptides that are present in most mucosal secretions include the cathelicidin LL-37 (Bals et al. 1998; Bals 2000) and secretory phospholipase A2 (Aho et al. 1997; Cole et al. 1999).

Most antimicrobial peptides and proteins are broad-spectrum microbicides that target Gram-positive and Gram-negative bacteria as well as fungi and some enveloped viruses (Schonwetter et al. 1995; Lehrer et al. 1999; Ganz 1999). How do these substances kill microbes or inhibit their multiplication? This section will be necessarily short as other chapters in this book delve deeper into the mechanism(s) of action of antimicrobial peptides and proteins. Nonetheless, some are enzymes that disrupt essential microbial structures, as exemplified by lysozyme and its ability to hydrolyze peptidoglycan, a key structural component of the bacterial cell wall. Others bind essential nutrients, denying them to microbes. Lactoferrin, an iron-binding protein, is representative of this host defense strategy. Still others, especially small antimicrobial proteins and peptides, act by disrupting microbial or viral membranes (Gazit et al. 1996; Mangoni et al. 1996; Ludtke et al. 1996; Shai 1999; Matsuzaki 2001; Weiss et al. 2002; Chen et al. 2003). Although antimicrobial peptides can be evolutionarily and structurally diverse, common properties typically include amphipathicity (spatial separation of polar and nonpolar residues) and cationicity (a net positive charge at physiological pH), attributes that assist peptide binding and insertion into microbial membranes. An emerging concept is that certain antimicrobial peptides have multiple roles in host defense that supersede their bacteriostatic or bactericidal capacities.

The sections that follow discuss antimicrobial proteins and peptides that are present in the cervicovaginal mucus, and detail the roles of these molecules that together contribute to mucosal innate host defense of the vagina and cervix. For certain molecules, the scant research in host defense of the female genital tract has been supplemented by parallel studies in other tissues. The antimicrobial peptides and proteins are listed in descending order of their aggregate prevalence as quantified in human vaginal fluid and cervical mucus plugs (Valore et al. 2002; Hein et al. 2002).

2.2
Calprotectin

Calprotectin, also called leukocyte protein L1, calgranulin A/calgranulin B, MRP-8/MRP-14 and S100A8/S100A9, is an abundant heterodimeric protein of neutrophils, monocytes, and keratinocytes, and can account for approximately 40% of the cytoplasmic protein content of human neutrophils (Brandtzaeg et al. 1995). It is present in normal vaginal fluid at approximately 34 µg ml^{-1}, and can be found at higher concentrations in inflammatory fluids, where it may be released from neutrophils by an as yet incompletely characterized secretory mechanism. At realistic concentrations, calprotectin inhibits the growth of fungi and bacteria in vitro (Steinbakk et al. 1990), in part by sequestering zinc (Clohessy and Golden 1996; Sohnle et al. 2000). Recently developed knockout mice deficient in MRP-14, the larger component of the calprotectin heterodimer, lacked an obvious phenotype, suggesting that the contribution of this protein complex to host defense may be subtle (Manitz et al. 2003; Hobbs et al. 2003), or that other molecules are compensatory. Another study revealed that MRP-8, the smaller component of calprotectin, was an activator of HIV-1 in latently infected monocytic cells (Hashemi et al. 2001). Furthermore, samples of cervicovaginal lavage fluid that contained higher concentrations of MRP-8 also exhibited greater activation of HIV in latently infected monocytic cells. While the mechanism is not known, insights into the role of mucosal polypeptides that increase the probability of transmission and infection of HIV-1 would be critical in the development of effective antiretroviral treatments and preventatives.

2.3
Secretory Leukocyte Protease Inhibitor (SLPI)

Secretory leukocyte protease inhibitor (SLPI) is a 12-kDa protein found in epithelial secretions but also produced by macrophages. It is especially abundant in cervical mucus plugs of pregnancy, where it reaches concentrations in the 100–1,000 µg ml^{-1} range (Hein et al. 2002). SLPI is approximately 10- to 100-fold less abundant in normal cervical and vaginal secretions. SLPI is weakly antibacterial and antifungal (Hiemstra et al. 1996; Tomee et al. 1998) but adequate SLPI concentrations and appropriate conditions for its activity may not be present in most mucosal fluids aside from cervical mucus plugs. The high molar ratio of SLPI:elastase in the cervical mucus plug suggests that anti-inflammatory and antimicrobial properties of SLPI play an important role during normal pregnancy (Helmig et al. 1995). In a South African study, increased SLPI concentrations in vaginal fluid were correlated with reduced rates of perinatal HIV-1 transmission (Pillay et al. 2001). The same study reported no significant correlation between lysozyme or lactoferrin and HIV-1 transmission. Draper and colleagues reported similarly that SLPI was decreased in women with BV and STDs, and suggested that decreased SLPI levels may increase susceptibility to HIV-1 infection (Draper et al. 2000).

A belief that HIV-1 is rarely transmitted through salivary secretions provided impetus to study endogenous inhibitors of HIV-1 produced by the oral mucosa. Several laboratories have reported an inhibitory activity of SLPI against HIV-1 (reviewed in Shugars et al. 1999). SLPI ($>$100 ng ml^{-1}) was shown to block HIV-1 internalization by inhibition of viral entry or capsid uncoating (McNeely et al. 1995, 1997). Conversely, studies from another group revealed that while saliva inhibited HIV-1 infection, up to 1,000 μg ml^{-1} of recombinant SLPI was not active against HIV-1 in vitro (Turpin et al. 1996), leaving the antiretroviral activity of SLPI in question. The potential anti-HIV-1 activity of SLPI might be best realized in concert with the other antimicrobial peptides and proteins of cervicovaginal fluids.

2.4
Lysozyme

Lysozyme (muramidase), is a 14.6-kDa enzyme whose bacteriolytic properties derive from cleavage of the bond linking N-acetyl muramic acid and N-acetyl-D-glucosamine residues in peptidoglycan, the macromolecule that confers shape and rigidity on cell walls of bacteria. Lysozyme also displays nonenzymatic antimicrobial activity that stems from its highly cationic nature, which presumably allows it to disrupt microbial membranes or activate autolytic bacterial enzymes (Wecke et al. 1982; Laible and Germaine 1985; Ibrahim et al. 2001). However, lysozyme alone is inactive against many pathogens, either because it cannot reach the peptidoglycan layer or because the structure of peptidoglycan is modified to make it less sensitive to the enzymatic action of lysozyme. Such resistance may be overcome by synergy with other antimicrobial components of epithelial secretions (Ellison and Giehl 1991; Singh et al. 2000).

The concentration of lysozyme in vaginal fluid is relatively low (~13 μg ml^{-1}; Valore et al. 2002); however, it is a particularly plentiful protein of cervical mucus plugs, reaching roughly 1 mg/g of plug (Hein et al. 2002). Lower lysozyme and lactoferrin levels in cervical mucus during preterm labor were shown to be correlated with chorioamnionitis, suggesting a reduction of this defense mechanism (Chimura et al. 1993). Lee-Huang and colleagues reported that lysozyme purified from human milk and neutrophils reduced HIV-1 production in chronically infected T lymphocytes and monocytes (Lee-Huang et al. 1999). While the antiretroviral activity of lysozyme was modest at best, lysozyme in the cervical mucus plug might reach concentrations sufficient for in vivo activity against HIV-1. The group has since discovered that octadecapeptide and nonapeptide motifs in the C-terminal region of lysozyme confer anti-HIV-1 activity at EC_{50} concentrations in the low-to-mid nanomolar range (Lee-Huang et al. 2005).

2.5
Lactoferrin

Lactoferrin, an 80-kDa homolog of the iron-carrier protein transferrin, is an abundant constituent of certain epithelia as well as the specific granules of neutrophils. Although in cervical mucus plugs the concentration is approximately 100 μg g^{-1} of plug (Hein et al. 2002), the concentration of lactoferrin in vaginal fluid is comparably low (~1 μg ml^{-1}; Valore et al. 2002). Lactoferrin exerts antibacterial activity through at least two distinct mechanisms (Bullen and Armstrong 1979; Arnold et al. 1982): by sequestration of iron, an essential element for microbes, and directly, independent of iron. Unlike transferrin, lactoferrin binds iron under acidic conditions in areas of microbial infection and inflammation. The iron-independent activity may be due to a highly cationic segment of lactoferrin (Chapple et al. 1998) that may disrupt microbial membranes. There is evidence that lysozyme and lactoferrin act synergistically against Gram-negative bacteria (Ellison and Giehl 1991). In the digestive tract, lactoferrin from milk may be cleaved by pepsin or other proteases to free an antimicrobial peptide segment, called lactoferricin (Bellamy et al. 1992), although this peptide has not yet been identified in urogenital tissues. Lactoferrin was reported to directly and indirectly inhibit several viruses, including HIV-1, HSV, and human cytomegalovirus (Hasegawa et al. 1994; Harmsen et al. 1995; Swart et al. 1998; Andersen et al. 2001). The mechanism of its action against HIV-1 occurred in an early phase of infection, most likely during adsorption of virus to the target cell (Puddu et al. 1998). In vitro, lactoferrin bound strongly to the V3 loop of HIV-1 gp120, causing inhibition of viral fusion and entry (Swart et al. 1998).

2.6
Defensins

Defensins comprise the most widely studied family of antimicrobial peptides, and thus will occupy a significant segment of this review. Nearly 100 different defensins are expressed by the leukocytes and epithelial cells of various birds and mammals (Lehrer et al. 1999; Ganz 1999, 2001, 2005; Hughes 1999; Ouellette and Bevins 2001; Lehrer and Ganz 2002; Bevins 2004; Selsted and Ouellette 2005). These peptides can be divided into three subfamilies: α-defensins, β-defensins, and θ-defensins (Tang et al. 1999). All of these defensins derive from an ancestral gene that existed before the evolutionary divergence of reptiles and birds (Liu et al. 1997; Zhao et al. 2001), contain six cysteines, and have largely β-sheet structures that are stabilized by three intramolecular disulfide bonds. Most defensins are active against a broad range of microbes and viruses. Although α- and β-defensins differ in the spacing and connectivity of their cysteines, they have similar topology (Zimmermann et al. 1995). By contrast, despite originating from α-defensin precursors, θ-defensins are structurally unique. The following subsections describe human α-, β-, and θ-defensins in the context of mucosal immunity.

2.6.1
α-Defensins

α-Defensins are one of the principal antibacterial components of human neutrophil granules (Ganz et al. 1985) and Paneth cells of the small intestinal crypts of Lieberkühn (Jones and Bevins 1992, 1993). The α-defensins have 29- to 35-residue β-sheet structures with six cysteines that form three intramolecular disulfide bonds by pairing cysteine residues 1 and 6, 2 and 4, and 3 and 5 (Ganz et al. 1985). Four α-defensins (HNP1–4) are stored in azurophil granules of the neutrophil, where they constitute roughly 30% of the granules' total protein (Ganz and Lehrer 1997). At much lower concentrations, three α-defensins (HNP1–3) have also been identified as components of primary T lymphocytes, natural killer (NK) cells, and monocytic cell lines (Agerberth et al. 2000). Neutrophil defensins are synthesized in bone marrow promyelocytes as 93- to 94-amino acid preprodefensins that undergo proteolytic processing by sequential removal of the signal sequence and propiece, so that the azurophil granules contain the mature active peptides almost exclusively (Valore and Ganz 1992; Harwig et al. 1992). During phagocytosis of bacteria by neutrophils, defensins are discharged into phagocytic vacuoles of neutrophils where they reach very high concentrations (roughly milligram per milliliter) (Joiner et al. 1989), and presumably contribute to the microbicidal milieu in the phagocytic vacuoles. It was discovered of late that specific granules of neutrophils contain unprocessed prodefensins (proHNPs), which undergo a significant level of constitutive exocytosis (Faurschou et al. 2005). The release of prodefensins, and their presumed extracellular activation, may comprise the extracellular arm of HNP host defense.

Similar to proHNPs of the neutrophil, the two α-defensins of human Paneth cells, human defensin (HD)-5 and HD6, are stored in secretory granules as inactive precursor peptides (Jones and Bevins 1992, 1993; Ouellette 2005).

Paneth cell-derived trypsin activates HD5 in the small intestine by proteolytic cleavage (Ghosh et al. 2002). Paneth cell granules discharge their defensins into the narrow crypt lumen by cholinergic stimuli associated with digestion and by the presence of microbes (Satoh 1988; Satoh et al. 1992; Qu et al. 1996; Ayabe et al. 2000). High concentrations of defensins, calculated in the milligram per milliliter range, are attained in the crypt but even in the intestinal lumen defensin concentrations are sufficient to exert antimicrobial effects (Ayabe et al. 2000).

α-Defensins are important antimicrobial components of the female genital tract. In a comprehensive study by Quayle and colleagues, HD5 was immunolocalized to the upper half of stratified epithelium of ectocervix and vagina, but secretion was not apparent (Quayle et al. 1998). However, in the endocervix, HD5 was immunolocalized to both the apically oriented granules of the columnar epithelium as well as the surface of the tissue, which implies that this tissue may be the source of HD5 in cervical mucus (Quayle et al. 1998). In cervicovaginal lavage fluid, the expression of HD5 was found to be maximal during the secretory phase of the menstrual cycle (Quayle et al. 1998), suggesting hormonal modulation of HD5 expression. Surprisingly, there was little change in the concentration of most other antimicrobial peptides and proteins in cervicovaginal fluid during the menstrual cycle (Valore et al. 2002). Healthy, nongravid women expressed low levels of HNP1–3 in their vaginal fluid ($\sim$0.35 μg ml^{-1}; Valore et al. 2002), which may indicate that subclinical inflammation is a constitutive mechanism of host defense. Women with intermediate and frank BV during pregnancy expressed up to 200-fold greater concentrations of α-defensin (Balu et al. 2002). High levels of HNPs in vaginal fluid at 24–29 weeks gestation might also be a predictor of preterm birth (Balu et al. 2003).

In environmentally favorable conditions, most defensins at micromolar concentrations display a remarkably broad spectrum of antimicrobial activity against Gram-positive and Gram-negative bacteria, fungi, and yeast (Selsted et al. 1984, 1993; Ganz et al. 1985; Selsted and Harwig 1987; Alcouloumre et al. 1993), mycobacteria (Ogata et al. 1992), and protozoa (Aley et al. 1994). In the presence of components of plasma and mucus, including salt, other monovalent and especially divalent cations (Lehrer et al. 1985b, 1988), and defensin-binding proteins (Panyutich and Ganz 1991; Panyutich et al. 1994, 1995), higher concentrations of defensins are required for microbial killing. Microbes also become more resistant to defensins when microbial metabolism is inhibited by the lack of nutrients or the presence of metabolic toxins (Lehrer et al. 1988). Countering these inhibitory influences, defensins reach high concentrations in several common biological contexts, including the phagocytic vacuoles of neutrophils (Joiner et al. 1989; Ganz and Lehrer

1994), small intestinal crypts (Ayabe et al. 2000), and intercellular spaces between keratinocytes (Shi et al. 1999; Oren et al. 2003). Nevertheless, even lower concentrations of defensins, such as those that prevail in the lumen of the small intestine, appear to have significant effects on intestinal flora and pathogens (Salzman et al. 2003), as illustrated by the ability of transgenic HD5 to protect mice against intestinal infection with *Salmonella typhimurium*.

Whereas the ability of defensins to kill bacteria and fungi is well documented, comparably few studies have examined their antiviral effects. Certain enveloped viruses can be inactivated by α-defensins purified from rabbit (Lehrer et al. 1985a) or human (Daher et al. 1986) leukocytes. These viruses include HSV-1, HSV-2, cytomegalovirus, vesicular stomatitis virus, and influenza virus A/WSN. In contrast, two nonenveloped viruses, echovirus type 11 and reovirus type 3, were not inactivated. The inhibitory effect of HNP1 on HSV-1 was temperature-dependent, and inhibited by the presence of serum. Rabbit α-defensin NP-1 was recently shown to be an entry inhibitor, preventing an early step in HSV-1 infection—either viral adherence or subsequent membrane fusion (Sinha et al. 2003). Stimulated by the global epidemic of AIDS, studies have shown that α-defensins are active against HIV-1. Nakashima and colleagues reported that rabbit, rat, and guinea pig defensins can inhibit HIV-1-induced cytopathogenicity of a CD4[+] human T cell line (Nakashima et al. 1993). Another group noted structural and functional similarities between the looped motifs of α-defensins and peptides derived from HIV-1 gp41 (Monell and Strand 1994). It was proposed that these loops are requisite for viral fusion and infectivity. Several studies extended these findings to confirm that human α-defensins are active against HIV-1 (Zhang et al. 2002), and that the mechanism of action is twofold: direct inactivation of HIV-1 virions and inhibition of HIV-1 replication through intracellular interference with PKC activity (Mackewicz et al. 2003; Chang et al. 2003; Chang et al. 2005).

2.6.2
β-Defensins

β-Defensins were first identified in the tracheal epithelium (Diamond et al. 1991) and granulocytes (Selsted et al. 1993) of cows. Although β-defensins are slightly larger than their α-defensin counterparts and have a different disulfide connectivity (pairing of cysteines 1–5, 2–4, and 3–6), the shapes of their peptide backbones are very similar to α-defensins (Zimmermann et al. 1995). Human β-defensins are predominantly produced by epithelia (Bensch et al. 1995; Harder et al. 1997, 2001; Lehrer and Ganz 2002). Although some defensins are expressed constitutively, others are produced in response to infection. HBD2–4 are induced upon simulation by inflammatory mediators from mast cells, epithelial cells, keratinocytes, monocytes, and dendritic cells (Duits et al. 2002; Niyonsaba et al. 2002). Human β-defensin-1 (HBD1) is expressed constitutively by barrier epithelial cells and keratinocytes, and may provide a ubiquitous protective blanket at those sites. Recently, several groups have shown that HBD1 is also up-regulated by proinflammatory processes in monocytes/macrophages and dendritic cells (Duits et al. 2002), as well as in human skin by transactivation of epidermal growth factor (Sorensen et al. 2005). Among tissues presenting the highest expression levels of HBD1 are those of the urogenital tract: kidney, vagina, endocervix, and ectocervix (Valore et al. 1998). The biological concentration of HBD1 in the female genital tract could be sufficient to contribute to antimicrobial host defense (Hein et al. 2002).

While microbes can induce the host's production of antimicrobial peptides, epithelial cells produce relatively scant amounts of β-defensins in *direct* response to a microbe, and thus require other receptors and signals for their activation. Among molecules that function as detectors of infection, Toll-like receptors (TLRs) recognize conserved molecular patterns that are components of microbial and viral pathogens. Examples include TLR4, which recognizes lipopolysaccharide (LPS) from Gram-negative bacteria, and TLR2, which is triggered by peptidoglycan of Gram-positive bacteria. Pattern recognition is an important component of the host's innate ability to discriminate between self and non-self. Early studies of LPS induction of HBD2 in human tracheobronchial epithelium revealed that HBD2 utilized CD14, which may have subsequently formed a complex with a Toll-like receptor (TLR) to induce NF-κB nuclear translocation and antimicrobial peptide gene activation (Becker et al. 2000). Similarly, mouse β-defensin-2, a peptide with similarities to human β-defensins, was also found to be LPS-inducible (Morrison et al. 1999). In other studies, the expression of HBD2 in airway epithelia was found to be mediated through TLR2, activated by either lipoteichoic acid, a component of peptidoglycan (X. Wang et al. 2003), or by lipopeptide derived from mycobacteria (Hertz et al. 2003). Likewise, HEK293 cells transfected with TLR2 also produced HBD2 in response to bacterial lipoproteins (Birchler et al. 2001). Live rhinovirus-16, a respiratory virus for the common cold, but not UV-inactivated virus, increased expression of HBD2 and HBD3 in bronchial epithelial cells (Duits et al. 2003). The investigators reported that these cells expressed TLR3, a receptor that recognizes dsRNA, and that dsRNA alone was sufficient to cause an up-regulation of β-defensins, implicating TLR3 in rhinoviral infection. Conversely, β-defensins can also act as endogenous ligands for TLRs. In mice, murine β-defensin-2 (MBD-2) acts as a ligand for TLR4 on immature dendritic cells, inducing dendritic cell maturation and the up-regulation of costimulatory molecules (Biragyn et al. 2002). Indeed,

C. albicans, LPS, and peptidoglycan have all been shown to up-regulate HBD2 in vaginal epithelial cells (Pivarcsi et al. 2005). *T. vaginalis* activates cells through TLR4 (Zariffard et al. 2004) and inhibits proinflammatory cytokine production in macrophages by suppressing NF-κB activation (Chang et al. 2004), suggesting an innate host defense component to this infection.

β-defensins were also recently shown to inhibit HIV-1 infection. Quinones-Mateu and colleagues (Quinones-Mateu et al. 2003) revealed that HIV-1 induced HBD2 and HBD3 expression in human oral epithelial cells. Moreover, expression of β-defensins down-modulated the expression of CXCR4, a chemokine (co)receptor required for HIV-1 entry. In vaginal fluid from healthy donors, HBD2 was present at less than 1 μg ml^{-1} (Valore et al. 2002), concentrations below those required for anti-HIV-1 activity in vitro. HBD2 levels were below the level of detection in cervical mucus plugs (Hein et al. 2002). It is not known whether HIV-1 can induce HBD2 or HBD3 in vaginal or cervical epithelium.

2.6.3
θ-Defensins

θ-Defensins were first isolated from the leukocytes and bone marrow of the rhesus monkey, *Macacca mulatta* (Tang et al. 1999; Leonova et al. 2001; Tran et al. 2002). These arginine-rich, circular 18-residue peptides arose from two precursor peptides, each of which contributes nine residues (three cysteines) to the mature θ-defensin (Tang et al. 1999; Leonova et al. 2001; Tran et al. 2002). They have a β-sheet conformation (Trabi et al. 2001) and represent the first truly circular peptides of vertebrate origin. Their intracellular cyclization occurs via binary, posttranslational head-to-tail ligations utilizing unknown cellular machinery. These splices join the backbones of the two nonapeptide precursors to make the resulting peptide circular, and oxidation of the six cysteines form an internal tri-disulfide ladder that renders it tetracyclic. Because the incorporated nine residue segments can be derived from identical or different genes, the θ-defensin family can generate molecular diversity through a unique posttranslational mechanism (Leonova et al. 2001; Tran et al. 2002). Whereas α- and β-defensin peptides are produced by both human and non-human primates (Tang et al. 1999), humans are not known to produce mature θ-defensin peptides (Nguyen et al. 2003). Thus, although humans express mRNA encoded by θ-defensin (DEFT) genes, the termination codon within the signal sequence prematurely arrests its translation (Cole et al. 2002). DEFT genes cluster on chromosome 8p23 (Nguyen et al. 2003), a locus that also contains the human α- and β-defensin genes (Liu et al. 1997).

Having cloned the human θ-defensin cDNA and also identified several DEFT genes by surveying the human genome database, it was a logical step to deduce the sequences of the nine residue precursor elements that were incorporated into the θ-defensins produced by our primate ancestors before the mutational event silenced translation of the precursors. These ancestral peptides, called retrocyclins, were then recreated by solid phase synthesis (Cole et al. 2002). While θ-defensin peptides are modest antibacterials (Tang et al. 1999; Leonova et al. 2001; Cole et al. 2002; Tran et al. 2002), they remarkably protected primary and transformed T lymphocytes from in vitro infection by both X4 and R5 strains of HIV-1 (Cole et al. 2002), and are more active in vitro than other α-defensins tested (unpublished data). Studies with pseudotyped luciferase reporter viruses indicated that retrocyclins act by preventing the entry of HIV-1 (Münk et al. 2003). Furthermore, the antiviral properties of retrocyclin and other θ-defensins might be intimately linked to an ability to bind carbohydrate epitopes displayed by viral and cell-surface glycoproteins involved in viral entry (W. Wang et al. 2003). Together, these studies suggest that retrocyclins inhibit HIV-1 entry either by blocking a postbinding conformational change in the viral envelope glycoproteins gp120 or gp41 that occurs on the pathway to fusion, preventing the insertion of the gp41 amino terminal fusion peptide into the target cell membrane, or preventing gp41 hairpin conversion. RTD-1, the rhesus monkey homolog of retrocyclin, has been shown to bind parallel to lipid membrane surfaces (Weiss et al. 2002), forming stable lipid–peptide domains that may lead to membrane perturbation (Abuja et al. 2004). Similar binding by retrocyclin may alter cell surface properties in a manner that prevents the fusion of cellular and viral membranes. Others have reported that the antiviral activities of retrocyclins and other θ-defensins extend to herpes simplex virus (HSV), also acting through the inhibition of HSV adhesion and entry (Sinha et al. 2003; Yasin et al. 2004).

X4 strains utilize the chemokine co-receptor CXCR4 along with the primary receptor CD4 to aid entry into receptive cells, while R5 strains instead utilize the co-receptor CCR5. As retrocyclins protect against both tropisms, they may be better therapeutic candidates than peptides that protect only against one. Indeed, as R5 strains are purported to be the primary tropism involved during sexual transmission of HIV-1, it will be important to design and develop compounds that are active against this HIV-1 tropism. Human retrocyclins were silenced after the orangutan and hominid lineages had diverged, approximately 7.5–10 million years ago (Nguyen et al. 2003). Given retrocyclin's antiviral properties against HIV-1, the evolutionary loss of retrocyclin may have contributed to HIV-1 susceptibility in modern humans. If so, then topical administration of retrocyclins to the vaginal mucosa could restore antiretroviral effector molecules that were never relinquished by many nonhuman primates.

2.7
Cathelicidin (LL-37)

Cathelicidins are a family of antimicrobial peptides with each member having a common precursor cathelin sequence similar the thiol protease inhibitor cystatin, yet containing a C-terminal active region of very diverse sequence. While animals such as pigs and cows are endowed with multiple cathelicidin genes (Zanetti et al. 1995), humans have only one, called hCAP-18/LL-37, so named because its precursor is a human cationic protein of 18 kD, whose mature form has 37 amino acids with two leucines at its N-terminus (Agerberth et al. 1995; Cowland et al. 1995; Larrick et al. 1995). LL-37 has a broad spectrum of activity against bacteria and fungi (Turner et al. 1998), even in the presence of salt in complex media. Cathelicidins are abundant components of mammalian neutrophils but are also found in seminal plasma and other secretions. In the neutrophil, cathelicidins are found in granule types that are predominantly destined for exocytosis (Cowland et al. 1995), suggesting that this peptide family may have an important role in host defense of extracellular spaces. In the female genital tract, LL-37 is present in vaginal fluid at 1 µg ml^{-1} or less (Valore et al. 2002), and in the inflamed and dysplastic cervix, hCAP18 is expressed in the superficial epithelial layers in a band-like pattern (Frohm et al. 1999). Following sexual intercourse, processing of hCAP18 into the antimicrobial peptide ALL-38 (LL-37 with an N-terminal alanine residue) was mediated by the prostate-derived protease gastricsin present in seminal plasma (but not vaginal fluid) (Sorensen et al. 2003). Furthermore, processing of hCAP18 to ALL-38 was pH dependent, while gastricsin was inactive at neutral or basic pH, processing occurred in the presence of acidic vaginal fluid ex vivo, or low pH buffer in vitro (Sorensen et al. 2003). The enzymatic activation of hCAP18 represents a novel mechanism to prevent infection following sexual intercourse (Sorensen et al. 2003).

2.8
Histones

As early as 1893, Vaughan et al. (1893) and Kossel (1896) realized that complexes of nucleic acids and histones were active against Gram-positive bacteria. This finding was later confirmed by several groups (Miller et al. 1942; Negroni and Fischer 1944; Hirsch 1958; Park et al. 2000; Wang et al. 2002). Histones, and the related salt-like nucleins called protamines, are well endowed with the two basic amino acids, arginine and lysine. As with many cationic antimicrobial (poly)peptides, the antibacterial activities of histones are likely related to electrostatic attraction to anionic microbial surfaces and subsequent cell permeabilization. In a classic paper, Hirsch described properties of histones that are remarkably similar to those of cationic antimicrobial (poly)peptides discovered decades later: histones are more active at acidic pH, and are sensitive to salt, serum, and polyanionic substances (Hirsch 1958).

Similar to other mucosal secretions, vaginal fluids from healthy individuals contain histone proteins. Valore and colleagues reported that vaginal fluid from one donor contained histones (Valore et al. 2002). Using a proteomic approach, we have identified histones and histone fragments in all vaginal fluid samples tested (unpublished observations). Why are histones present in mucosal secretions? Do these proteins confer an active function, or are they just byproducts of cellular decay? Until recently, the latter was most plausible explanation. However, a very recent study by Brinkmann and colleagues provided an alternative mechanism behind the presence of extracellular histones (Brinkmann et al. 2004). They reported that activated neutrophils release NETs (neutrophil extracellular traps), long elaborations of chromatin and neutrophil elastase, that are independent of apoptosis or necrosis. NETs bound and inactivated both Gram-positive and Gram-negative bacteria and prevented their dispersal. Moreover, NETs were abundant in experimental dysentery and in spontaneous human appendicitis. It is not known whether inflammatory cells in the cervicovaginal mucosa elaborate NETs and their associated histones as a host defense mechanism, or if histones are released simply as a result of cellular damage.

3
Conclusion

There has been mounting interest in exploring the role of antimicrobial peptides and proteins in the innate host defense of vaginal and cervical mucosae. These mucosal surfaces are blanketed by a mixture of antimicrobial peptides and proteins perhaps because such a mixture provides a broad antimicrobial spectrum and decreases the likelihood that microbial resistance will emerge. Fortunately, little emerging resistance to antimicrobial peptides has been reported even upon repeat exposure of microbes under laboratory conditions (reviewed in Zasloff 2002), and though microbes must have frequently encountered antimicrobial peptides throughout evolution, these natural antibiotics still remain highly effective in vitro against many microbial targets.

It is curious that humans have retained multiple versions of the same family of antimicrobial peptides, as demonstrated with the defensins. One conjecture is that defensin sequences have diverged and generated different spectra of activity that are partially overlapping. Thus, microbes and viruses that are resistant to one type of defensin might still preserve sensitivity to another. Because of overlapping functions of defensins, pathogens encountering a number of defensin types would be less likely to subvert the host's defenses. Molecular diversity is well demonstrated by the antimicrobial activity of β-defensins. *S. aureus* is strikingly insensitive to HBD1 and HBD2, yet even multidrug-resistant strains are susceptible to HBD3 (Harder et al. 2001). Interestingly, low concentrations of HBD3 are also active at physiological concentrations of salt, a property that is not exhibited by HBD1 or HBD2. Another possibility is that by interacting in an additive or synergistic manner, individual peptides and proteins with modest intrinsic antimicrobial activity could produce considerably greater effects when combined. The abundance and multiplicity of antimicrobial peptides and proteins in the fluid covering the vagina and cervix suggests that overall protection may be best achieved from the combined effects of multiple host-derived molecules.

Acknowledgements This work was supported by grants R01AI052017, R01HL070876, R01AI060753, and U19AI065430 from the National Institutes of Health. The author expresses his appreciation to Dr. Ole Sørensen for critically reviewing this chapter.

References

Abuja PM, Zenz A, Trabi M, Craik DJ, Lohner K (2004) The cyclic antimicrobial peptide RTD-1 induces stabilized lipid-peptide domains more efficiently than its open-chain analogue. FEBS Lett 566:301–306

Agerberth B, Gunne H, Odeberg J, Kogner P, Boman HG, Gudmundsson GH (1995) FALL-39, a putative human peptide antibiotic, is cysteine-free and expressed in bone marrow and testis. Proc Natl Acad Sci U S A 92:195–199

Agerberth B, Charo J, Werr J, Olsson B, Idali F, Lindbom L, Kiessling R, Jornvall H, Wigzell H, Gudmundsson GH (2000) The human antimicrobial and chemotactic peptides LL-37 and alpha-defensins are expressed by specific lymphocyte and monocyte populations. Blood 96:3086–3093

Agnew KJ, Hillier SL (1995) The effect of treatment regimens for vaginitis and cervicitis on vaginal colonization by lactobacilli. Sex Transm Dis 22:269–273

Aho HJ, Grenman R, Sipila J, Peuravuori H, Hartikainen J, Nevalainen TJ (1997) Group II phospholipase A2 in nasal fluid, mucosa and paranasal sinuses. Acta Otolaryngol (Stockh) 117:860–863

Alcouloumre MS, Ghannoum MA, Ibrahim AS, Selsted ME, Edwards JEJ (1993) Fungicidal properties of defensin NP-1 and activity against Cryptococcus neoformans in vitro. Antimicrob Agents Chemother 37:2628–2632

Aley SB, Zimmerman M, Hetsko M, Selsted ME, Gillin FD (1994) Killing of Giardia lamblia by cryptdins and cationic neutrophil peptides. Infect Immun 62:5397–5403

Andersen JH, Osbakk SA, Vorland LH, Traavik T, Gutteberg TJ (2001) Lactoferrin and cyclic lactoferricin inhibit the entry of human cytomegalovirus into human fibroblasts. Antiviral Res 51:141–149

Arnold RR, Russell JE, Champion WJ, Brewer M, Gauthier JJ (1982) Bactericidal activity of human lactoferrin: differentiation from the stasis of iron deprivation. Infect Immun 35:792–799

Aroutcheva A, Gariti D, Simon M, Shott S, Faro J, Simoes JA, Gurguis A, Faro S (2001) Defense factors of vaginal lactobacilli. Am J Obstet Gynecol 185:375–379

Ayabe T, Satchell DP, Wilson CL, Parks WC, Selsted ME, Ouellette AJ (2000) Secretion of microbicidal α-defensins by intestinal Paneth cells in response to bacteria. Nat Immunol 1:113–118

Bals R (2000) Epithelial antimicrobial peptides in host defense against infection. Respir Res 1:141–150

Bals R, Wang X, Zasloff M, Wilson JM (1998) The peptide antibiotic LL-37/hCAP-18 is expressed in epithelia of the human lung where it has broad antimicrobial activity at the airway surface. Proc Natl Acad Sci U S A 95:9541–9546

Balu RB, Savitz DA, Ananth CV, Hartmann KE, Miller WC, Thorp JM, Heine RP (2002) Bacterial vaginosis and vaginal fluid defensins during pregnancy. Am J Obstet Gynecol 187:1267–1271

Balu RB, Savitz DA, Ananth CV, Hartmann KE, Miller WC, Thorp JM, Heine RP (2003) Bacterial vaginosis, vaginal fluid neutrophil defensins, and preterm birth. Obstet Gynecol 101:862–868

Barousse MM, Steele C, Dunlap K, Espinosa T, Boikov D, Sobel JD, Fidel PL Jr (2001) Growth inhibition of Candida albicans by human vaginal epithelial cells. J Infect Dis 184:1489–1493

Becker MN, Diamond G, Verghese MW, Randell SH (2000) CD14-dependent lipopolysaccharide-induced beta-defensin-2 expression in human tracheo-bronchial epithelium. J Biol Chem 275:29731–29736

Bellamy W, Takase M, Wakabayashi H, Kawase K, Tomita M (1992) Antibacterial spectrum of lactoferricin B, a potent bactericidal peptide derived from the N-terminal region of bovine lactoferrin. J Appl Bacteriol 73:472–479

Bensch KW, Raida M, Magert HJ, Schulz-Knappe P, Forssmann WG (1995) hBD-1: a novel beta-defensin from human plasma. FEBS Lett 368:331–335

Bevins CL (2004) The Paneth cell and the innate immune response. Curr Opin Gastroenterol 20:572–580

Biragyn A, Ruffini PA, Leifer CA, Klyushnenkova E, Shakhov A, Chertov O, Shirakawa AK, Farber JM, Segal DM, Oppenheim JJ, Kwak LW (2002) Toll-like receptor 4-dependent activation of dendritic cells by beta-defensin 2. Science 298:1025–1029

Birchler T, Seibl R, Buchner K, Loeliger S, Seger R, Hossle JP, Aguzzi A, Lauener RP (2001) Human Toll-like receptor 2 mediates induction of the antimicrobial peptide human beta-defensin 2 in response to bacterial lipoprotein. Eur J Immunol 31:3131–3137

Brandtzaeg P, Gabrielsen TO, Dale I, Muller F, Steinbakk M, Fagerhol MK (1995) The leucocyte protein L1 (calprotectin): a putative nonspecific defence factor at epithelial surfaces. Adv Exp Med Biol 371A:201–206

Brinkmann V, Reichard U, Goosmann C, Fauler B, Uhlemann Y, Weiss DS, Weinrauch Y, Zychlinsky A (2004) Neutrophil extracellular traps kill bacteria. Science 303:1532–1535

Brown D Jr (2004) Clinical variability of bacterial vaginosis and trichomoniasis. J Reprod Med 49:781–786

Brunham RC, Rey-Ladino J (2005) Immunology of Chlamydia infection: implications for a Chlamydia trachomatis vaccine. Nat Rev Immunol 5:149–161

Bullen JJ, Armstrong JA (1979) The role of lactoferrin in the bactericidal function of polymorphonuclear leucocytes. Immunology 36:781–791

Cates W Jr (1999) Estimates of the incidence and prevalence of sexually transmitted diseases in the United States. American Social Health Association Panel. Sex Transm Dis 26: S2–S7

Chang JH, Ryang YS, Morio T, Lee SK, Chang EJ (2004) Trichomonas vaginalis inhibits proinflammatory cytokine production in macrophages by suppressing NF-kappaB activation. Mol Cells 18:177–185

Chang TL, Francois F, Mosoian A, Klotman ME (2003) CAF-mediated human immunodeficiency virus (HIV) type 1 transcriptional inhibition is distinct from alpha-defensin-1 HIV inhibition. J Virol 77:6777–6784

Chang TL, Vargas J Jr, DelPortillo A, Klotman ME (2005) Dual role of alpha-defensin-1 in anti-HIV-1 innate immunity. J Clin Invest 115:765–773

Chapple DS, Joannou CL, Mason DJ, Shergill JK, Odell EW, Gant V, Evans RW (1998) A helical region on human lactoferrin. Its role in antibacterial pathogenesis. Adv Exp Med Biol 443:215–220

Chen FY, Lee MT, Huang HW (2003) Evidence for membrane thinning effect as the mechanism for peptide-induced pore formation. Biophys J 84:3751–3758

Cherpes TL, Meyn LA, Krohn MA, Lurie JG, Hillier SL (2003) Association between acquisition of herpes simplex virus type 2 in women and bacterial vaginosis. Clin Infect Dis 37:319–325

Chimura T, Hirayama T, Takase M (1993) Lysozyme in cervical mucus of patients with chorioamnionitis. Jpn J Antibiot 46:726–729

Cibley LJ, Cibley LJ (1991) Cytolytic vaginosis. Am J Obstet Gynecol 165:1245–1249

Clohessy PA, Golden BE (1996) The mechanism of calprotectin's candidastatic activity appears to involve zinc chelation. Biochem Soc Trans 24:309S

Cole AM, Dewan P, Ganz T (1999) Innate antimicrobial activity of nasal secretions. Infect Immun 67:3267–3275

Cole AM, Hong T, Boo LM, Nguyen T, Zhao C, Bristol G, Zack JA, Waring AJ, Yang OO, Lehrer RI (2002) Retrocyclin: a primate peptide that protects cells from infection by T- and M-tropic strains of HIV-1. Proc Natl Acad Sci U S A 99:1813–1818

Cowland JB, Johnsen AH, Borregaard N (1995) hCAP-18, a cathelin/pro-bactenecin-like protein of human neutrophil specific granules. FEBS Lett 368:173–176

Daher KA, Selsted ME, Lehrer RI (1986) Direct inactivation of viruses by human granulocyte defensins. J Virol 60:1068–1074

Diamond G, Zasloff M, Eck H, Brasseur M, Maloy WL, Bevins CL (1991) Tracheal antimicrobial peptide, a cysteine-rich peptide from mammalian tracheal mucosa: peptide isolation and cloning of a cDNA. Proc Natl Acad Sci U S A 88:3952–3956

Draper DL, Landers DV, Krohn MA, Hillier SL, Wiesenfeld HC, Heine RP (2000) Levels of vaginal secretory leukocyte protease inhibitor are decreased in women with lower reproductive tract infections. Am J Obstet Gynecol 183:1243–1248

Duits LA, Ravensbergen B, Rademaker M, Hiemstra PS, Nibbering PH (2002) Expression of beta-defensin 1 and 2 mRNA by human monocytes, macrophages and dendritic cells. Immunology 106:517–525

Duits LA, Nibbering PH, van Strijen E, Vos JB, Mannesse-Lazeroms SP, van Sterkenburg MA, Hiemstra PS (2003) Rhinovirus increases human beta-defensin-2 and -3 mRNA expression in cultured bronchial epithelial cells. FEMS Immunol Med Microbiol 38:59–64

Edwards JL, Apicella MA (2004) The molecular mechanisms used by Neisseria gonorrhoeae to initiate infection differ between men and women. Clin Microbiol Rev 17:965–981

Edwards L (2004) The diagnosis and treatment of infectious vaginitis. Dermatol Ther 17:102–110

Ellison RT3, Giehl TJ (1991) Killing of Gram-negative bacteria by lactoferrin and lysozyme. J Clin Invest 88:1080–1091

Faurschou M, Kamp S, Cowland JB, Udby L, Johnsen AH, Calafat J, Winther H, Borregaard N (2005) Prodefensins are matrix proteins of specific granules in human neutrophils. J Leukoc Biol 78:785–793

Fidel PL Jr (2004) History and new insights into host defense against vaginal candidiasis. Trends Microbiol 12:220–227

Fidel PL Jr (2005) Immunity in vaginal candidiasis. Curr Opin Infect Dis 18:107–111

Fidel PL Jr, Barousse M, Espinosa T, Ficarra M, Sturtevant J, Martin DH, Quayle AJ, Dunlap K (2004) An intravaginal live Candida challenge in humans leads to new hypotheses for the immunopathogenesis of vulvovaginal candidiasis. Infect Immun 72:2939–2946

Fleming A (1922) On a remarkable bacteriolytic element found in tissues and secretions. Proc R Soc Lond [Biol] 93:306–317

Frohm NM, Sandstedt B, Sorensen O, Weber G, Borregaard N, Stahle-Backdahl M (1999) The human cationic antimicrobial protein (hCAP18), a peptide antibiotic, is widely expressed in human squamous epithelia and colocalizes with interleukin-6. Infect Immun 67:2561–2566

Ganz T (1999) Defensins and host defense. Science 286:420–421

Ganz T (2001) Defensins in the urinary tract and other tissues. J Infect Dis 183 [Suppl 1]:S41–S42

Ganz T (2005) Defensins and other antimicrobial peptides: a historical perspective and an update. Comb Chem High Throughput Screen 8:209–217

Ganz T, Lehrer RI (1994) Defensins. Curr Opin Immunol 6:584–589

Ganz T, Lehrer RI (1997) Antimicrobial peptides of leukocytes. Curr Opin Hematol 4:53–58

Ganz T, Selsted ME, Szklarek D, Harwig SS, Daher K, Bainton DF, Lehrer RI (1985) Defensins. Natural peptide antibiotics of human neutrophils. J Clin Invest 76:1427–1435

Gazit E, Miller IR, Biggin PC, Sansom MS, Shai Y (1996) Structure and orientation of the mammalian antibacterial peptide cecropin P1 within phospholipid membranes. J Mol Biol 258:860–870

Ghosh D, Porter E, Shen B, Lee SK, Wilk D, Drazba J, Yadav SP, Crabb JW, Ganz T, Bevins CL (2002) Paneth cell trypsin is the processing enzyme for human defensin-5. Nat Immunol 3:583–590

Ghosh SK, Zhao J, Philogene MC, Alzaharani A, Rane S, Banerjee A (2004) Pathogenic consequences of Neisseria gonorrhoeae pilin glycan variation. Microbes Infect 6:693–701

Grouard G, Clark EA (1997) Role of dendritic and follicular dendritic cells in HIV infection and pathogenesis. Curr Opin Immunol 9:563–567

Harder J, Bartels J, Christophers E, Schroeder J-M (1997) A peptide antibiotic from human skin. Nature 387:861–862

Harder J, Bartels J, Christophers E, Schroder JM (2001) Isolation and characterization of human beta -defensin-3, a novel human inducible peptide antibiotic. J Biol Chem 276:5707–5713

Harmsen MC, Swart PJ, de Bethune MP, Pauwels R, De Clercq E, The TH, Meijer DK (1995) Antiviral effects of plasma and milk proteins: lactoferrin shows potent activity against both human immunodeficiency virus and human cytomegalovirus replication in vitro. J Infect Dis 172:380–388

Harwig SS, Park AS, Lehrer RI (1992) Characterization of defensin precursors in mature human neutrophils. Blood 79:1532–1537

Hasegawa K, Motsuchi W, Tanaka S, Dosako S (1994) Inhibition with lactoferrin of in vitro infection with human herpes virus. Jpn J Med Sci Biol 47:73–85

Hashemi FB, Mollenhauer J, Madsen LD, Sha BE, Nacken W, Moyer MB, Sorg C, Spear GT (2001) Myeloid-related protein (MRP)-8 from cervico-vaginal secretions activates HIV replication. AIDS 15:441–449

Hein M, Valore EV, Helmig RB, Uldbjerg N, Ganz T (2002) Antimicrobial factors in the cervical mucus plug. Am J Obstet Gynecol 187:137–144

Helmig R, Uldbjerg N, Ohlsson K (1995) Secretory leukocyte protease inhibitor in the cervical mucus and in the fetal membranes. Eur J Obstet Gynecol Reprod Biol 59:95–101

Hertz CJ, Wu Q, Porter EM, Zhang YJ, Weismuller KH, Godowski PJ, Ganz T, Randell SH, Modlin RL (2003) Activation of Toll-like receptor 2 on human tracheo-bronchial epithelial cells induces the antimicrobial peptide human beta defensin-2. J Immunol 171:6820–6826

Hiemstra PS, Maassen RJ, Stolk J, Heinzel-Wieland R, Steffens GJ, Dijkman JH (1996) Antibacterial activity of antileukoprotease. Infect Immun 64:4520–4524

Hillier SL (1999) Normal vaginal flora. In: Holmes KK et al (eds) Sexually transmitted diseases. McGraw-Hill, New York, pp 191–204

Hirsch JG (1958) Bactericidal action of histone. J Exp Med 108:925–944

Hobbs JA, May R, Tanousis K, McNeill E, Mathies M, Gebhardt C, Henderson R, Robinson MJ, Hogg N (2003) Myeloid cell function in MRP-14 (S100A9) null mice. Mol Cell Biol 23:2564–2576

Horowitz BJ, Mardh PA, Nagy E, Rank EL (1994) Vaginal lactobacillosis. Am J Obstet Gynecol 170:857–861

Hughes AL (1999) Evolutionary diversification of the mammalian defensins. Cell Mol Life Sci 56:94–103

Ibrahim HR, Matsuzaki T, Aoki T (2001) Genetic evidence that antibacterial activity of lysozyme is independent of its catalytic function. FEBS Lett 506:27–32

Joiner KA, Ganz T, Albert J, Rotrosen D (1989) The opsonizing ligand on Salmonella typhimurium influences incorporation of specific, but not azurophil, granule constituents into neutrophil phagosomes. J Cell Biol 109:2771–2782

Jones DE, Bevins CL (1992) Paneth cells of the human small intestine express an antimicrobial peptide gene. J Biol Chem 267:23216–23225

Jones DE, Bevins CL (1993) Defensin-6 mRNA in human Paneth cells: implications for antimicrobial peptides in host defense of the human bowel. FEBS Lett 315:187–192

Kossel A (1896) Uber die basichen Stoffe des Zellkerns. Z Ohysiol Chem 22:176–190

Laible NJ, Germaine GR (1985) Bactericidal activity of human lysozyme, muramidase-inactive lysozyme, and cationic polypeptides against Streptococcus sanguis and Streptococcus faecalis: inhibition by chitin oligosaccharides. Infect Immun 48:720–728

Larrick JW, Hirata M, Balint RF, Lee J, Zhong J, Wright SC (1995) Human CAP18: a novel antimicrobial lipopolysaccharide-binding protein. Infect Immun 63:1291–1297

Lee-Huang S, Huang PL, Sun Y, Huang PL, Kung HF, Blithe DL, Chen HC (1999) Lysozyme and RNases as anti-HIV components in beta-core preparations of human chorionic gonadotropin. Proc Natl Acad Sci U S A 96:2678–2681

Lee-Huang S, Maiorov V, Huang PL, Ng A, Lee HC, Chang YT, Kallenbach N, Huang PL, Chen HC (2005) Structural and functional modeling of human lysozyme reveals a unique nonapeptide, HL9, with anti-HIV activity. Biochemistry 44:4648–4655

Lehrer RI, Ganz T (2002) Defensins of vertebrate animals. Curr Opin Immunol 14:96–102

Lehrer RI, Daher K, Ganz T, Selsted ME (1985a) Direct inactivation of viruses by MCP-1 and MCP-2, natural peptide antibiotics from rabbit leukocytes. J Virol 54:467–472

Lehrer RI, Szklarek D, Ganz T, Selsted ME (1985b) Correlation of binding of rabbit granulocyte peptides to Candida albicans with candidacidal activity. Infect Immun 49:207–211

Lehrer RI, Ganz T, Szklarek D, Selsted ME (1988) Modulation of the in vitro candidacidal activity of human neutrophil defensins by target cell metabolism and divalent cations. J Clin Invest 81:1829–1835

Lehrer RI, Bevins CL, Ganz T (1999) Defensins and other antimicrobial peptides. In: Ogra PL, Mestecky J, Lamm ME, Strober W, Bienenstock J, McGhee JR (eds) Mucosal immunology. Academic Press, San Diego, pp 89–99

Leonova L, Kokryakov VN, Aleshina GM, Hong T, Nguyen T, Zhao C, Waring AJ, Lehrer RI (2001) Circular minidefensins and posttranslational generation of molecular diversity. J Leukoc Biol 70:461–464

Liu L, Zhao C, Heng HHQ, Ganz T (1997) The human β-defensin-1 and α-defensins are encoded by adjacent genes: two peptide families with differing disulfide topology share a common ancestry. Genomics 43:316–320

Ludtke SJ, He K, Heller WT, Harroun TA, Yang L, Huang HW (1996) Membrane pores induced by magainin. Biochemistry 35:13723–13728

Mackewicz CE, Yuan J, Tran P, Diaz L, Mack E, Selsted ME, Levy JA (2003) Alpha-defensins can have anti-HIV activity but are not CD8 cell anti-HIV factors. AIDS 17:F23–F32

Mangoni ME, Aumelas A, Charnet P, Roumestand C, Chiche L, Despaux E, Grassy G, Calas B, Chavanieu A (1996) Change in membrane permeability induced by protegrin 1: implication of disulphide bridges for pore formation. FEBS Lett 383:93–98

Manitz MP, Horst B, Seeliger S, Strey A, Skryabin BV, Gunzer M, Frings W, Schonlau F, Roth J, Sorg C, Nacken W (2003) Loss of S100A9 (MRP14) Results in reduced interleukin-8-induced CD11b surface expression, a polarized microfilament system, and diminished responsiveness to chemoattractants in vitro. Mol Cell Biol 23:1034–1043

Matsuzaki K (2001) Why and how are peptide-lipid interactions utilized for self defence? Biochem Soc Trans 29:598–601

McCray PB, Bentley L (1997) Human airway epithelia express a β-defensin. Am J Respir Cell Mol Biol 16:343–349

McGregor JA, French JI (2000) Bacterial vaginosis in pregnancy. Obstet Gynecol Surv 55:S1–S19

McNeely TB, Dealy M, Dripps DJ, Orenstein JM, Eisenberg SP, Wahl SM (1995) Secretory leukocyte protease inhibitor: a human saliva protein exhibiting anti-human immunodeficiency virus 1 activity in vitro. J Clin Invest 96:456–464

McNeely TB, Shugars DC, Rosendahl M, Tucker C, Eisenberg SP, Wahl SM (1997) Inhibition of human immunodeficiency virus type 1 infectivity by secretory leukocyte protease inhibitor occurs prior to viral reverse transcription. Blood 90:1141–1149

Meredith SD, Raphael GD, Baraniuk JN, Banks SM, Kaliner MA (1989) The pathophysiology of rhinitis. III. The control of IgG secretion. J Allergy Clin Immunol 84:920–930

Miller BF, Abrams R, Dorfman A, Klein M (1942) Antibacterial properties of protamine and histone. Science 96:428–430

Monell CR, Strand M (1994) Structural and functional similarities between synthetic HIV gp41 peptides and defensins. Clin Immunol Immunopathol 71:315–324

Morrison GM, Davidson DJ, Dorin JR (1999) A novel mouse beta defensin, Defb2, which is upregulated in the airways by lipopolysaccharide. FEBS Lett 442:112–116

Münk C, Wei G, Yang OO, Waring AJ, Wang W, Hong T, Lehrer RI, Landau NR, Cole AM (2003) The theta-defensin, retrocyclin, inhibits HIV-1 entry. AIDS Res Hum Retroviruses 19:875–881

Nakashima H, Yamamoto N, Masuda M, Fujii N (1993) Defensins inhibit HIV replication in vitro (letter). AIDS 7:1129

Negroni P, Fischer I (1944) Antibiotic action of protamines and histones. Rev Soc Argentina Biol 20:307–314

Nguyen TX, Cole AM, Lehrer RI (2003) Evolution of primate theta-defensins: a serpentine path to a sweet tooth. Peptides 24:1647–1654

Niyonsaba F, Iwabuchi K, Matsuda H, Ogawa H, Nagaoka I (2002) Epithelial cell-derived human beta-defensin-2 acts as a chemotaxin for mast cells through a pertussis toxin-sensitive and phospholipase C-dependent pathway. Int Immunol 14:421–426

Ogata K, Linzer BA, Zuberi RI, Ganz T, Lehrer RI, Catanzaro A (1992) Activity of defensins from human neutrophilic granulocytes against *Mycobacterium avium-Mycobacterium intracellulare*. Infect Immun 60:4720–4725

Oren A, Ganz T, Liu L, Meerloo T (2003) In human epidermis, [beta]-defensin 2 is packaged in lamellar bodies. Exp Mol Pathol 74:180–182

Ouellette AJ (2005) Paneth cell alpha-defensins: peptide mediators of innate immunity in the small intestine. Springer Semin Immunopathol 27:133–146

Ouellette AJ, Bevins CL (2001) Paneth cell defensins and innate immunity of the small bowel. Inflamm Bowel Dis 7:43–50

Panyutich A, Ganz T (1991) Activated alpha 2-macroglobulin is a principal defensin-binding protein. Am J Respir Cell Mol Biol 5:101–106

Panyutich AV, Szold O, Poon PH, Tseng Y, Ganz T (1994) Identification of defensin binding to C1 complement. FEBS Lett 356:169–173

Panyutich AV, Hiemstra PS, Van Wetering S, Ganz T (1995) Human neutrophil defensin and serpins form complexes and inactivate each other. Am J Respir Cell Mol Biol 12:351–357

Park CB, Yi KS, Matsuzaki K, Kim MS, Kim SC (2000) Structure-activity analysis of buforin II, a histone H2A-derived antimicrobial peptide: the proline hinge is responsible for the cell-penetrating ability of buforin II. Proc Natl Acad Sci U S A 97:8245–8250

Patterson BK, Landay A, Andersson J, Brown C, Behbahani H, Jiyamapa D, Burki Z, Stanislawski D, Czerniewski MA, Garcia P (1998) Repertoire of chemokine receptor expression in the female genital tract: implications for human immunodeficiency virus transmission. Am J Pathol 153:481–490

Patton DL, Thwin SS, Meier A, Hooton TM, Stapleton AE, Eschenbach DA (2000) Epithelial cell layer thickness and immune cell populations in the normal human vagina at different stages of the menstrual cycle. Am J Obstet Gynecol 183:967–973

Pillay K, Coutsoudis A, Gadzi-Naqvi AK, Kuhn L, Coovadia HM, Janoff EN (2001) Secretory leukocyte protease inhibitor in vaginal fluids and perinatal human immunodeficiency virus type 1 transmission. J Infect Dis 183:653–656

Pivarcsi A, Nagy I, Koreck A, Kis K, Kenderessy-Szabo A, Szell M, Dobozy A, Kemeny L (2005) Microbial compounds induce the expression of pro-inflammatory cytokines, chemokines and human beta-defensin-2 in vaginal epithelial cells. Microbes Infect 7:1117–1127

Puddu P, Borghi P, Gessani S, Valenti P, Belardelli F, Seganti L (1998) Antiviral effect of bovine lactoferrin saturated with metal ions on early steps of human immunodeficiency virus type 1 infection. Int J Biochem Cell Biol 30:1055–1062

Qu XD, Lloyd KC, Walsh JH, Lehrer RI (1996) Secretion of type II phospholipase A2 and cryptdin by rat small intestinal Paneth cells. Infect Immun 64:5161–5165

Quayle AJ (2002) The innate and early immune response to pathogen challenge in the female genital tract and the pivotal role of epithelial cells. J Reprod Immunol 57:61–79

Quayle AJ, Porter EM, Nussbaum AA, Wang YM, Brabec C, Yip KP, Mok SC (1998) Gene expression, immunolocalization, and secretion of human defensin-5 in human female reproductive tract. Am J Pathol 152:1247–1258

Quinones-Mateu ME, Lederman MM, Feng Z, Chakraborty B, Weber J, Rangel HR, Marotta ML, Mirza M, Jiang B, Kiser P, Medvik K, Sieg SF, Weinberg A (2003) Human epithelial beta-defensins 2 and 3 inhibit HIV-1 replication. AIDS 17: F39–F48

Reid G (2001) Probiotic agents to protect the urogenital tract against infection. Am J Clin Nutr 73:437S–443S

Salzman NH, Ghosh D, Huttner KM, Paterson Y, Bevins CL (2003) Protection against enteric salmonellosis in transgenic mice expressing a human intestinal defensin. Nature 422:522–526

Satoh Y (1988) Effect of live and heat-killed bacteria on the secretory activity of Paneth cells in germ-free mice. Cell Tissue Res 251:87–93

Satoh Y, Ishikawa K, Oomori Y, Takeda S, Ono K (1992) Bethanechol and a G-protein activator, NaF/AlCl3, induce secretory response in Paneth cells of mouse intestine. Cell Tissue Res 269:213–220

Schonwetter BS, Stolzenberg ED, Zasloff MA (1995) Epithelial antibiotics induced at sites of inflammation. Science 267:1645–1648

Schwebke JR, Burgess D (2004) Trichomoniasis. Clin Microbiol Rev 17:794–803

Selsted ME, Harwig SS (1987) Purification, primary structure, and antimicrobial activities of a guinea pig neutrophil defensin. Infect Immun 55:2281–2286

Selsted ME, Ouellette AJ (2005) Mammalian defensins in the antimicrobial immune response. Nat Immunol 6:551–557

Selsted ME, Szklarek D, Lehrer RI (1984) Purification and antibacterial activity of antimicrobial peptides of rabbit granulocytes. Infect Immun 45:150–154

Selsted ME, Tang YQ, Morris WL, McGuire PA, Novotny MJ, Smith W, Henschen AH, Cullor JS (1993) Purification, primary structures, and antibacterial activities of beta-defensins, a new family of antimicrobial peptides from bovine neutrophils. J Biol Chem 268:6641–6648

Shafer WM, Veal WL, Lee EH, Zarantonelli L, Balthazar JT, Rouquette C (2001) Genetic organization and regulation of antimicrobial efflux systems possessed by Neisseria gonorrhoeae and Neisseria meningitidis. J Mol Microbiol Biotechnol 3:219–224

Shai Y (1999) Mechanism of the binding, insertion and destabilization of phospholipid bilayer membranes by alpha-helical antimicrobial and cell non-selective membrane-lytic peptides. Biochim Biophys Acta 1462:55–70

Shi J, Zhang G, Wu H, Ross C, Blecha F, Ganz T (1999) Porcine epithelial beta-defensin 1 is expressed in the dorsal tongue at antimicrobial concentrations. Infect Immun 67:3121–3127

Shugars DC, Alexander AL, Fu K, Freel SA (1999) Endogenous salivary inhibitors of human immunodeficiency virus. Arch Oral Biol 44:445–453

Singh PK, Jia HP, Wiles K, Hesselberth J, Liu L, Conway BD, Greenberg EP, Valore EV, Welsh MJ, Ganz T, Tack BF, McCray PBJ (1998) Production of β-defensins by human airway epithelia. Proc Natl Acad Sci U S A 95:14961–14966

Singh PK, Tack BF, McCray PB Jr, Welsh MJ (2000) Synergistic and additive killing by antimicrobial factors found in human airway surface liquid. Am J Physiol Lung Cell Mol Physiol 279:L799–L805

Sinha S, Cheshenko N, Lehrer RI, Herold BC (2003) NP-1, a rabbit alpha-defensin, prevents the entry and intercellular spread of herpes simplex virus type 2. Antimicrob Agents Chemother 47:494–500

Sobel JD (1992) Pathogenesis and treatment of recurrent vulvovaginal candidiasis. Clin Infect Dis 14 [Suppl 1]:S148–S153

Sobel JD (2004) Current trends and challenges in candidiasis. Oncology (Huntingt) 18:7–8

Sohnle PG, Hunter MJ, Hahn B, Chazin WJ (2000) Zinc-reversible antimicrobial activity of recombinant calprotectin (migration inhibitory factor-related proteins 8 and 14). J Infect Dis 182:1272–1275

Sorensen OE, Gram L, Johnsen AH, Andersson E, Bangsboll S, Tjabringa GS, Hiemstra PS, Malm J, Egesten A, Borregaard N (2003) Processing of seminal plasma hCAP-18 to ALL-38 by gastricsin: a novel mechanism of generating antimicrobial peptides in vagina. J Biol Chem 278:28540–28546

Sorensen OE, Thapa DR, Rosenthal A, Liu L, Roberts AA, Ganz T (2005) Differential regulation of beta-defensin expression in human skin by microbial stimuli. J Immunol 174:4870–4879

Steinbakk M, Naess-Andresen CF, Lingaas E, Dale I, Brandtzaeg P, Fagerhol MK (1990) Antimicrobial actions of calcium binding leucocyte L1 protein, calprotectin. Lancet 336:763–765

Steinman RM, Inaba K (1999) Myeloid dendritic cells. J Leukoc Biol 66:205–208

Swart PJ, Kuipers EM, Smit C, Van Der Strate BW, Harmsen MC, Meijer DK (1998) Lactoferrin. Antiviral activity of lactoferrin. Adv Exp Med Biol 443:205–213

Taha TE, Hoover DR, Dallabetta GA, Kumwenda NI, Mtimavalye LA, Yang LP, Liomba GN, Broadhead RL, Chiphangwi JD, Miotti PG (1998) Bacterial vaginosis and disturbances of vaginal flora: association with increased acquisition of HIV. AIDS 12:1699–1706

Taha TE, Gray RH, Kumwenda NI, Hoover DR, Mtimavalye LA, Liomba GN, Chiphangwi JD, Dallabetta GA, Miotti PG (1999) HIV infection and disturbances of vaginal flora during pregnancy. J Acquir Immune Defic Syndr Hum Retrovirol 20:52–59

Tang YQ, Yuan J, Osapay G, Osapay K, Tran D, Miller CJ, Ouellette AJ, Selsted ME (1999) A cyclic antimicrobial peptide produced in primate leukocytes by the ligation of two truncated α-defensins. Science 286:498–502

Tomee JF, Koeter GH, Hiemstra PS, Kauffman HF (1998) Secretory leukoprotease inhibitor: a native antimicrobial protein presenting a new therapeutic option? Thorax 53:114–116

Trabi M, Schirra HJ, Craik DJ (2001) Three-dimensional structure of RTD-1, a cyclic antimicrobial defensin from Rhesus macaque leukocytes. Biochemistry 40:4211–4221

Tran D, Tran PA, Tang YQ, Yuan J, Cole T, Selsted ME (2002) Homodimeric thetadefensins from Rhesus macaque leukocytes—Isolation, synthesis, antimicrobial activities, and bacterial binding properties of the cyclic peptides. J Biol Chem 277:3079–3084

Turner J, Cho Y, Dinh NN, Waring AJ, Lehrer RI (1998) Activities of LL-37, a cathelin-associated antimicrobial peptide of human neutrophils. Antimicrob Agents Chemother 42:2206–2214

Turpin JA, Schaeffer CA, Bu M, Graham L, Buckheit RW Jr, Clanton D, Rice WG (1996) Human immunodeficiency virus type-1 (HIV-1) replication is unaffected by human secretory leukocyte protease inhibitor. Antiviral Res 29:269–277

UNAIDS/World Health Organization (2004) Global estimates of HIV/AIDS epidemic. www unaids org [accessed June 13, 2005]

Valore EV, Ganz T (1992) Posttranslational processing of defensins in immature human myeloid cells. Blood 79:1538–1544

Valore EV, Park CH, Quayle AJ, Wiles KR, McCray PB, Ganz T (1998) Human β-defensin-1: an antimicrobial peptide of urogenital tissues. J Clin Invest 101:1633–1642

Valore EV, Park CH, Igreti SL, Ganz T (2002) Antimicrobial components of vaginal fluid. Am J Obstet Gynecol 187:561–568

Vaughan VC, Novy FG, McClintock CT (1893) The germicidal properties of nucleins. Med News 62:536–538

Wang SA, Papp JR, Stamm WE, Peeling RW, Martin DH, Holmes KK (2005) Evaluation of antimicrobial resistance and treatment failures for Chlamydia trachomatis: a meeting report. J Infect Dis 191:917–923

Wang W, Cole AM, Hong T, Waring AJ, Lehrer RI (2003) Retrocyclin, an Antiretroviral θ-defensin, is a lectin. J Immunol 170:4708–4716

Wang X, Zhang Z, Louboutin JP, Moser C, Weiner DJ, Wilson JM (2003) Airway epithelia regulate expression of human beta-defensin 2 through Toll-like receptor 2. FASEB J 17:1727–1729

Wang YQ, Griffiths WJ, Jornvall H, Agerberth B, Johansson J (2002) Antibacterial peptides in stimulated human granulocytes—characterization of ubiquitinated histone H1A. Eur J Biochem 269:512–518

Wasserheit JN (1992) Epidemiological synergy. Interrelationships between human immunodeficiency virus infection and other sexually transmitted diseases. Sex Transm Dis 19:61–77

Wecke J, Lahav M, Ginsburg I, Giesbrecht P (1982) Cell wall degradation of Staphylococcus aureus by lysozyme. Arch Microbiol 131:116–123

Weiss TM, Yang L, Ding L, Wang WC, Waring AJ, Lehrer RI, Huang HW (2002) Two states of a cyclic antimicrobial peptide theta-defensin in lipid bilayers. Biophys J 82:7A

Wiesenfeld HC, Hillier SL, Krohn MA, Landers DV, Sweet RL (2003) Bacterial vaginosis is a strong predictor of Neisseria gonorrhoeae and Chlamydia trachomatis infection. Clin Infect Dis 36:663–668

Yasin B, Wang W, Pang M, Cheshenko N, Hong T, Waring AJ, Herold BC, Wagar EA, Lehrer RI (2004) Theta defensins protect cells from infection by herpes simplex virus by inhibiting viral adhesion and entry. J Virol 78:5147–5156

Zanetti M, Gennaro R, Romeo D (1995) Cathelicidins: a novel protein family with a common proregion and a variable C-terminal antimicrobial domain. FEBS Lett 374:1–5

Zariffard MR, Harwani S, Novak RM, Graham PJ, Ji X, Spear GT (2004) Trichomonas vaginalis infection activates cells through toll-like receptor 4. Clin Immunol 111:103–107

Zasloff M (2002) Antimicrobial peptides of multicellular organisms. Nature 415:389–395

Zhang L, He T, Talal A, Wang G, Frankel SS, Ho DD (1998) In vivo distribution of the human immunodeficiency virus/simian immunodeficiency virus coreceptors: CXCR4, CCR3, and CCR5. J Virol 72:5035–5045

Zhang L, Yu W, He T, Yu J, Caffrey RE, Dalmasso EA, Fu S, Pham T, Mei J, Ho JJ, Zhang W, Lopez P, Ho DD (2002) Contribution of human alpha-defensin-1, -2 and -3 to the anti-HIV-1 activity of CD8 antiviral factor. Science 298:995–1000

Zhao C, Nguyen T, Liu L, Sacco RE, Brogden KA, Lehrer RI (2001) Gallinacin-3, an Inducible Epithelial β-Defensin in the Chicken. Infect Immun 69:2684–2691

Zimmermann GR, Legault P, Selsted ME, Pardi A (1995) Solution structure of bovine neutrophil beta-defensin-12: the peptide fold of the beta-defensins is identical to that of the classical defensins. Biochemistry 34:13663–13671

CTMI (2006) 306:231–250

Molecular Mechanisms of Bacterial Resistance to Antimicrobial Peptides

D. Kraus · A. Peschel (✉)

Cellular and Molecular Microbiology Division, Medical Microbiology and Hygiene Institute, University of Tübingen, Elfriede-Aulhorn-Str. 6, 72076 Tübingen, Germany
andreas.peschel@uni-tuebingen.de

Abstract Cationic antimicrobial peptides (CAMPs) are integral compounds of the antimicrobial arsenals in virtually all kinds of organisms, with important roles in microbial ecology and higher organisms' host defense. Many bacteria have developed countermeasures to limit the efficacy of CAMPs such as defensins, cathelicidins, kinocidins, or bacteriocins. The best-studied bacterial CAMP resistance mechanisms involve electrostatic repulsion of CAMPs by modification of cell envelope molecules, proteolytic cleavage of CAMPs, production of CAMP-trapping proteins, or extrusion of CAMPs by energy-dependent efflux pumps. The repertoire of CAMPs produced by a given host organism and the efficiency of microbial CAMP resistance mechanisms appear to be crucial in host–pathogen interactions, governing the composition of commensal microbial communities and the virulence of bacterial pathogens. However, all CAMP resistance mechanisms have limitations and bacteria have never succeeded in becoming fully insensitive to a broad range of CAMPs. CAMPs or conserved CAMP resistance factors are discussed as new mediators and targets, respectively, of novel and sustainable anti-infective strategies.

1
Introduction

One of nature's most ancient strategies for combating unwelcome bacteria is the production of membrane-damaging antimicrobial peptides (Zasloff 2002; Hancock and Chapple 1999). Such molecules are produced by certain bacterial or archaeal strains (bacteriocins) (Riley and Wertz 2002), by plants (plant defensins) (Lay and Anderson 2005), by protozoons (Leippe and Herbst 2004), and by virtually all classes of animals (Zasloff 2002). In order to equip these molecules with a high affinity for bacterial membranes most of them have cationic properties and are referred to as CAMPs (cationic antimicrobial peptides). The antimicrobial activity of CAMPs depends on an ionic milieu comparable to the conditions found in mammalian body fluids (Dorschner et al. 2006). CAMPs include linear, usually α-helical peptides such as the amphibian magainin, the murine CRAMP, and the human LL-37 (Nizet and Gallo 2003), disulfide bridge-stabilized peptides with β-sheet structures such as the α-, β-, and θ-defensins (Ganz 2003; Lehrer 2004), and large chemokines or chemokine-derived molecules with antimicrobial activity named kinocidins (Yang et al. 2003; Dürr and Peschel 2002), to mention only a few typical classes of vertebrate CAMPs. Bacteriocins often contain unusual modifications such as thioether bridges (Guder et al. 2000).

CAMPs have been shown to play crucial roles in microbial ecology and in higher organisms' host defense. However, microorganisms have also found many ways to limit the efficacy of CAMPs (Groisman 1994; Ernst et al. 2001; Peschel 2002; Nizet 2005). Bacteriocin-producing bacteria are resistant to the produced peptides, which enable them to survive while competing microorganisms are inhibited (Riley and Wertz 2002). Bacterial commensals and pathogens of higher organisms, on the other hand, use CAMP resistance mechanisms as a prerequisite to invade and colonize host tissues (Peschel 2002). Unlike antibiotic resistance genes, most CAMP resistance genes are usually not found on plasmids, transposons, or other laterally transferable genetic elements but on the bacterial chromosome in the vicinity of housekeeping genes. At least some of them are considered to have appeared rather early in evolution and seem to be integral parts of the genomes of bacteria whose habitats involve the frequent exposure to CAMPs. As another consequence of their long presence in bacteria, some of the cell wall modifications leading to CAMP resistance affect other bacterial functions such as the attachment and activity of cell wall proteins (Peschel et al. 2000), biofilm formation (Gross et al. 2001), or interaction with epithelial cells (Weidenmaier et al. 2004). Extensive research activities have led to a very large number of studies on bacterial CAMP resistance (Table 1; Fig. 1). This review focuses on established molecular principles of CAMP resistance rather than giving a complete overview on all publications concerning this topic. The ecological aspects of CAMP resistance along with their relevance in microbial biofilm formation and biofilm-associated infections are discussed elsewhere (Otto 2005).

As one would expect, CAMPs seem to be subjected to a very rapid and active evolution (Maxwell et al. 2003), probably as a means to react to the equally fast evolving bacterial resistance mechanisms (Patil et al. 2004). Accordingly, the various mammalian genera are highly variable in the sequences and structures of produced antimicrobial peptides. It can be assumed that the pattern of antimicrobial molecules of a given species is one of the factors that govern the spectrum of its commensal and pathogenic microorganisms. For instance, the production of antiretroviral θ-defensins is discussed as a crucial factor determining resistance (in monkeys) or susceptibility (humans) to HIV (Nguyen et al. 2003). The human gut is particularly rich in bacterial colonizers, which is probably the reason why specialized cells in the crypts of Lieberkühn produce an extraordinarily large spectrum of antimicrobial peptides ranging from CRS peptides (mice) (Hornef et al. 2004) to various α-defensins (most mammalian species including humans) (Lehrer 2004; Ganz 2003). Elucidating the basis of microbial CAMP resistance mechanisms will be crucial for understanding, monitoring, and interfering with bacterial colonization and infection.

2
How Widespread Is CAMP Resistance?

Considering the fact that probably each bacterial species encounters CAMP-producing competing microorganisms or host cells, one would expect that most bacteria have evolved at least some strategies to evade CAMP-mediated killing. In fact, increasing research activities have clearly demonstrated that this is true for many microbial habitats. Skin bacteria such as staphylococci, oral bacteria such as streptococci, and intestinal bacteria such as salmonellae have been described to resist high concentrations of locally produced CAMPs (Peschel 2002; Ernst et al. 2001; Nizet 2005) (Table 1). Soil bacteria such as *Bacillus subtilis* also have CAMP resistance mechanisms, probably as a means

Table 1 Mechanisms and prevalence of bacterial CAMP resistance

Resistance mechanism	Species	Reference
Proteolytic cleavage		
PgtE	*Salmonella enterica*	(Guina et al. 2000)
OmpT	*Escherichia coli*	(Stumpe et al. 1998)
Aureolysin, serin protease V8	*Staphylococcus aureus*	(Sieprawska-Lupa et al. 2004)
Unidentified proteases	*Pseudomonas aeruginosa, Enterococcus faecalis, Proteus mirabilis, Porphyromonas gingivalis, Prevotella* spp.	(Schmidtchen et al. 2002)
Production of external CAMP-binding molecules		
SIC protein, M1 protein	*Streptococcus pyogenes*	(Frick et al. 2003; Nizet 2005)
Staphylokinase	*Staphylococcus aureus*	(Jin et al. 2004)
CAMP-specific drug exporters		
MtrCDE	*Neisseria gonorrhoeae*	(Shafer et al. 1998)
EpiFEG	*Staphylococcus epidermidis*; many antibiotic producers	(Peschel and Götz 1996; Jack et al. 1998)
RosA/B	*Yersinia* spp.	(Bengoechea and Skurnik 2000)
Alteration of the electrostatic properties of the bacterial cell surface		
Modification of lipid A with aminoarabinose	*Salmonella enterica*, many Gram-negative spp.	(Ernst et al. 2001b; Miller et al. 2005)
Alanylation of teichoic acids	*Staphylococcus aureus*; many Gram-positive bacteria	(Peschel et al. 1999; Abachin et al. 2002; Poyart et al. 2003; Perego et al. 1995)
Lysinylation of phospholipids	*Staphylococcus aureus*; many Gram-positive and Gram-negative bacteria	(Peschel et al. 2001; Staubitz and Peschel 2002; Ratledge and Wilkinson 1988)
Further mechanisms		
Additional fatty acid in lipid A	*Salmonella enterica*, many Gram-negative spp.	(Guo et al. 1998; Miller et al. 2005)
Modification of mycolic acid	*Mycobacterium tuberculosis*	(Gao et al. 2003)
Reduced cytoplasmic membrane potential	*Staphylococcus aureus*	(Yeaman et al. 1998)

Table 1 (continued)

Resistance mechanism	Species	Reference
Slime and capsule polymers, biofilm formation	*Klebsiella pneumoniae, Staphylococcus epidermidis,* many other bacteria	(Campos et al. 2004; Otto 2005)
Inhibition of CAMP production	*Shigella* spp.	(Islam et al. 2001)

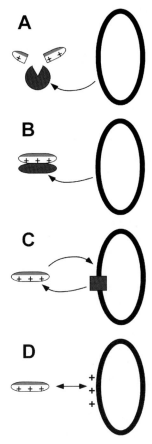

Fig. 1A–D Mechanisms of bacterial CAMP resistance by proteolytic cleavage of CAMPs (**A**), CAMP-trapping molecules (**B**), CAMP extruding transport proteins (**C**), or electrostatic repulsion of CAMPs (**D**)

to achieve protection against bacteriocins frequently produced by other soil microorganisms and fungi (Staubitz and Peschel 2002; Cao and Helmann 2004). The available bacterial genome sequences reveal the presence of CAMP resistance genes in the majority of microbial species, indicating that CAMP resistance is in fact a very widespread bacterial trait (Weidenmaier et al. 2003; Miller et al. 2005). Extensive investigations in some prototype species such as *Salmonella enterica* and *Staphylococcus aureus* have revealed the presence of several resistance mechanisms in one bacterial species, which seem to complement each other in order to achieve high-level resistance to a broad spectrum of CAMPs (Ernst et al. 2001; Peschel 2002). However, different isolates of one particular bacterial species may vary widely in their susceptibility to CAMPs (Midorikawa et al. 2003; Joly et al. 2004) indicating that the various mechanisms may be differently expressed or functional in different clones.

3
Proteolytic Cleavage of CAMPs

The most straightforward way for a bacterial species to inactivate antimicrobial peptides is the production of peptidases and proteases that cleave CAMPs (Fig. 1A). Such enzymes have been described in Gram-negative and Gram-positive bacteria. *S. enterica* produces the outer membrane protease PgtE, which is capable of cleaving the cathelicidin LL-37 and other alpha helical CAMPs (Guina et al. 2000). *S. aureus* expresses several proteases; the metalloprotease aureolysin and the serine protease V8 can cleave LL-37 and the in vitro resistance to LL-37 has been associated with aureolysin production (Sieprawska-Lupa et al. 2004). Many other bacterial species including *Pseudomonas aeruginosa*, *Enterococcus faecalis*, *Proteus mirabilis*, *Porphyromonas gingivalis*, and *Prevotella* spp. also produce proteases that cleave linear CAMPs (Schmidtchen et al. 2002).

4
Production of External CAMP-Binding Molecules

Some bacterial species express secreted or surface-anchored proteins that bind certain CAMPs with a very high affinity and thereby prevent their access to the cytoplasmic membrane (Fig. 1B). *Streptococcus pyogenes* secretes the SIC protein, which binds and thereby inactivates LL-37 (Frick et al. 2003). Its production has been correlated with the invasiveness of *S. pyogenes* strains.

A similar approach is used by *S. aureus* in order to achieve resistance to α-defensins. The fibrinolytic exoprotein staphylokinase does not only bind to plasminogen, but also has a high affinity for human α-defensins (Jin et al. 2004). Staphylokinase thereby contributes significantly to α-defensin resistance and staphylokinase production correlates with the in vitro resistance of *S. aureus* isolates to defensins.

The *S. pyogenes* M proteins are covalently attached to the peptidoglycan. Several functions have been assigned to the various M protein domains, which are highly variable in structure and size (Bisno et al. 2003). One of the M protein serotypes, the globally disseminated M1 clone, seems to play a critical role in resistance to LL-37 by binding this peptide with the hypervariable M1 C-terminus (Nizet 2005). Inactivation of the M1 gene or its heterologous expression leads to reduced susceptibility and increased resistance to LL-37 respectively.

5
CAMP-Specific Drug Exporters

Protection against small amphiphatic drugs is often mediated by extrusion of the molecules by energy-dependent export proteins in the cytoplasmic membrane. Many of these resistance factors have a broad substrate spectrum and are referred to as multiple drug resistance exporters (van Veen and Konings 1997). As CAMPs also have amphiphatic, membrane-damaging properties, it is not surprising that some of the known bacterial MDRs confer resistance to certain CAMPs (Fig. 1C). The *Neisseria gonorrhoeae* MtrCDE MDR, a member of the resistance/nodulation/division (RND) class of microbial efflux pumps, contributes to resistance to the small porcine β-sheet CAMP protegrin 1 and to the α-helical human peptide LL-37 (Shafer et al. 1998). Attenuated virulence of *mtr*-mutated *N. gonorrhoeae* suggests a considerable role of the MtrCDE system in evasion of CAMP-mediated killing (Jerse et al. 2003). The QacA efflux pump contributes to *S. aureus* resistance to platelet-derived CAMPs (tPMPs, thrombin-induced microbicidal proteins) (Kupferwasser et al. 1999). However, this mechanism appears to be independent of the transport function of QacA and to result from another activity of this membrane protein.

Bacterial producers of lanthionine-containing bacteriocins usually bear ABC transporters that provide resistance against the produced antimicrobial peptide. These systems always seem to be very specific for the produced peptide and do not protect against a larger spectrum of CAMPs (Riley and Wertz 2002; Peschel and Götz 1996).

6
CAMP Resistance by Altering the Electrostatic Properties of the Bacterial Cell Surface

CAMPs share positive net charges with most antimicrobial molecules and enzymes such as lysozyme, secretory group IIA phospholipase A_2 (PLA2), RNase 7, and myeloperoxidase. These cationic properties are in a striking contrast to the generally anionic net charge of the molecules forming the bacterial cell envelopes such as peptidoglycan, most phospholipids, lipid A (Gram-negatives) and teichoic acids (Gram-positives) (Weidenmaier et al. 2003). In contrast, the outer leaflets of human cell membranes are usually composed of uncharged or zwitterionic lipids such as phosphatidylcholine and sphingolipids (Devaux and Morris 2004), which are unfavorable for binding and integration of CAMPs. It is assumed that host defense factors have evolved cationic properties in order to impart a high and selective affinity for bacterial cell surface molecules (Weidenmaier et al. 2003). Most of the anionic bacterial cell envelope molecules are very ancient and invariable, and it seems to be impossible for microorganisms to replace these molecules with different structures that would be less favorable for interactions with CAMPs. However, many bacteria are able to modify their cell surfaces in order to reduce their negative net charge and thus acquire protection against inactivation by CAMPs (Fig. 1D). Detailed studies of this phenomenon are again available for *S. aureus* and *S. enterica*.

The teichoic acids of staphylococci and other Gram-positive bacteria are composed of alternating glycerolphosphate or ribitolphosphate groups and are substituted with N-acetylglucosamine or D-alanine (Neuhaus and Baddiley 2003). These polymers are anchored to the cytoplasmic membrane (lipoteichoic acids) or connected to the peptidoglycan (wall teichoic acids). The great number of phosphate groups impart polyanionic properties on teichoic acids. D-alanine incorporation introduces positively charged amino groups into teichoic acids, leading to a partial neutralization of the polymers (Peschel et al. 1999). This modification limits the interactions of CAMPs with the staphylococcal cell wall and decreases the susceptibility to a broad variety of cationic host factors ranging from defensins (Peschel et al. 1999) and PLA2 (Koprivnjak et al. 2002) to myeloperoxidase (Collins et al. 2002). In addition to staphylococci, this resistance mechanism has also been described in *Listeria monocytogenes* (Abachin et al. 2002), *Streptococcus agalactiae* (Poyart et al. 2003), *S. pyogenes* (Kristian et al. 2005), and *B. subtilis* (Wecke et al. 1997). The *dltABCD* operon responsible for D-alanine transfer into teichoic acids occurs in the genomes of most bacteria of the low G+C branch of Gram-positive bacteria, indicating that teichoic acid alanylation represents a very widespread CAMP resistance mechanism (Weidenmaier et al. 2003).

Most of the bacterial phospholipids such as phosphatidylglycerol, cardiolipin, and others share anionic properties with cell wall polymers (Huijbregts et al. 2000). Many bacterial species, however, including staphylococci, enterococci, listeriae and *P. aeruginosa*, are able to modify a considerable amount of phosphatidylglycerol with L-lysine (Ratledge and Wilkinson 1988), which leads again to neutralization of the cell surface net charge and, consequently, to reduced binding of CAMPs and other cationic host defense molecules. MprF, a novel membrane enzyme, is responsible for the synthesis of lysylphosphatidylglycerol (Staubitz et al. 2004; Oku et al. 2004), and its inactivation leads to a considerably increased susceptibility of *S. aureus* to a large variety of CAMPs (Peschel et al. 2001; Kristian et al. 2003a; Koprivnjak et al. 2002; Weidenmaier et al. 2005). *mprF* homologs are found in the genomes of many Gram-positive and Gram-negative bacteria, among them many human, animal, and plant pathogens, and even in some archaeal species, suggesting that these bacteria employ very similar mechanisms to achieve protection against CAMPs (Weidenmaier et al. 2003).

Many Gram-negative bacteria have similar CAMP resistance strategies. Modifications of lipid A, the conserved integral membrane part of the lipopolysaccharide, are responsible for CAMP resistance in *S. enterica* and *P. aeruginosa* (Ernst et al. 1999, 2001). The anionic character of lipid A can be reduced, for instance, by incorporation of cationic aminoarabinose (Nummila et al. 1995; Gunn et al. 1998). Many Gram-negative species bear the *pmr* genes responsible for amioarabinose transfer into lipid A in their genomes (Miller et al. 2005), suggesting that this modification is a widespread trait in Gram-negative bacteria.

Other cell wall modifications such as synthesis of the neutral phospholipid phosphatidylethanolamine (Cao and Helmann 2004), the neutralization of peptidoglycan muropeptides by iso-D-glutamate amidation (Gustafson et al. 1994), and the transfer of positively charged ethanolamine into lipopolysaccharide (Nummila et al. 1995) may also have the purpose of reducing the efficacy of CAMPs. Obviously, many of the bacterial mechanisms of CAMP resistance reflect the same molecular strategy, even though the modified target molecules and the involved genes are unrelated.

7
Further Bacterial Mechanisms of CAMP Resistance

In order to kill bacteria, CAMPs need to integrate into bacterial membranes, diffuse laterally, and form complexes with other CAMP molecules, which leads to pore formation and efflux of protons and small molecules (Sahl et al. 2004). Membrane fluidity is thus a critical aspect in CAMP-mediated killing and, in some cases, changes in the composition of lipid fatty acids have been implicated in CAMP resistance. Introduction of an additional fatty acid into the lipid A of *S. enterica* mediated by the PagA protein reduces the susceptibility to LL-37 and protegrin PG1 (Guo et al. 1998). Related genes that may play similar roles are found in several Gram-negative pathogens' genomes (Miller et al. 2005) and increased acylation of lipid A has been implicated in adaptation of *P. aeruginosa* during persistent lung infection in cystic fibrosis patients (Ernst et al. 1999). The occurrence of shorter acyl chains in mycolic acids of a *Mycobacterium tuberculosis kasB* mutant leads to increased susceptibility to defensins and lysozyme (Gao et al. 2003). Mycolic acids form an outer membrane-like shield on the mycobacterial surface and the altered acyl chains of the mutant increase the permeability for several antibiotics and CAMPs. *S. aureus* resistance to platelet microbicidal proteins has also been associated with changes in the composition of lipid fatty acids and concomitantly altered membrane fluidity (Bayer et al. 2000). Inactivation of the major cold shock gene *cspA* leads to susceptibility of *S. aureus* to a cathepsin G-derived CAMP for unclear reasons (Katzif et al. 2003). Since mutation of *cspA* also led to deficiency in the yellow membrane carotinoid staphyloxanthine, altered composition and fluidity of the cytoplasmic membrane may be the reason for CAMP susceptibility in this mutant.

CAMPs need a certain threshold membrane potential to integrate into lipid bilayers. Bacterial cytoplasmic membranes usually have a strong potential since they contain the respiratory chain generating a proton-motive force. Eukaryotic cytoplasmic membranes, in contrast, are much less energized, which is one of the factors for the relative insensitivity of eukaryotic cells for CAMPs. Spontaneous mutations in genes encoding bacterial respiratory chain components often lead to small colony phenotypes since these mutants show a strongly attenuated growth behavior. *S. aureus* small colony variants (SCVs), however, have a better capacity to persist in human cells and they are often responsible for recurrent infections (Proctor et al. 1998). SCVs have a lower membrane potential and they are less susceptible to many CAMPs (Yeaman et al. 1998). Accordingly, the SCV phenotype can be regarded as a CAMP resistance mechanism and CAMP resistance may be one of the reasons for the increased ability of SCVs to persist in host tissues.

In some instances, capsular polymers have been shown to contribute to CAMP resistance (Campos et al. 2004; Vuong et al. 2004). Bacterial capsules are usually considered as an antiopsonic and antiphagocytotic virulence factor. They do usually not represent a major diffusion barrier for small molecules such as CAMPs. In some cases, however, the extracellular slime matrix of capsules and biofilms have been shown to provide protection against certain CAMPs (Campos et al. 2004; Vuong et al. 2004). This phenomenon may depend on the net charge of capsule polymers, as the exopolymer PIA involved in CAMP resistance in *Staphylococcus epidermidis* has a positive net charge and may thus contribute to repulsion of cationic antimicrobial molecules. Slime polymers and the special metabolic adaptations of bacteria in biofilms seem to play important roles in evasion of CAMP-mediated killing. Their relevance in CAMP resistance is reviewed in detail elsewhere in this book (Otto 2005).

Another elegant method of CAMP resistance is used by *Shigella* species, which inhibit expression of LL-37 and β-defensin 1 in human colonic epithelia cells. This event involves *Shigella* plasmid DNA (Islam et al. 2001). The underlying mechanisms are not yet understood.

8
Regulation of CAMP Resistance Mechanisms

Most of the regulatory mechanisms involved in resistance of bacteria against CAMPs are not well understood yet, both in terms of regulating signals and of regulatory proteins. However, there are some well-characterized regulatory pathways in Gram-negative bacteria, which have been shown to play crucial roles in CAMP resistance. The PhoP/PhoQ two-component system plays a key role in the virulence of *S. enterica*, *P. aeruginosa*, and *Yersinia pseudotuberculosis* (Groisman 2001). PhoP/PhoQ-controlled genes such as PagP are necessary for lipid A modification leading to CAMP resistance, as shown in *S. enterica* (Guo et al. 1998). PhoP/PhoQ responds to changes in the magnesium and calcium ion concentrations (García Véscovi et al. 1996), and it has been shown to be activated by the presence of subinhibitory concentrations of CAMPs in *S. enterica* (Bader et al. 2003). The sensor kinase PhoQ directly recognizes CAMPs, thereby displacing PhoQ-bound divalent cations and leading to activation of the response regulator PhoP (Bader et al. 2005). A second two-component regulatory system, PmrA/PmrB, responds to extracellular iron (Wosten et al. 2000), and it is also controlled by PhoP/PhoQ in *S. enterica* (Groisman 2001). It confers resistance to several CAMPs by transcriptional activation of two loci, *pmrE* and *pmrHFIJKLM*, which are required for the biosynthesis of a lipid A variant with 4-aminoarabinose modification (Gunn et al. 2000). This modification leads to a reduction of the anionic character of the bacterial lipid A and, consequently, to CAMP resistance, as discussed above. A related system seems to respond directly to CAMP exposure in *P. aeruginosa* (McPhee et al. 2003).

Much less is known about the regulation mechanisms and stimuli involved in CAMP resistance of Gram-positive bacteria. The global virulence regulatory system *agr* of *S. aureus* is involved in the regulation of the *dlt*-operon, responsible for the alanylation of techoic acids (Dunman et al. 2001). Another two-component regulation system, DltRS, controls expression of the *dlt* operon in *S. agalactiae* (Poyart et al. 2003). Inactivation of regulatory genes has led to increased CAMP susceptibility in *S. pyogenes* (Nizet et al. 2001) and *L. monocytogenes* (Cotter et al. 2002), but the genes controlled by these regulators have remained unknown.

9
Role of CAMP Resistance in Host Colonization and Infection

In addition to obligate pathogens, several bacterial commensals or opportunistic pathogens have been shown to resist high concentrations of CAMPs (Sahly et al. 2003; Shelburne et al. 2005; Brissette et al. 2004; Nishimura et al. 2004). This ability is generally considered as a prerequisite for the colonization of human epithelia whose secretions in the airway as well as in the gastrointestinal and genitourinary tracts contain high amounts of CAMPs such as β-defensins and LL-37. Only a few animal studies have addressed the role of CAMP resistance in bacterial colonization. *S. aureus* colonizes the anterior nares in 30%–40% of the human population (Peacock et al. 2001), which is one of the crucial risk factors for developing severe wound and skin infections or life-threatening systemic infections such as endocarditis and sepsis (von Eiff et al. 2001; Wertheim et al. 2004). A CAMP-susceptible *S. aureus* *dlt*A mutant has recently been shown to have a strongly reduced capacity to colonize the nares of cotton rats, which represent a good model of human nasal colonization (Weidenmaier et al. 2004). However, since the *dlt*A mutation leads to altered teichoic acids and since teichoic acid structure is critical in *S. aureus* binding to nasal epithelial cells, it is not yet clear whether the abrogated capacity of this mutant to colonize cotton rat nares is a result of reduced binding to epithelial cells, increased killing by nasal CAMPs, or both. Further in vivo studies will be necessary to elucidate the relevance of CAMP resistance in colonization.

The importance of CAMP resistance in localized infections of various organ systems has been demonstrated for many different pathogens and in many animal models. Skin infections caused by *S. pyogenes* (Nizet et al. 2001), *S. aureus* abscess-like tissue cage infections (Kristian et al. 2003b), *Legionella pneumophila* lung infections (Edelstein et al. 2003), *S. enterica* gastrointestinal infections (Gunn et al. 2000), and *N. gonorrhoeae* genital tract infections (Jerse et al. 2003), to name but a few examples, are strongly affected if CAMP susceptible mutants are used. Increased bacterial killing by CAMPs produced by epithelial cells of infected organs or released by phagocytes upon contact with bacteria is most probably the reason for the observed virulence attenuation. In line with this notion, CAMP-susceptible bacterial mutants are inactivated faster and more efficiently by CAMP-producing phagocytes (Collins et al. 2002; Kristian et al. 2003a, 2005).

CAMP-susceptible *S. agalactiae* and *S. aureus* mutants are also less virulent in blood stream infections studied in mouse sepsis or rabbit endocarditis models (Poyart et al. 2003; Collins et al. 2002; Weidenmaier et al. 2005). Depending on the particular pathogen and the animal model used, alleviated killing by blood phagocytes, inactivation by microbicidal proteins released by activated platelets, or both seems to be the reason for the reduced virulence of CAMP-susceptible mutants under these conditions.

10
Perspectives

The production of CAMPs is a very ancient and still successful strategy of to inhibit microorganisms. Considering the short half-life of the effectiveness of modern antibiotics it seems to be a mystery how CAMPs remained so efficient during evolution. Even the great variety of bacterial CAMP resistance mechanisms has not led to microorganisms with complete resistance to all kinds of CAMPs. It seems that evolution has always found new ways to circumvent the microbial CAMP resistance mechanisms, for instance by rendering CAMPs protease-resistant or by combining two or more antimicrobial mechanisms in one molecule, as shown for the highly versatile bacteriocin nisin (Pag and Sahl 2002). The extraordinary success of CAMPs may be based on the fact that bacteria cannot completely change the composition and properties of their cytoplasmic membrane. The high metabolic costs of becoming resistant to CAMPs, for instance by the energy-consuming, extensive modifications of the cell envelope, may be another reason why it is so difficult for bacteria to develop totally efficient CAMP resistance mechanisms. Nevertheless, some CAMP resistance mechanisms seem to date back to a very early origin, as *mprF*-related genes, for instance, are found in both bacterial and some archaeal genomes (Staubitz and Peschel 2002).

The amazing effectiveness of CAMPs suggest a use of such molecules in antimicrobial therapy. In fact, several CAMPs have yielded promising results in clinical trials (Andres and Dimarcq 2004). The lactococcal bacteriocin nisin has been used as a food preservative for decades (Pag and Sahl 2002) and daptomycin, a noncationic membrane-damaging antimicrobial lipopeptide with activity against multidrug-resistant staphylococci and enterococci has recently been approved for the use in human infections (Steenbergen et al. 2005), underscoring the therapeutic potential of membrane-active antimicrobial compounds such as CAMPs. On the other hand, highly conserved bacterial CAMP resistance proteins such as MprF or DltABCD may represent interesting new targets for novel anti-infective compounds that would not kill the bacteria but render them susceptible to innate antimicrobial host molecules (Weidenmaier et al. 2003). A deeper understanding of CAMPs and CAMP resistance mechanisms will help to exploit both innate human host defenses and bacterial evasion strategies.

Acknowledgements We thank co-workers and collaborators for help and support. Our research is supported by grants from the German Research Council, the German Ministry of Education and Research, the European Union, and the Interdisciplinary Center for Clinical Research of the University of Tübingen.

References

Abachin E, Poyart C, Pellegrini E, Milohanic E, Fiedler F, Berche P, Trieu-Cuot P (2002) Formation of D-alanyl-lipoteichoic acid is required for adhesion and virulence of *Listeria monocytogenes*. Mol Microbiol 43:1–14

Andres E, Dimarcq JL (2004) Cationic antimicrobial peptides: update of clinical development. J Intern Med 255:519–520

Bader MW, Navarre WW, Shiau W, Nikaido H, Frye JG, McClelland M, Fang FC, Miller SI (2003) Regulation of *Salmonella typhimurium* virulence gene expression by cationic antimicrobial peptides. Mol Microbiol 50:219–230

Bader MW, Sanowar S, Daley ME, Schneider AR, Cho U, Xu W, Klevit RE, Le Moual H, Miller SI (2005) Recognition of antimicrobial peptides by a bacterial sensor kinase. Cell 122:461–472

Bayer AS, Prasad R, Chandra J, Koul A, Smriti M, Varma A, Skurray RA, Firth N, Brown MH, Koo S-P, Yeaman MR (2000) In vitro resistance of *Staphylococcus aureus* to thrombin-induced platelet microbicidal proteins is associated with alterations in cytoplasmic membrane fluidity. Infect Immun 68:3548–3553

Bengoechea JA, Skurnik M (2000) Temperature-regulated efflux pump/potassium antiporter system mediates resistance to cationic antimicrobial peptides in *Yersinia*. Mol Microbiol 37:67–80

Bisno AL, Brito MO, Collins CM (2003) Molecular basis of group A streptococcal virulence. Lancet Infect Dis 3:191–200

Brissette CA, Simonson LG, Lukehart SA (2004) Resistance to human beta-defensins is common among oral treponemes. Oral Microbiol Immunol 19:403–407

Campos MA, Vargas MA, Regueiro V, Llompart CM, Alberti S, Bengoechea JA (2004) Capsule polysaccharide mediates bacterial resistance to antimicrobial peptides. Infect Immun 72:7107–7114

Cao M, Helmann JD (2004) The *Bacillus subtilis* extracytoplasmic-function sigmaX factor regulates modification of the cell envelope and resistance to cationic antimicrobial peptides. J Bacteriol 186:1136–1146

Collins LV, Kristian SA, Weidenmaier C, Faigle M, Van Kessel KP, Van Strijp JA, Gotz F, Neumeister B, Peschel A (2002) *Staphylococcus aureus* strains lacking D-alanine modifications of teichoic acids are highly susceptible to human neutrophil killing and are virulence attenuated in mice. J Infect Dis 186:214–219

Cotter PD, Guinane CM, Hill C (2002) The LisRK signal transduction system determines the sensitivity of *Listeria monocytogenes* to nisin and cephalosporins. Antimicrob Agents Chemother 46:2784–2790

Devaux PF, Morris R (2004) Transmembrane asymmetry and lateral domains in biological membranes. Traffic 5:241–246

Dorschner RA, Lopez-Garcia B, Peschel A, Kraus D, Morikawa K, Nizet V, Gallo RL (2006) The mammalian ionic environment dictates microbial susceptibility to antimicrobial defense peptides. FASEB J 20:35–42

Dunman PM, Murphy E, Haney S, Palacios D, Tucker-Kellogg G, Wu S, Brown EL, Zagursky RJ, Shlaes D, Projan SJ (2001) Transcription profiling-based identification of *Staphylococcus aureus* genes regulated by the *agr* and/or *sarA* loci. J Bacteriol 183:7341–7353

Dürr M, Peschel A (2002) Chemokines meet defensins—the merging concepts of chemoattractants and antimicrobial peptides in host defense. Infect Immun 70:6515–6517

Edelstein PH, Hu B, Higa F, Edelstein MA (2003) lvgA, a novel *Legionella pneumophila* virulence factor. Infect Immun 71:2394–2403

Ernst RK, Yi EC, Guo L, Lim KB, Burns JL, Hackett M, Miller SI (1999) Specific lipopolysaccharide found in cystic fibrosis airway *Pseudomonas aeruginosa*. Science 286:1561–1565

Ernst RK, Guina T, Miller SI (2001) *Salmonella typhimurium* outer membrane remodeling: role in resistance to host innate immunity. Microbes Infect 3:1327–1334

Frick IM, Akesson P, Rasmussen M, Schmidtchen A, Bjorck L (2003) SIC, a secreted protein of *Streptococcus pyogenes* that inactivates antibacterial peptides. J Biol Chem 278:16561–16566

Ganz T (2003) Defensins: antimicrobial peptides of innate immunity. Nat Rev Immunol 3:710–720

Gao LY, Laval F, Lawson EH, Groger RK, Woodruff A, Morisaki JH, Cox JS, Daffe M, Brown EJ (2003) Requirement for *kasB* in *Mycobacterium* mycolic acid biosynthesis, cell wall impermeability and intracellular survival: implications for therapy. Mol Microbiol 49:1547–1563

García Véscovi E, Soncini FC, Groisman EA (1996) Mg^{2+} as an extracellular signal: environmental regulation of *Salmonella* virulence. Cell 84:165–174

Groisman EA (1994) How bacteria resist killing by host defense peptides. Trends Microbiol Sci 2:444–448

Groisman EA (2001) The pleiotropic two-component regulatory system PhoP-PhoQ. J Bacteriol 183:1835–1842

Gross M, Cramton S, Goetz F, Peschel A (2001) Key role of teichoic acid net charge in *Staphylococcus aureus* colonization of artificial surfaces. Infect Immun 69:3423–3426

Guder A, Wiedemann I, Sahl HG (2000) Posttranslationally modified bacteriocins—the lantibiotics. Biopolymers 55:62–73

Guina T, Yi EC, Wang H, Hackett M, Miller SI (2000) A PhoP-regulated outer membrane protease of *Salmonella enterica* serovar typhimurium promotes resistance to alpha-helical antimicrobial peptides. J Bacteriol 182:4077–4086

Gunn JS, Lim KB, Krueger J, Kim K, Guo L, Hackett M, Miller SI (1998) PmrA-PmrB-regulated genes necessary for 4-aminoarabinose lipid A modification and polymyxin resistance. Mol Microbiol 27:1171–1182

Gunn JS, Ryan SS, Van Velkinburgh JC, Ernst RK, Miller SI (2000) Genetic and functional analysis of a PmrA-PmrB-regulated locus necessary for lipopolysaccharide modification, antimicrobial peptide resistance, and oral virulence of *Salmonella enterica* serovar Typhimurium. Infect Immun 68:6139–6146

Guo L, Lim KB, Poduje CM, Daniel M, Gunn JS, Hackett M, Miller SI (1998) Lipid A acylation and bacterial resistance against vertebrate antimicrobial peptides. Cell 95:189–198

Gustafson J, Strassle A, Hachler H, Kayser FH, Berger-Bachi B (1994) The *femC* locus of *Staphylococcus aureus* required for methicillin resistance includes the glutamine synthetase operon. J Bacteriol 176:1460–1467

Hancock RE, Chapple DS (1999) Peptide antibiotics. Antimicrob Agents Chemother 43:1317–1323

Hornef MW, Putsep K, Karlsson J, Refai E, Andersson M (2004) Increased diversity of intestinal antimicrobial peptides by covalent dimer formation. Nat Immunol 5:836–843

Huijbregts RP, de Kroon AI, de Kruijff B (2000) Topology and transport of membrane lipids in bacteria. Biochim Biophys Acta 1469:43–61

Islam D, Bandholtz L, Nilsson J, Wigzell H, Christensson B, Agerberth B, Gudmundsson G (2001) Downregulation of bactericidal peptides in enteric infections: a novel immune escape mechanism with bacterial DNA as a potential regulator. Nat Med 7:180–185

Jack RW, Bierbaum G, Sahl H-G (1998) Lantibiotics and related peptides. Springer, Berlin Heidelberg New York

Jerse AE, Sharma ND, Simms AN, Crow ET, Snyder LA, Shafer WM (2003) A gonococcal efflux pump system enhances bacterial survival in a female mouse model of genital tract infection. Infect Immun 71:5576–5582

Jin T, Bokarewa M, Foster T, Mitchell J, Higgins J, Tarkowski A (2004) *Staphylococcus aureus* resists human defensins by production of staphylokinase, a novel bacterial evasion mechanism. J Immunol 172:1169–1176

Joly S, Maze C, McCray PB Jr, Guthmiller JM (2004) Human beta-defensins 2 and 3 demonstrate strain-selective activity against oral microorganisms. J Clin Microbiol 42:1024–1029

Katzif S, Danavall D, Bowers S, Balthazar JT, Shafer WM (2003) The major cold shock gene, cspA, is involved in the susceptibility of *Staphylococcus aureus* to an antimicrobial peptide of human cathepsin G. Infect Immun 71:4304–4312

Koprivnjak T, Peschel A, Gelb MH, Liang NS, Weiss JP (2002) Role of charge properties of bacterial envelope in bactericidal action of human Group IIA phospholipase A2 against *Staphylococcus aureus*. J Biol Chem 277:47636–47644

Kristian SA, Durr M, Van Strijp JA, Neumeister B, Peschel A (2003a) MprF-mediated lysinylation of phospholipids in *Staphylococcus aureus* leads to protection against oxygen-independent neutrophil killing. Infect Immun 71:546–549

Kristian SA, Lauth X, Nizet V, Goetz F, Neumeister B, Peschel A, Landmann R (2003b) Alanylation of teichoic acids protects *Staphylococcus aureus* against Toll-like receptor 2-dependent host defense in a mouse tissue cage infection model. J Infect Dis 188:414–423

Kristian S, Datta V, Weidenmaier C, Kansal R, Fedtke I, Peschel A, Gallo R, Nizet V (2005) D-alanylation of teichoic acids promotes group A streptococcus antimicrobial peptide resistance, neutrophil survival, and epithelial cell invasion. J Bacteriol 187:6719–6725

Kupferwasser LI, Skurray RA, Brown MH, Firth N, Yeaman MR, Bayer AS (1999) Plasmid-mediated resistance to thrombin-induced platelet microbicidal protein in staphylococci: role of the *qacA* locus. Antimicrob Agents Chemother 43:2395–2399

Lay FT, Anderson MA (2005) Defensins—components of the innate immune system in plants. Curr Protein Pept Sci 6:85–101

Lehrer RI (2004) Primate defensins. Nat Rev Microbiol 2:727–738

Leippe M, Herbst R (2004) Ancient weapons for attack and defense: the pore-forming polypeptides of pathogenic enteric and free-living amoeboid protozoa. J Eukaryot Microbiol 51:516–521

Maxwell AI, Morrison GM, Dorin JR (2003) Rapid sequence divergence in mammalian beta-defensins by adaptive evolution. Mol Immunol 40:413–421

McPhee JB, Lewenza S, Hancock RE (2003) Cationic antimicrobial peptides activate a two-component regulatory system, PmrA-PmrB, that regulates resistance to polymyxin B and cationic antimicrobial peptides in *Pseudomonas aeruginosa*. Mol Microbiol 50:205–217

Midorikawa K, Ouhara K, Komatsuzawa H, Kawai T, Yamada S, Fujiwara T, Yamazaki K, Sayama K, Taubman MA, Kurihara H, Hashimoto K, Sugai M (2003) *Staphylococcus aureus* susceptibility to innate antimicrobial peptides, beta-defensins and CAP18, expressed by human keratinocytes. Infect Immun 71:3730–3739

Miller SI, Ernst RK, Bader MW (2005) LPS, TLR4 and infectious disease diversity. Nat Rev Microbiol 3:36–46

Neuhaus FC, Baddiley J (2003) A continuum of anionic charge: structures and functions of D-alanyl-teichoic acids in Gram-positive bacteria. Microbiol Mol Biol Rev 67:686–723

Nguyen TX, Cole AM, Lehrer RI (2003) Evolution of primate theta-defensins: a serpentine path to a sweet tooth. Peptides 24:1647–1654

Nishimura E, Eto A, Kato M, Hashizume S, Imai S, Nisizawa T, Hanada N (2004) Oral streptococci exhibit diverse susceptibility to human beta-defensin-2: antimicrobial effects of hBD-2 on oral streptococci. Curr Microbiol 48:85–87

Nizet V (2005) Antimicrobial peptide resistance in human bacterial pathogens. In: Gallo RL (ed) Antimicrobial peptides in human health and disease. Horizon Bioscience, Norfolk, UK, pp 277–304

Nizet V, Gallo RL (2003) Cathelicidins and innate defense against invasive bacterial infection. Scand J Infect Dis 35:670–676

Nizet V, Ohtake T, Lauth X, Trowbridge J, Rudisill J, Dorschner RA, Pestonjamasp V, Piraino J, Huttner K, Gallo RL (2001) Innate antimicrobial peptide protects the skin from invasive bacterial infection. Nature 414:454–457

Nummila K, Kilpeläinen I, Zähringer U, Vaara M, Helander IM (1995) Lipopolysaccharides of polymyxin B-resistant mutants of *Escherichia coli* are extensively substituted by 2-aminoethyl pyrophosphate and contain aminoarabinose in lipid A. Mol Microbiol 16:271–278

Oku Y, Kurokawa K, Ichihashi N, Sekimizu K (2004) Characterization of the *Staphylococcus aureus mprF* gene, involved in lysinylation of phosphatidylglycerol. Microbiology 150:45–51

Otto M (2005) Bacterial evasion of antimicrobial peptides by biofilm formation. In: Shafer WM (ed) Antimicrobial peptides and human disease. Curr Top Microbiol Immunol (in press)

Pag U, Sahl HG (2002) Multiple activities in lantibiotics—models for the design of novel antibiotics? Curr Pharm Des 8:815–833

Patil A, Hughes AL, Zhang G (2004) Rapid evolution and diversification of mammalian alpha-defensins as revealed by comparative analysis of rodent and primate genes. Physiol Genomics 20:1–11

Peacock SJ, de Silva I, Lowy FD (2001) What determines nasal carriage of *Staphylococcus aureus*? Trends Microbiol 9:605–610

Perego M, Glaser P, Minutello A, Strauch MA, Leopold K, Fischer W (1995) Incorporation of D-alanine into lipoteichoic acid and wall teichoic acid in *Bacillus subtilis*. J Biol Chem 270:15598–15606

Peschel A (2002) How do bacteria resist human antimicrobial peptides? Trends Microbiol 10:179–186

Peschel A, Götz F (1996) Analysis of the *Staphylococcus epidermidis* genes *epiF*, *E*, and *G* involved in epidermin immunity. J Bacteriol 178:531–536

Peschel A, Otto M, Jack RW, Kalbacher H, Jung G, Götz F (1999) Inactivation of the *dlt* operon in *Staphylococcus aureus* confers sensitivity to defensins, protegrins and other antimicrobial peptides. J Biol Chem 274:8405–8410

Peschel A, Vuong C, Otto M, Götz F (2000) The D-alanine residues of *Staphylococcus aureus* teichoic acids alter the susceptibility to vancomycin and the activity of autolysins. Antimicrob Agents Chemother 44:2845–2847

Peschel A, Jack RW, Otto M, Collins LV, Staubitz P, Nicholson G, Kalbacher H, Nieuwenhuizen WF, Jung G, Tarkowski A, van Kessel KPM, van Strijp JAG (2001) *Staphylococcus aureus* resistance to human defensins and evasion of neutrophil killing via the novel virulence factor MprF is based on modification of membrane lipids with L-lysine. J Exp Med 193:1067–1076

Poyart C, Pellegrini E, Marceau M, Baptista M, Jaubert F, Lamy MC, Trieu-Cuot P (2003) Attenuated virulence of *Streptococcus agalactiae* deficient in D-alanyl-lipoteichoic acid is due to an increased susceptibility to defensins and phagocytic cells. Mol Microbiol 49:1615–1625

Proctor RA, Kahl B, von Eiff C, Vaudaux PE, Lew DP, Peters G (1998) Staphylococcal small colony variants have novel mechanisms for antibiotic resistance. Clin Infect Dis 27 [Suppl 1]:68–74

Ratledge C, Wilkinson SG (1988) Microbial lipids. Academic Press, London

Riley MA, Wertz JE (2002) Bacteriocins: evolution, ecology, and application. Annu Rev Microbiol 56:117–137

Sahl HG, Pag U, Bonness S, Wagner S, Antcheva N, Tossi A (2004) Mammalian defensins: structures and mechanism of antibiotic activity. J Leukoc Biol 77:466–475

Sahly H, Schubert S, Harder J, Rautenberg P, Ullmann U, Schroder J, Podschun R (2003) *Burkholderia* is highly resistant to human Beta-defensin 3. Antimicrob Agents Chemother 47:1739–1741

Schmidtchen A, Frick IM, Andersson E, Tapper H, Bjorck L (2002) Proteinases of common pathogenic bacteria degrade and inactivate the antibacterial peptide LL-37. Mol Microbiol 46:157–168

Shafer WM, Qu X-D, Waring AJ, Lehrer RI (1998) Modulation of *Neisseria gonorrhoeae* susceptibility to vertebrate antibacterial peptides due to a member of the resistance/nodulation/division efflux pump family. Proc Natl Acad Sci U S A 95:1829–1833

Shelburne CE, Coulter WA, Olguin D, Lantz MS, Lopatin DE (2005) Induction of β-defensin resistance in the oral anaerobe *Porphyromonas gingivalis*. Antimicrob Agents Chemother 49:183–187

Sieprawska-Lupa M, Mydel P, Krawczyk K, Wojcik K, Puklo M, Lupa B, Suder P, Silberring J, Reed M, Pohl J, Shafer W, McAleese F, Foster T, Travis J, Potempa J (2004) Degradation of human antimicrobial peptide LL-37 by *Staphylococcus aureus*-derived proteinases. Antimicrob Agents Chemother 48:4673–4679

Staubitz P, Peschel A (2002) MprF-mediated lysinylation of phospholipids in *Bacillus subtilis*—protection against bacteriocins in terrestrial environments? Microbiology 148:3331–3332

Staubitz P, Neumann H, Schneider T, Wiedemann I, Peschel A (2004) MprF-mediated biosynthesis of lysylphosphatidylglycerol, an important determinant in staphylococcal defensin resistance. FEMS Microbiol Lett 231:67–71

Steenbergen JN, Alder J, Thorne GM, Tally FP (2005) Daptomycin: a lipopeptide antibiotic for the treatment of serious Gram-positive infections. J Antimicrob Chemother 53:283–288

Stumpe S, Schmid R, Stephens DL, Georgiou G, Bakker EP (1998) Identification of OmpT as the protease that hydrolyzes the antimicrobial peptide protamine before it enters growing cells of *Escherichia coli*. J Bacteriol 180:4002–4006

Van Veen HW, Konings WN (1997) Drug efflux proteins in multidrug resistant bacteria. Biol Chem 378:769–777

Von Eiff C, Becker K, Machka K, Stammer H, Peters G (2001) Nasal carriage as a source of *Staphylococcus aureus* bacteremia. Study group. N Eng J Med 344:11–16

Vuong C, Voyich JM, Fischer ER, Braughton KR, Whitney AR, DeLeo FR, Otto M (2004) Polysaccharide intercellular adhesin (PIA) protects *Staphylococcus epidermidis* against major components of the human innate immune system. Cell Microbiol 6:269–275

Wecke J, Madela K, Fischer W (1997) The absence of D-alanine from lipoteichoic acid and wall teichoic acid alters surface charge, enhances autolysis and increases susceptibility to methicillin in *Bacillus subtilis*. Microbiology 143:2953–2960

Weidenmaier C, Kristian SA, Peschel A (2003) Bacterial resistance to antimicrobial host defenses—an emerging target for novel antiinfective strategies? Curr Drug Targets 4:643–649

Weidenmaier C, Kokai-Kun JF, Kristian SA, Chanturyia T, Kalbacher H, Gross M, Nicholson G, Neumeister B, Mond JJ, Peschel A (2004) Role of teichoic acids in *Staphylococcus aureus* nasal colonization, a major risk factor in nosocomial infections. Nat Med 10:243–245

Weidenmaier C, Peschel A, Kempf VA, Lucindo N, Yeaman MR, Bayer AS (2005) DltABCD- and MprF-mediated cell envelope modifications of *Staphylococcus aureus* confer resistance to platelet microbicidal proteins and contribute to virulence in a rabbit endocarditis model. Infect Immun 73:8033–8038

Wertheim HF, Vos MC, Ott A, van Belkum A, Voss A, Kluytmans JA, van Keulen PH, Vandenbroucke-Grauls CM, Meester MH, Verbrugh HA (2004) Risk and outcome of nosocomial *Staphylococcus aureus* bacteraemia in nasal carriers versus non-carriers. Lancet 364:703–705

Wosten MM, Kox LF, Chamnongpol S, Soncini FC, Groisman EA (2000) A signal transduction system that responds to extracellular iron. Cell 103:113–125

Yang D, Chen Q, Hoover DM, Staley P, Tucker KD, Lubkowski J, Oppenheim JJ (2003) Many chemokines including CCL20/MIP-3alpha display antimicrobial activity. J Leukoc Biol 74:448–455

Yeaman MR, Bayer AS, Koo S-P, Foss W, Sullam PM (1998) Platelet microbicidal proteins and neutrophil defensin disrupt the *Staphylococcus aureus* cytoplasmic membrane by distinct mechanisms of action. J Clin Invest 101:178–187

Zasloff M (2002) Antimicrobial peptides of multicellular organisms. Nature 415:389–395

CTMI (2006) 306:251–258

Bacterial Evasion of Antimicrobial Peptides by Biofilm Formation

M. Otto (✉)

Laboratory of Human Bacterial Pathogenesis, Rocky Mountain Laboratories,
National Institute of Allergy and Infectious Diseases,
The National Institutes of Health, 903 S. 4th Street, Hamilton, MT 59840, USA
motto@niaid.nih.gov

Abstract Biofilm formation is a main virulence determinant in many bacterial infections. It significantly increases bacterial resistance to antibiotics and innate host defense. In general, the specific physiology of biofilms and the barrier function of the extracellular biofilm matrix determine resistance to antibacterials. However, resistance to antimicrobial peptides appears to be mainly based on the interaction with biofilm and capsule exopolymers. These polymers may work by electrostatic repulsion and/or sequestration of antibacterial substances. As biofilm polymers play an eminent role in biofilm structuring and resistance, their destruction by dedicated enzymes is a promising attempt to prevent colonization and develop treatment for biofilm-associated infections.

1
Introduction

The formation of surface-attached cellular agglomerations, so-called biofilms, contributes significantly to bacterial resistance to antibiotics and innate host defense [1]. For these reasons, biofilm formation by pathogenic bacteria has an enormous impact on the outcome of many bacterial infections. According to several estimates, biofilm-associated infections cost the public health system in the United States alone several billion dollars per year. A majority (65%) of infections treated by physicians in the developed world are characterized by the involvement of biofilms [1]. Prominent biofilm-associated diseases in humans are *Pseudomonas aeruginosa* colonization of the lungs of cystic fibrosis patients, the colonization of indwelling medical devices by staphylococci, and dental plaque formation by oral streptococci and other bacteria in mixed-species biofilms.

During human infection, neutrophils as the cornerstone of innate host defense ingest and kill invading bacteria. They eliminate bacteria by a combination of mechanisms, including reactive oxygen species and antimicrobial peptides (AMPs). Furthermore, many epithelial cell types also secrete specific AMPs to kill colonizing bacteria without the need of neutrophil uptake [2]. Although neutrophils can efficiently penetrate biofilms [3], there is increasing evidence that biofilms protect from neutrophil killing and phagocytosis [4]. However, the molecular basis of biofilm resistance is not yet entirely clear. Specifically, our knowledge about the interaction of AMPs with biofilms is relatively limited. Here, I will summarize some of the more recent discoveries on the mechanisms that bacteria use to achieve resistance to AMPs during biofilm formation.

2
The Basis of Biofilm Resistance

Bacteria in biofilms can reach resistance levels that are approximately 10–1,000 times higher than during planktonic growth [1]. The difference is due in part to the production of the biofilm-typical exopolysaccharide (EPS) structure, which is believed to decrease the penetration of antibacterial substances to their target [5]. While this is the case for a series of antibiotics, e.g., ciprofloxacin, several other antibacterial substances have been shown to break through the exopolysaccharide layer. For example, rifampicin and vancomycin can penetrate the extracellular matrix of *Staphylococcus epidermidis* [6]. However, biofilm resistance is also based on the slow growth and low metabolic activity of biofilms. In *S. epidermidis*, genome-wide analysis of biofilm-specific gene expression has shown that basic cellular processes are down-regulated and aerobic energy production is shifted toward anaerobic fermentation [7]. These results may explain why the activities of antibiotics that predominantly inhibit processes of metabolically active cells are impaired against bacteria in biofilms. Another explanation of biofilm resistance is heterogeneity. It has been hypothesized that a small subpopulation of cells in a biofilm may survive increased concentrations of an antibacterial substance due to a specific physiological status. While most other cells are killed, the survival of such "persister" cells prevents the colony from being erased entirely [8]. All these mechanisms probably work together to achieve resistance to a variety of substances in manifold physiological situations and contribute to the fact that biofilm-associated infections are very difficult to eradicate.

3
Polysaccharide Intercellular Adhesin: A Widespread Biofilm Exopolysaccharide Component with Protective Properties

The status of low metabolic activity and the persister model have been used to explain biofilm resistance to antibiotics in general. In contrast, research on biofilm resistance to AMPs has only been initiated recently and up to now was predominantly focused on resistance mediated by exopolysaccharides and other polymeric substances. Exopolysaccharide (EPS) is produced by a variety of bacteria and is frequently an essential component of the extracellular matrix of biofilm-forming bacteria [5]. However, our knowledge about specific EPS molecules is limited. Specifically, we lack understanding about how the structural features of EPS molecules influence bacterial pathogenesis.

More recently, Vuong et al. in my group found that the major biofilm exopolysaccharide of *S. epidermidis* and *S. aureus*, polysaccharide intercellular adhesin (PIA), protects from AMPs of human skin and neutrophil specific granules, LL-37 and human β-defensin 3 [9]. Most likely, resistance is based in part on the cationic character of PIA. The positive net charge of PIA is very unusual for a bacterial exopolysaccharide and likely causes electrostatic repulsion of the commonly cationic AMPs. However, PIA also protected from the unusual anionic peptide dermcidin, which is secreted by human epithelia. Therefore, at least part of the mechanism by which PIA protects from AMPs is probably not due to electrostatic repulsion.

Chemically, PIA is a homopolymer of N-acetylglucosamine (GlcNAc) subunits [10]. A certain percentage of GlcNAc residues are de-acetylated (~10%–20%). Deacetylation produces free amino groups that become positively charged at neutral to acid pH (pK ~6.9), resulting in a positive net charge of the otherwise neutral PIA molecule. My laboratory has recently demonstrated that the surface-attached protein IcaB gene is responsible for the deacetylation of GlcNAc residues in PIA [11]. Of note, in contrast to wild-type PIA, nondeacetylated PIA in an isogenic *icaB* mutant failed to attach to the bacterial cell surface. Furthermore, the *icaB* mutant was not able to colonize surfaces that are relevant to *S. epidermidis* infection, namely human epithelial cells and plastic surfaces. Resistance to AMPs and neutrophil phagocytosis in the *icaB* mutant was reduced to a level found in a mutant devoid of any PIA. Remarkably, as a consequence of impaired biofilm formation and resistance to innate host defense, the mutant failed to cause infection of medical devices in an animal infection model.

Importantly, the biosynthetic machinery for PIA and the molecule itself are being found in an increasing number of pathogenic, biofilm-forming microorganisms, such as *Escherichia coli*, *Yersinia pestis*, *Actinobacillus actinomycetemcomitans*, and *Bordetella pertussis* [12–14]. Thus, PIA-mediated resistance to AMPs is likely of widespread importance.

4
Pseudomonas Aeruginosa Exopolymers

P. aeruginosa is the most common pathogen found in the lung of cystic fibrosis patients [15]. There is circumstantial, yet convincing evidence that *P. aeruginosa* forms biofilms in infected patients [16]. *P. aeruginosa* produces an extracellular matrix, of which the anionic β-D-mannuronate and α-L-guluronate polymer alginate is the best characterized component. Although contrasting reports exist about the role of alginate in biofilm-associated infection [17, 18], it might play an important role in the resistance of *P. aeruginosa* to antibacterials. Interestingly, in a very recent report it has been shown that alginate induces helix formation and peptide–peptide association in AMPs, in a way similar to AMP interaction with the cytoplasmic membrane [19]. By promoting peptide aggregation, this mechanism is believed to result in the removal of the AMP from the cytoplasmic membrane, the primary target of AMP action.

In general, negatively charged or neutral bacterial exopolymers are found far more frequently than cationic polymers such as PIA and may also significantly contribute to specific resistance to antibacterial substances [5]. For example, periplasmic glucans of *P. aeruginosa* have been shown to bind the antibiotic tobramycin [20]. This suggests that EPS molecules might prevent the accumulation of antimicrobial substance at its target by a sequestration mechanism. Such a mechanism may also be the basis of PIA-mediated resistance to the anionic AMP dermcidin, which has been described above. EPS might thus work by electrostatic repulsion or by sequestration of AMPs, or possibly even combine the two mechanisms against peptides with opposite charges.

5
Poly-γ-Glutamic Acid: A Protective Capsule in *Bacillus Anthracis* and Coagulase-Negative Staphylococci

Capsules are extracellular protective layers that surround various bacteria and can be composed of chemically diverse molecules, such as polypeptides or polysaccharides. Possibly, they may also contribute to the formation of the extracellular biofilm matrix. Whereas some capsules have been shown to consist mainly of one chemically distinct polymer, such as the poly-γ-glutamic acid (PGA) capsule of *Bacillus anthracis*, others are composed of a variety of not always completely identified substances. There are many reports that capsules protect from phagocytosis, for example in *E. coli*, *S. aureus*, and *B. anthracis*, to name but a few [21–23]. It has also been shown that encapsulated strains are usually more virulent than their nonencapsulated counterparts and it is believed that capsule mediates increased resistance to human innate host defense [24]. However, it has not been investigated if capsules also provide resistance to AMPs.

Kocianova et al. in my group have shown recently that *S. epidermidis* and some other coagulase-negative staphylococci produce a PGA capsule-like protective extracellular layer [25]. PGA in *S. epidermidis* protected the bacteria from neutrophil phagocytosis. Importantly, it also provided protection to representative AMPs of human skin and neutrophil-specific granules. Remarkably, we observed a significant level of protection, although the PGA production level was approximately 1 million times lower than reported for *Bacillus* species. Most likely, therefore, the much thicker PGA capsule of *B. anthracis* may provide even more pronounced protection from AMPs, which remains to be demonstrated. Further, the *cap* genes that encode production of PGA are also found in other pathogenic microorganisms, for example *Leptospira interrogans* and *Fusobacterium nucleatum* [26, 27]. Although PGA production in these cases has not been demonstrated yet, presence of the *cap* genes suggests that PGA is produced and might also be involved in AMP resistance.

Interestingly, the *cap* gene locus is up-regulated in *S. epidermidis* biofilms [7]. PGA production might thus contribute to the biofilm-specific AMP resistance in *S. epidermidis*, although PGA itself, in contrast to PIA, did not influence biofilm formation in this bacterium [25]. With regard to protection from innate host defense, the biological roles of biofilm exopolysaccharides, such as PIA, and those of capsule substances are likely very similar. However, it should be stressed that polymers with an additional function in biofilm formation also act by promoting the formation of an extended, three-dimensional biofilm structure. This structure further contributes to resistance as it may decrease penetration of antibacterial substances to cells in deeper biofilm layers.

6
Conclusion and Outlook

In conclusion, resistance to AMPs in biofilms appears to be based, at least in part, on the interaction with specific extracellular biofilm polymers. The polymers may act via different mechanisms, but all aim at removing the AMP from its target. However, only in cases in which these polymers represent biofilm-specific molecules can we speak of a truly biofilm-inherent resistance. Many of the extracellular polymers are also produced in different physiological situations. They might constitute a much more general mechanism of resistance against AMPs, which is not directly related to biofilms.

Among the predominant biofilm-forming pathogens in humans, oral streptococci are the ones about which we know least with regard to exopolysaccharide structure and function, and the mechanisms of resistance to AMPs. The situation in these bacteria is complicated by the fact that they usually produce glycosyltransferases, leading to a constant change in EPS composition [28]. It will be an important task of future research to elucidate the roles of EPS in these pathogens.

Based on its role in biofilm formation and resistance to innate host defense, EPS is a prime target for therapeutic maneuvers aimed at treating biofilm-associated infections. The enzymatic destruction of EPS would leave bacteria prone to being killed by conventional antibiotics. Recently, an enzyme has been discovered in the dental pathogen *A. actinomycetemcomitans*, which degrades PIA from several biofim-forming pathogens [29, 30]. The use of this and possibly other EPS-destroying enzymes might be an important new approach to target biofilm-associated infections.

References

1. Costerton JW, Stewart PS, Greenberg EP (1999) Bacterial biofilms: a common cause of persistent infections. Science 284:1318–1322
2. Hancock RE, Diamond G (2000) The role of cationic antimicrobial peptides in innate host defences. Trends Microbiol 8:402–410
3. Leid JG, Shirtliff ME, Costerton JW, Stoodley AP (2002) Human leukocytes adhere to, penetrate, and respond to *Staphylococcus aureus* biofilms. Infect Immun 70:6339–6345
4. Jesaitis AJ, Franklin MJ, Berglund D, Sasaki M, Lord CI, Bleazard JB, Duffy JE, Beyenal H, Lewandowski Z (2003) Compromised host defense on *Pseudomonas aeruginosa* biofilms: characterization of neutrophil and biofilm interactions. J Immunol 171:4329–4339
5. Sutherland I (2001) Biofilm exopolysaccharides: a strong and sticky framework. Microbiology 147:3–9

6. Dunne WM Jr, Mason EO Jr, Kaplan SL (1993) Diffusion of rifampin and vancomycin through a *Staphylococcus epidermidis* biofilm. Antimicrob Agents Chemother 37:2522–2526

7. Yao Y, Sturdevant DE, Otto M (2005) Genomewide analysis of gene expression in *Staphylococcus epidermidis* biofilms: insights into the pathophysiology of *S. epidermidis* biofilms and the role of phenol-soluble modulins in formation of biofilms. J Infect Dis 191:289–298

8. Keren I, Kaldalu N, Spoering A, Wang Y, Lewis K (2004) Persister cells and tolerance to antimicrobials. FEMS Microbiol Lett 230:13–18

9. Vuong C, Voyich JM, Fischer ER, Braughton KR, Whitney AR, DeLeo FR, Otto M (2004) Polysaccharide intercellular adhesin (PIA) protects *Staphylococcus epidermidis* against major components of the human innate immune system. Cell Microbiol 6:269–275

10. Mack D, Fischer W, Krokotsch A, Leopold K, Hartmann R, Egge H, Laufs R (1996) The intercellular adhesin involved in biofilm accumulation of *Staphylococcus epidermidis* is a linear beta-1,6-linked glucosaminoglycan: purification and structural analysis. J Bacteriol 178:175–183

11. Vuong C, Kocianova S, Voyich JM, Yao Y, Fischer ER, DeLeo FR, Otto M (2004) A crucial role for exopolysaccharide modification in bacterial biofilm formation, immune evasion, and virulence. J Biol Chem 279:54881–54886

12. Kaplan JB, Velliyagounder K, Ragunath C, Rohde H, Mack D, Knobloch JK, Ramasubbu N (2004) Genes involved in the synthesis and degradation of matrix polysaccharide in *Actinobacillus actinomycetemcomitans* and *Actinobacillus pleuropneumoniae* biofilms. J Bacteriol 186:8213–8220

13. Wang X, Preston JFI, Romeo T (2004) The *pgaABCD* locus of *Escherichia coli* promotes the synthesis of a polysaccharide adhesin required for biofilm formation. J Bacteriol 186:2724–2734

14. Darby C, Hsu JW, Ghori N, Falkow S (2002) *Caenorhabditis elegans*: plague bacteria biofilm blocks food intake. Nature 417:243–244

15. Garau J, Gomez L (2003) *Pseudomonas aeruginosa* pneumonia. Curr Opin Infect Dis 16:135–143

16. Singh PK, Schaefer AL, Parsek MR, Moninger TO, Welsh MJ, Greenberg EP (2000) Quorum-sensing signals indicate that cystic fibrosis lungs are infected with bacterial biofilms. Nature 407:762–764

17. Stapper AP, Narasimhan G, Ohman DE, Barakat J, Hentzer M, Molin S, Kharazmi A, Hoiby N, Mathee K (2004) Alginate production affects *Pseudomonas aeruginosa* biofilm development and architecture, but is not essential for biofilm formation. J Med Microbiol 53:679–690

18. Wozniak DJ, Wyckoff TJ, Starkey M, Keyser R, Azadi P, O'Toole GA, Parsek MR (2003) Alginate is not a significant component of the extracellular polysaccharide matrix of PA14 and PAO1 *Pseudomonas aeruginosa* biofilms. Proc Natl Acad Sci U S A 100:7907–7912

19. Chan C, Burrows LL, Deber CM (2004) Helix induction in antimicrobial peptides by alginate in biofilms. J Biol Chem 279:38749–38754

20. Mah TF, Pitts B, Pellock B, Walker GC, Stewart PS, O'Toole GA (2003) A genetic basis for *Pseudomonas aeruginosa* biofilm antibiotic resistance. Nature 426:306–310

21. Horwitz MA, Silverstein SC (1980) Influence of the *Escherichia coli* capsule on complement fixation and on phagocytosis and killing by human phagocytes. J Clin Invest 65:82–94
22. O'Riordan K, Lee JC (2004) *Staphylococcus aureus* capsular polysaccharides. Clin Microbiol Rev 17:218–234
23. Little SF, Ivins BE (1999) Molecular pathogenesis of *Bacillus anthracis* infection. Microbes Infect 1:131–139
24. Hornef MW, Wick MJ, Rhen M, Normark S (2002) Bacterial strategies for overcoming host innate and adaptive immune responses. Nat Immunol 3:1033–1040
25. Kocianova S, Vuong C, Yao Y, Voyich JM, Fischer ER, DeLeo FR, Otto M (2005) Key role of poly-γ-DL-glutamic acid in immune evasion and virulence of *Staphylococcus epidermidis*. J Clin Invest 115:688–694
26. Ren SX, Fu G, Jiang XG, Zeng R, Miao YG, Xu H, Zhang YX, Xiong H, Lu G, Lu LF et al (2003) Unique physiological and pathogenic features of Leptospira interrogans revealed by whole-genome sequencing. Nature 422:888–893
27. Kapatral V, Anderson I, Ivanova N, Reznik G, Los T, Lykidis A, Bhattacharyya A, Bartman A, Gardner W, Grechkin G et al (2002) Genome sequence and analysis of the oral bacterium *Fusobacterium nucleatum* strain ATCC 25586. J Bacteriol 184:2005–2018.
28. Neu TR (1996) Significance of bacterial surface-active compounds in interaction of bacteria with interfaces. Microbiol Rev 60:151–166
29. Itoh Y, Wang X, Hinnebusch BJ, Preston JF 3rd, Romeo T (2005) Depolymerization of beta-1,6-N-acetyl-D-glucosamine disrupts the integrity of diverse bacterial biofilms. J Bacteriol 187:382–387
30. Kaplan JB, Meyenhofer MF, Fine DH (2003) Biofilm growth and detachment of *Actinobacillus actinomycetemcomitans*. J Bacteriol 185:1399–1404

Subject Index

Current Topics in Microbiology and Immunology

Volumes published since 1989 (and still available)

Vol. 262: **Oldstone, Michael B.A. (Ed.):** Arenaviruses I. 2002. 30 figs. XVIII, 197 pp. ISBN 3-540-42244-7

Vol. 263: **Oldstone, Michael B. A. (Ed.):** Arenaviruses II. 2002. 49 figs. XVIII, 268 pp. ISBN 3-540-42705-8

Vol. 264/I: **Hacker, Jörg; Kaper, James B. (Eds.):** Pathogenicity Islands and the Evolution of Microbes. 2002. 34 figs. XVIII, 232 pp. ISBN 3-540-42681-7

Vol. 264/II: **Hacker, Jörg; Kaper, James B. (Eds.):** Pathogenicity Islands and the Evolution of Microbes. 2002. 24 figs. XVIII, 228 pp. ISBN 3-540-42682-5

Vol. 265: **Dietzschold, Bernhard; Richt, Jürgen A. (Eds.):** Protective and Pathological Immune Responses in the CNS. 2002. 21 figs. X, 278 pp. ISBN 3-540-42668X

Vol. 266: **Cooper, Koproski (Eds.):** The Interface Between Innate and Acquired Immunity, 2002. 15 figs. XIV, 116 pp. ISBN 3-540-42894-X

Vol. 267: **Mackenzie, John S.; Barrett, Alan D. T.; Deubel, Vincent (Eds.):** Japanese Encephalitis and West Nile Viruses. 2002. 66 figs. X, 418 pp. ISBN 3-540-42783X

Vol. 268: **Zwickl, Peter; Baumeister, Wolfgang (Eds.):** The Proteasome-Ubiquitin Protein Degradation Pathway. 2002. 17 figs. X, 213 pp. ISBN 3-540-43096-2

Vol. 269: **Koszinowski, Ulrich H.; Hengel, Hartmut (Eds.):** Viral Proteins Counteracting Host Defenses. 2002. 47 figs. XII, 325 pp. ISBN 3-540-43261-2

Vol. 270: **Beutler, Bruce; Wagner, Hermann (Eds.):** Toll-Like Receptor Family Members and Their Ligands. 2002. 31 figs. X, 192 pp. ISBN 3-540-43560-3

Vol. 271: **Koehler, Theresa M. (Ed.):** Anthrax. 2002. 14 figs. X, 169 pp. ISBN 3-540-43497-6

Vol. 272: **Doerfler, Walter; Böhm, Petra (Eds.):** Adenoviruses: Model and Vectors in Virus-Host Interactions. Virion and Structure, Viral Replication, Host Cell Interactions. 2003. 63 figs., approx. 280 pp. ISBN 3-540-00154-9

Vol. 273: **Doerfler, Walter; Böhm, Petra (Eds.):** Adenoviruses: Model and Vectors in VirusHost Interactions. Immune System, Oncogenesis, Gene Therapy. 2004. 35 figs., approx. 280 pp. ISBN 3-540-06851-1

Vol. 274: **Workman, Jerry L. (Ed.):** Protein Complexes that Modify Chromatin. 2003. 38 figs., XII, 296 pp. ISBN 3-540-44208-1

Vol. 275: **Fan, Hung (Ed.):** Jaagsiekte Sheep Retrovirus and Lung Cancer. 2003. 63 figs., XII, 252 pp. ISBN 3-540-44096-3

Vol. 276: **Steinkasserer, Alexander (Ed.):** Dendritic Cells and Virus Infection. 2003. 24 figs., X, 296 pp. ISBN 3-540-44290-1

Vol. 277: **Rethwilm, Axel (Ed.):** Foamy Viruses. 2003. 40 figs., X, 214 pp. ISBN 3-540-44388-6

Vol. 278: **Salomon, Daniel R.; Wilson, Carolyn (Eds.):** Xenotransplantation. 2003. 22 figs., IX, 254 pp. ISBN 3-540-00210-3

Vol. 279: **Thomas, George; Sabatini, David; Hall, Michael N. (Eds.):** TOR. 2004. 49 figs., X, 364 pp. ISBN 3-540-00534X

Vol. 280: **Heber-Katz, Ellen (Ed.):** Regeneration: Stem Cells and Beyond. 2004. 42 figs., XII, 194 pp. ISBN 3-540-02238-4

Vol. 281: **Young, John A. T. (Ed.):** Cellular Factors Involved in Early Steps of Retroviral Replication. 2003. 21 figs., IX, 240 pp. ISBN 3-540-00844-6

Vol. 282: **Stenmark, Harald (Ed.):** Phosphoinositides in Subcellular Targeting and Enzyme Activation. 2003. 20 figs., X, 210 pp. ISBN 3-540-00950-7

Vol. 283: **Kawaoka, Yoshihiro (Ed.):** Biology of Negative Strand RNA Viruses: The Power of Reverse Genetics. 2004. 24 figs., IX, 350 pp. ISBN 3-540-40661-1

Vol. 284: **Harris, David (Ed.):** Mad Cow Disease and Related Spongiform Encephalopathies. 2004. 34 figs., IX, 219 pp. ISBN 3-540-20107-6

Vol. 285: **Marsh, Mark (Ed.):** Membrane Trafficking in Viral Replication. 2004. 19 figs., IX, 259 pp. ISBN 3-540-21430-5

Vol. 286: **Madshus, Inger H. (Ed.):** Signalling from Internalized Growth Factor Receptors. 2004. 19 figs., IX, 187 pp. ISBN 3-540-21038-5

Vol. 287: **Enjuanes, Luis (Ed.):** Coronavirus Replication and Reverse Genetics. 2005. 49 figs., XI, 257 pp. ISBN 3-540-21494-1

Vol. 288: **Mahy, Brain W. J. (Ed.):** Foot-and-Mouth-Disease Virus. 2005. 16 figs., IX, 178 pp. ISBN 3-540-22419X

Vol. 289: **Griffin, Diane E. (Ed.):** Role of Apoptosis in Infection. 2005. 40 figs., IX, 294 pp. ISBN 3-540-23006-8

Vol. 290: **Singh, Harinder; Grosschedl, Rudolf (Eds.):** Molecular Analysis of B Lymphocyte Development and Activation. 2005. 28 figs., XI, 255 pp. ISBN 3-540-23090-4

Vol. 291: **Boquet, Patrice; Lemichez Emmanuel (Eds.)** Bacterial Virulence Factors and Rho GTPases. 2005. 28 figs., IX, 196 pp. ISBN 3-540-23865-4

Vol. 292: **Fu, Zhen F (Ed.):** The World of Rhabdoviruses. 2005. 27 figs., X, 210 pp. ISBN 3-540-24011-X

Vol. 293: **Kyewski, Bruno; Suri-Payer, Elisabeth (Eds.):** CD4+CD25+ Regulatory T Cells: Origin, Function and Therapeutic Potential. 2005. 22 figs., XII, 332 pp. ISBN 3-540-24444-1

Vol. 294: **Caligaris-Cappio, Federico, Dalla Favera, Ricardo (Eds.):** Chronic Lymphocytic Leukemia. 2005. 25 figs., VIII, 187 pp. ISBN 3-540-25279-7

Vol. 295: **Sullivan, David J.; Krishna Sanjeew (Eds.):** Malaria: Drugs, Disease and Post-genomic Biology. 2005. 40 figs., XI, 446 pp. ISBN 3-540-25363-7

Vol. 296: **Oldstone, Michael B. A. (Ed.):** Molecular Mimicry: Infection Induced Autoimmune Disease. 2005. 28 figs., VIII, 167 pp. ISBN 3-540-25597-4

Vol. 297: **Langhorne, Jean (Ed.):** Immunology and Immunopathogenesis of Malaria. 2005. 8 figs., XII, 236 pp. ISBN 3-540-25718-7

Vol. 298: **Vivier, Eric; Colonna, Marco (Eds.):** Immunobiology of Natural Killer Cell Receptors. 2005. 27 figs., VIII, 286 pp. ISBN 3-540-26083-8

Vol. 299: **Domingo, Esteban (Ed.):** Quasispecies: Concept and Implications. 2006. 44 figs., XII, 401 pp. ISBN 3-540-26395-0

Vol. 300: **Wiertz, Emmanuel J.H.J.; Kikkert, Marjolein (Eds.):** Dislocation and Degradation of Proteins from the Endoplasmic Reticulum. 2006. 19 figs., VIII, 168 pp. ISBN 3-540-28006-5

Vol. 301: **Doerfler, Walter; Böhm, Petra (Eds.):** DNA Methylation: Basic Mechanisms. 2006. 24 figs., VIII, 324 pp. ISBN 3-540-29114-8

Vol. 302: **Robert N. Eisenman (Ed.):** The Myc/Max/Mad Transcription Factor Network. 2006. 28 figs. XII, 278 pp. ISBN 3-540-23968-5

Vol. 303: **Thomas E. Lane (Ed.):** Chemokines and Viral Infection. 2006. 14 figs. XII, 154 pp. ISBN 3-540-29207-1

Vol. 304: **Stanley A. Plotkin (Ed.):** Mass Vaccination: Global Aspects -- Progress and Obstacles. 2006. 40 figs. IX, ... pp. ISBN 3-540-29382-5

Vol. 305: **Radbruch, Andreas; Lipsky, Peter E. (Eds.):** Current Concepts in Autoimmunity. 2006. 29 figs. IIX, 276 pp. ISBN 3-540-29713-8

Printing: Krips bv, Meppel
Binding: Stürtz, Würzburg